# GNU Octave

A high-level interactive language for numerical computations
Edition 3 for Octave version 3.0.2
August 2008

John W. Eaton
David Bateman
Søren Hauberg

A catalogue record for this book is available from the British Library.

First Printing, September 2008, for Octave version 3.

Published by Network Theory Limited.

Email: info@network-theory.co.uk

ISBN: 0-9546120-6-X

This book supersedes the previous edition for Octave v2 (ISBN 0-9541617-2-6).

Original cover design by David Nicholls.

Further information about this book is available from
http://www.network-theory.co.uk/octave/manual/

This book has an unconditional guarantee. If you are not fully satisfied with your purchase for any reason, please contact the publisher at the address above.

Copyright © 2008 Network Theory Ltd. Minor changes for publication. Edited sparse matrix chapter for clarity. Removed Finance, Control Theory, Quaternion and Dynamic Linking chapters, to be published separately.
Copyright © 1996, 1997, 1999, 2000, 2001, 2002, 2005, 2006, 2007 John W. Eaton.
This is the third edition of the Octave documentation, and is consistent with version 3.0.2 of Octave.
Permission is granted to make and distribute verbatim copies of this manual provided the copyright notice and this permission notice are preserved on all copies.
Permission is granted to copy and distribute modified versions of this manual under the conditions for verbatim copying, provided that the entire resulting derived work is distributed under the terms of a permission notice identical to this one.
Permission is granted to copy and distribute translations of this manual into another language, under the same conditions as for modified versions.
Portions of this document have been adapted from the gawk, readline, gcc, and C library manuals, published by the Free Software Foundation, Inc., 51 Franklin Street, Fifth Floor, Boston, MA 02110-1301–1307, USA.

# Table of Contents

Publisher's Preface .................................. 1

Preface ............................................. 3
    Acknowledgements ................................. 4
    How You Can Contribute to Octave ................. 6
    Distribution ..................................... 6

**1  A Brief Introduction to Octave ................ 9**
    1.1  Running Octave ............................... 9
    1.2  Simple Examples .............................. 9
        1.2.1  Creating a Matrix ...................... 9
        1.2.2  Matrix Arithmetic ..................... 10
        1.2.3  Solving Linear Equations .............. 10
        1.2.4  Integrating Differential Equations .... 10
        1.2.5  Producing Graphical Output ............ 11
        1.2.6  Editing What You Have Typed ........... 12
        1.2.7  Help and Documentation ................ 12
    1.3  Conventions ................................. 13
        1.3.1  Fonts ................................. 13
        1.3.2  Evaluation Notation ................... 13
        1.3.3  Printing Notation ..................... 14
        1.3.4  Error Messages ........................ 14
        1.3.5  Format of Descriptions ................ 14
            1.3.5.1  A Sample Function Description ..... 14
            1.3.5.2  A Sample Command Description ...... 15
            1.3.5.3  A Sample Variable Description ..... 15

**2  Getting Started ............................... 17**
    2.1  Invoking Octave from the Command Line ....... 17
        2.1.1  Command Line Options .................. 17
        2.1.2  Startup Files ......................... 20
    2.2  Quitting Octave ............................. 21
    2.3  Commands for Getting Help ................... 22
    2.4  Command Line Editing ........................ 23
    2.5  How Octave Reports Errors ................... 24
    2.6  Executable Octave Programs .................. 25
    2.7  Comments in Octave Programs ................. 26

## 3 Data Types ............................................. 29
### 3.1 Built-in Data Types ........................................ 29
#### 3.1.1 Numeric Objects ............................... 30
#### 3.1.2 Missing Data ................................. 30
#### 3.1.3 String Objects ................................ 30
#### 3.1.4 Data Structure Objects ......................... 31
#### 3.1.5 Cell Array Objects ............................. 31
### 3.2 Object Sizes .............................................. 31

## 4 Numeric Data Types .......................... 33
### 4.1 Matrices .................................................. 34
#### 4.1.1 Empty Matrices ................................ 37
### 4.2 Ranges ................................................... 38
### 4.3 Integer Data Types ........................................ 38
#### 4.3.1 Integer Arithmetic ............................. 41
### 4.4 Bit Manipulations ......................................... 41
### 4.5 Logical Values ............................................ 43
### 4.6 Predicates for Numeric Objects ............................ 44

## 5 Strings ............................................... 47
### 5.1 Creating Strings .......................................... 48
### 5.2 Comparing Strings ........................................ 51
### 5.3 Manipulating Strings ...................................... 52
### 5.4 String Conversions ........................................ 57
### 5.5 Character Class Functions ................................. 61

## 6 Data Containers .............................. 63
### 6.1 Data Structures ........................................... 63
#### 6.1.1 Structure Arrays ............................... 66
#### 6.1.2 Creating Structures ............................ 67
#### 6.1.3 Manipulating Structures ........................ 69
#### 6.1.4 Processing Data in Structures ................... 70
### 6.2 Cell Arrays ............................................... 71
#### 6.2.1 Creating Cell Array ............................ 72
#### 6.2.2 Indexing Cell Arrays ........................... 74
#### 6.2.3 Cell Arrays of Strings .......................... 75
#### 6.2.4 Processing Data in Cell Arrays .................. 76
### 6.3 Comma Separated Lists .................................... 78

## 7 Variables ........................................... 81
### 7.1 Global Variables .......................................... 82
### 7.2 Persistent Variables ....................................... 83
### 7.3 Status of Variables ........................................ 85
### 7.4 Summary of Built-in Variables ............................. 89
### 7.5 Defaults from the Environment ............................ 91

## 8 Expressions ... 93
- 8.1 Index Expressions ... 93
- 8.2 Calling Functions ... 95
  - 8.2.1 Call by Value ... 96
  - 8.2.2 Recursion ... 97
- 8.3 Arithmetic Operators ... 98
- 8.4 Comparison Operators ... 100
- 8.5 Boolean Expressions ... 101
  - 8.5.1 Element-by-element Boolean Operators ... 101
  - 8.5.2 Short-circuit Boolean Operators ... 102
- 8.6 Assignment Expressions ... 103
- 8.7 Increment Operators ... 106
- 8.8 Operator Precedence ... 106

## 9 Evaluation ... 109
- 9.1 Calling a Function by its Name ... 109
- 9.2 Evaluation in a Different Context ... 111

## 10 Statements ... 113
- 10.1 The `if` Statement ... 113
- 10.2 The `switch` Statement ... 115
  - 10.2.1 Notes for the C programmer ... 116
- 10.3 The `while` Statement ... 117
- 10.4 The `do-until` Statement ... 118
- 10.5 The `for` Statement ... 118
  - 10.5.1 Looping Over Structure Elements ... 120
- 10.6 The `break` Statement ... 120
- 10.7 The `continue` Statement ... 121
- 10.8 The `unwind_protect` Statement ... 122
- 10.9 The `try` Statement ... 123
- 10.10 Continuation Lines ... 123

## 11 Functions and Script Files ... 125
- 11.1 Defining Functions ... 125
- 11.2 Multiple Return Values ... 128
- 11.3 Variable-length Argument Lists ... 129
- 11.4 Variable-length Return Lists ... 131
- 11.5 Returning From a Function ... 132
- 11.6 Default Arguments ... 132
- 11.7 Function Files ... 133
  - 11.7.1 Manipulating the load path ... 134
  - 11.7.2 Subfunctions ... 136
  - 11.7.3 Overloading and Autoloading ... 136
  - 11.7.4 Function Locking ... 137
- 11.8 Script Files ... 139

11.9 Function Handles, Inline Functions, and Anonymous Functions ................................................. 140
    11.9.1 Function Handles ............................ 141
    11.9.2 Anonymous Functions ......................... 141
    11.9.3 Inline Functions ............................. 142
11.10 Commands .............................................. 143
11.11 Organization of Functions Distributed with Octave ...... 144

## 12 Errors and Warnings ....................... 147
12.1 Handling Errors ......................................... 147
    12.1.1 Raising Errors ............................... 147
    12.1.2 Catching Errors ............................. 149
12.2 Handling Warnings ...................................... 152
    12.2.1 Issuing Warnings ............................ 152
    12.2.2 Enabling and Disabling Warnings ............. 153

## 13 Debugging ................................... 157
13.1 Entering Debug Mode ................................... 157
13.2 Breakpoints ............................................. 158
13.3 Debug Mode ............................................ 159

## 14 Input and Output .......................... 161
14.1 Basic Input and Output ................................. 161
    14.1.1 Terminal Output ............................. 161
        14.1.1.1 Paging Screen Output ............... 164
    14.1.2 Terminal Input .............................. 166
    14.1.3 Simple File I/O ............................. 167
        14.1.3.1 Saving Data on Unexpected Exits ... 171
    14.1.4 Rational Approximations ..................... 172
14.2 C-Style I/O Functions ................................... 173
    14.2.1 Opening and Closing Files ................... 174
    14.2.2 Simple Output .............................. 175
    14.2.3 Line-Oriented Input ......................... 176
    14.2.4 Formatted Output ........................... 176
    14.2.5 Output Conversion for Matrices ............. 178
    14.2.6 Output Conversion Syntax ................... 178
    14.2.7 Table of Output Conversions ................ 179
    14.2.8 Integer Conversions ......................... 180
    14.2.9 Floating-Point Conversions .................. 181
    14.2.10 Other Output Conversions .................. 182
    14.2.11 Formatted Input ............................ 182
    14.2.12 Input Conversion Syntax .................... 183
    14.2.13 Table of Input Conversions ................. 184
    14.2.14 Numeric Input Conversions ................. 185
    14.2.15 String Input Conversions ................... 185
    14.2.16 Binary I/O ................................. 186

## 15 Plotting .................................................. 193
### 15.1 Plotting Basics ........................................... 193
- 15.1.1 Two-Dimensional Plots ........................ 193
- 15.1.2 Three-Dimensional Plotting .................... 207
- 15.1.3 Plot Annotations .............................. 211
- 15.1.4 Multiple Plots on One Page .................... 213
- 15.1.5 Multiple Plot Windows ......................... 213
- 15.1.6 Printing Plots ................................ 214
- 15.1.7 Test Plotting Functions ....................... 216

### 15.2 Advanced Plotting ........................................ 217
- 15.2.1 Graphics Objects .............................. 217
- 15.2.2 Graphics Object Properties .................... 222
  - 15.2.2.1 Root Figure Properties .............. 222
  - 15.2.2.2 Figure Properties ................... 222
  - 15.2.2.3 Axes Properties ..................... 223
  - 15.2.2.4 Line Properties ..................... 225
  - 15.2.2.5 Text Properties ..................... 226
  - 15.2.2.6 Image Properties .................... 229
  - 15.2.2.7 Patch Properties .................... 230
  - 15.2.2.8 Surface Properties .................. 230
- 15.2.3 Managing Default Properties ................... 230
- 15.2.4 Colors ........................................ 231
- 15.2.5 Line Styles ................................... 232
- 15.2.6 Marker Styles ................................. 232
- 15.2.7 Interaction with gnuplot ...................... 232

## 16 Matrix Manipulation ...................................... 235
### 16.1 Finding Elements and Checking Conditions ................. 235
### 16.2 Rearranging Matrices ..................................... 237
### 16.3 Applying a Function to an Array .......................... 244
### 16.4 Special Utility Matrices ................................. 244
### 16.5 Random Matrices .......................................... 246
### 16.6 Famous Matrices .......................................... 249

## 17 Arithmetic ................................................ 253
### 17.1 Utility Functions ........................................ 253
### 17.2 Complex Arithmetic ....................................... 258
### 17.3 Trigonometry ............................................. 259
### 17.4 Sums and Products ........................................ 262
### 17.5 Special Functions ........................................ 264
### 17.6 Coordinate Transformations ............................... 268
### 17.7 Mathematical Constants ................................... 269

14.2.17 Temporary Files ................................... 189
14.2.18 End of File and Errors ............................ 190
14.2.19 File Positioning .................................. 191

## 18 Linear Algebra ... 271
- 18.1 Techniques used for Linear Algebra ... 271
- 18.2 Basic Matrix Functions ... 271
- 18.3 Matrix Factorizations ... 276
- 18.4 Functions of a Matrix ... 282

## 19 Nonlinear Equations ... 283

## 20 Sparse Matrices ... 287
- 20.1 Basics ... 287
  - 20.1.1 Storage of Sparse Matrices ... 287
  - 20.1.2 Creating Sparse Matrices ... 288
  - 20.1.3 Sparse Matrix Properties ... 293
  - 20.1.4 Sparse Matrix Types ... 294
  - 20.1.5 Graphical Representations of Sparse Matrices ... 295
  - 20.1.6 Basic Operators and Functions on Sparse Matrices ... 297
    - 20.1.6.1 Sparse Functions ... 297
    - 20.1.6.2 The Return Types of Operators and Functions ... 298
    - 20.1.6.3 Mathematical Considerations ... 299
  - 20.1.7 Reordering ... 300
- 20.2 Linear Algebra on Sparse Matrices ... 308
- 20.3 Iterative Techniques applied to sparse matrices ... 312
- 20.4 Real Life Example of the use of Sparse Matrices ... 318

## 21 Numerical Integration ... 323
- 21.1 Functions of One Variable ... 323
- 21.2 Orthogonal Collocation ... 325
- 21.3 Functions of Multiple Variables ... 326

## 22 Differential Equations ... 329
- 22.1 Ordinary Differential Equations ... 329
- 22.2 Differential-Algebraic Equations ... 332

## 23 Optimization ... 339
- 23.1 Linear Programming ... 339
- 23.2 Quadratic Programming ... 345
- 23.3 Nonlinear Programming ... 346
- 23.4 Linear Least Squares ... 348

## 24 Statistics........................................ 351
- 24.1 Descriptive Statistics......................... 351
- 24.2 Basic Statistical Functions................... 355
- 24.3 Statistical Plots.............................. 358
- 24.4 Tests.......................................... 359
- 24.5 Models......................................... 367
- 24.6 Distributions.................................. 368
- 24.7 Random Number Generation ..................... 376

## 25 Sets............................................. 381
- 25.1 Set Operations................................ 381

## 26 Polynomial Manipulations.................... 385
- 26.1 Evaluating Polynomials........................ 385
- 26.2 Finding Roots................................. 386
- 26.3 Products of Polynomials....................... 387
- 26.4 Derivatives and Integrals..................... 389
- 26.5 Polynomial Interpolation...................... 390
- 26.6 Miscellaneous Functions....................... 392

## 27 Interpolation................................ 393
- 27.1 One-dimensional Interpolation................. 393
- 27.2 Multi-dimensional Interpolation............... 398

## 28 Geometry..................................... 403
- 28.1 Delaunay Triangulation........................ 403
  - 28.1.1 Plotting the Triangulation ............. 405
  - 28.1.2 Identifying points in Triangulation..... 406
- 28.2 Voronoi Diagrams.............................. 409
- 28.3 Convex Hull................................... 412
- 28.4 Interpolation on Scattered Data............... 413

## 29 Signal Processing............................ 415
- 29.1 Fast Fourier Transforms....................... 415
- 29.2 Filters and Windowing Functions............... 418

## 30 Image Processing............................. 427
- 30.1 Loading and Saving Images..................... 427
- 30.2 Displaying Images............................. 428
- 30.3 Representing Images........................... 430
- 30.4 Plotting on top of Images..................... 433
- 30.5 Color Conversion.............................. 434

## 31 Audio Processing .......................... 435
- 31.1 Audio Conversion Functions ............................. 435
- 31.2 Loading and Saving Audio Files ......................... 435

## 32 System Utilities ............................ 439
- 32.1 Timing Utilities ........................................ 439
- 32.2 Filesystem Utilities .................................... 448
- 32.3 File Archiving Utilities ................................ 454
- 32.4 Networking Utilities .................................... 456
- 32.5 Controlling Subprocesses ................................ 457
- 32.6 Process, Group, and User IDs ........................... 462
- 32.7 Environment Variables ................................... 462
- 32.8 Current Working Directory ............................... 463
- 32.9 Password Database Functions ............................. 464
- 32.10 Group Database Functions .............................. 465
- 32.11 System Information .................................... 466
- 32.12 Hashing Functions ..................................... 468

## 33 Packages .................................... 469
- 33.1 Installing and Removing Packages ....................... 469
- 33.2 Using Packages .......................................... 470
- 33.3 Administrating Packages ................................. 470
- 33.4 Creating Packages ....................................... 471
  - 33.4.1 The DESCRIPTION File ..................... 472
  - 33.4.2 The INDEX file ........................... 474
  - 33.4.3 PKG_ADD and PKG_DEL directives ........... 475

## Appendix A   Command Line Editing ........... 477
- A.0.1 Cursor Motion ................................ 477
- A.0.2 Killing and Yanking .......................... 478
- A.0.3 Commands For Changing Text ................... 479
- A.0.4 Letting Readline Type For You ................ 479
- A.0.5 Commands For Manipulating The History ....... 480
- A.0.6 Customizing readline ......................... 482
- A.0.7 Customizing the Prompt ....................... 483
- A.0.8 Diary and Echo Commands ...................... 484

## Appendix B   Test and Demo Functions ......... 487
- B.1 Test Functions ........................................... 487
- B.2 Demonstration Functions .................................. 492

## Appendix C     Tips and Standards ............... 495
    C.1   Writing Clean Octave Programs........................... 495
    C.2   Tips for Making Code Run Faster........................ 495
    C.3   Tips on Writing Comments............................... 496
    C.4   Conventional Headers for Octave Functions............... 497
    C.5   Tips for Documentation Strings.......................... 499

## Appendix D     Known Causes of Trouble ......... 505
    D.1   Actual Bugs We Haven't Fixed Yet....................... 505
    D.2   Reporting Bugs......................................... 505
    D.3   Have You Found a Bug?................................. 506
    D.4   Where to Report Bugs.................................. 506
    D.5   How to Report Bugs.................................... 507
    D.6   Sending Patches for Octave ............................. 509
    D.7   How To Get Help with Octave .......................... 510

## Appendix E     Installing Octave .................. 511
    E.1   Installation Problems ................................... 514

## Appendix F     Emacs Octave Support ........... 519
    F.1   Installing EOS.......................................... 519
    F.2   Using Octave Mode..................................... 519
    F.3   Running Octave From Within Emacs..................... 524
    F.4   Using the Emacs Info Reader for Octave.................. 526

## Appendix G     GNU GENERAL PUBLIC LICENSE
    ................................................ 527

## Books from the publisher ....................... 537

## Index............................................. 539

# Publisher's Preface

This manual documents the use of GNU Octave, an interactive environment for numerical computation.

GNU Octave is *free software*. The term "free software" has nothing to do with price—it is about freedom. It refers to your freedom to run, copy, distribute, study, change and improve the software. With GNU Octave you have all these freedoms.

GNU Octave is part of the GNU Project. The GNU Project was launched in 1984 to develop a complete Unix-like operating system which is free software: the GNU system. It was conceived as a way of bringing back the cooperative spirit that prevailed in the computing community in earlier days, by removing the obstacles to cooperation imposed by the owners of proprietary software.

The Free Software Foundation is a tax-exempt charity that raises funds for work on the GNU Project and is dedicated to promoting the freedom to modify and redistribute computer programs. You can support the GNU Project by becoming an associate member of the Free Software Foundation and paying regular membership dues. For more information, visit the website www.fsf.org.

Brian Gough<br>Publisher<br>August 2008

# Preface

Octave was originally intended to be companion software for an undergraduate-level textbook on chemical reactor design being written by James B. Rawlings of the University of Wisconsin-Madison and John G. Ekerdt of the University of Texas.

Clearly, Octave is now much more than just another 'courseware' package with limited utility beyond the classroom. Although our initial goals were somewhat vague, we knew that we wanted to create something that would enable students to solve realistic problems, and that they could use for many things other than chemical reactor design problems.

There are those who would say that we should be teaching the students Fortran instead, because that is the computer language of engineering, but every time we have tried that, the students have spent far too much time trying to figure out why their Fortran code crashes and not enough time learning about chemical engineering. With Octave, most students pick up the basics quickly, and are using it confidently in just a few hours.

Although it was originally intended to be used to teach reactor design, it has been used in several other undergraduate and graduate courses in the Chemical Engineering Department at the University of Texas, and the math department at the University of Texas has been using it for teaching differential equations and linear algebra as well. If you find it useful, please let us know. We are always interested to find out how Octave is being used in other places.

Virtually everyone thinks that the name Octave has something to do with music, but it is actually the name of a former professor of mine who wrote a famous textbook on chemical reaction engineering, and who was also well known for his ability to do quick 'back of the envelope' calculations. We hope that this software will make it possible for many people to do more ambitious computations just as easily.

Everyone is encouraged to share this software with others under the terms of the GNU General Public License (see Appendix G [Copying], page 527) as described at the beginning of this manual. You are also encouraged to help make Octave more useful by writing and contributing additional functions for it, and by reporting any problems you may have.

## Acknowledgements

Many people have already contributed to Octave's development. The following people have helped write parts of Octave or helped out in various other ways (listed alphabetically).

Ben Abbott
Muthiah Annamalai
Ben Barrowes
Heinz Bauschke
Don Bindner
Richard Bovey
Marco Caliari
Jean-Francois Cardoso
David Castelow
Albert Chin-A-Young
Michael Creel
Jorge Barros de Abreu
David M. Doolin
Dirk Eddelbuettel
Peter Ekberg
Ramon Garcia Fernandez
Castor Fu
Klaus Gebhardt
Michael Goffioul
Keith Goodman
Kai Habel
Kim Hansen
Yozo Hida
A. Scottedward Hodel
David Hoover
Cyril Humbert
Geoff Jacobsen
Steven G. Johnson
Mohamed Kamoun
Joel Keay
Aaron A. King
Heine Kolltveit
Piotr Krzyzanowski
Miroslaw Kwasniak
Claude Lacoursiere
Dirk Laurie
Timo Lindfors
David Livings
Emil Lucretiu
Jens-Uwe Mager
Makoto Matsumoto
G. D. McBain
Antoine Moreau

Andy Adler
Shai Ayal
Alexander Barth
Karl Berry
Jakub Bogusz
Marcus Brinkmann
Daniel Calvelo
Joao Cardoso
Vincent Cautaerts
J. D. Cole
Jeff Cunningham
Philippe Defert
Pascal A. Dupuis
Paul Eggert
Rolf Fabian
Torsten Finke
Eduardo Gallestey
Driss Ghaddab
Glenn Golden
Etienne Grossmann
William Hadisoeseno
Søren Hauberg
Ryan Hinton
Richard A. Holcombe
Kurt Hornik
Teemu Ikonen
Mats Jansson
Heikki Junes
Lute Kamstra
Mumit Khan
Arno J. Klaassen
Ken Kouno
Volker Kuhlmann
Rafael Laboissiere
Walter Landry
Maurice LeBrun
Benjamin Lindner
Erik de Castro Lopo
Hoxide Ma
Ricardo Marranita
Tatsuro Matsuoka
Christoph Mayer
Kai P. Mueller

Joel Andersson
Roger Banks
David Bateman
David Billinghurst
Moritz Borgmann
Remy Bruno
John C. Campbell
Larrie Carr
Clinton Chee
Martin Costabel
Martin Dalecki
Bill Denney
John W. Eaton
Stephen Eglen
Stephen Fegan
J.D. Munoz Frias
Walter Gautschi
Nicolo Giorgetti
Tomislav Goles
Peter Gustafson
Benjamin Hall
Daniel Heiserer
Roman Hodek
Tom Holroyd
Christopher Hulbert
Alan W. Irwin
Cai Jianming
Atsushi Kajita
Thomas Kasper
Paul Kienzle
Geoffrey Knauth
Oyvind Kristiansen
Tetsuro Kurita
Kai Labusch
Bill Lash
Friedrich Leisch
Ross Lippert
Massimo Lorenzin
James Macnicol
Orestes Mas
Laurent Mazet
Stefan Monnier
Victor Munoz

Carmen Navarrete
Rick Niles
Michael O'Brien
Luis F. Ortiz
Sylvain Pelissier
Jim Peterson
Hans Ekkehard Plesser
Ondrej Popp
James B. Rawlings
Michael Reifenberger
Matthew W. Roberts
Kevin Ruland
Juhani Saastamoinen
Michel D. Schmid
Ludwig Schwardt
Baylis Shanks
Julius Smith
Quentin H. Spencer
Russell Standish
John Swensen
Duncan Temple Lang
Utkarsh Upadhyay
James R. Van Zandt
Thomas Walter
Bob Weigel
Fook Fah Yap
Alex Zvoleff

Todd Neal
Takuji Nishimura
Thorsten Ohl
Scott Pakin
Per Persson
Danilo Piazzalunga
Tom Poage
Jef Poskanzer
Eric S. Raymond
Jason Riedy
Andrew Ross
Olli Saarela
Ben Sapp
Nicol N. Schraudolph
Daniel J. Sebald
Joseph P. Skudlarek
Shan G. Smith
Christoph Spiel
Doug Stewart
Ariel Tankus
Olaf Till
Stefan van der Walt
Gregory Vanuxem
Olaf Weber
Andreas Weingessel
Michael Zeising

Al Niessner
Eric Norum
Arno Onken
Gabriele Pannocchia
Primozz Peterlin
Nicholas Piper
Orion Poplawski
Francesco Potorti
Balint Reczey
Petter Risholm
Mark van Rossum
Toni Saarela
Alois Schloegl
Sebastian Schubert
Dmitri A. Sergatskov
John Smith
Joerg Specht
Richard Stallman
Thomas Stuart
Georg Thimm
Thomas Treichl
Peter Van Wieren
Ivana Varekova
Thomas Weber
Michael Weitzel
Federico Zenith

Special thanks to the following people and organizations for supporting the development of Octave:

- The United States Department of Energy, through grant number DE-FG02-04ER25635.
- Ashok Krishnamurthy, David Hudak, Juan Carlos Chaves, and Stanley C. Ahalt of the Ohio Supercomputer Center.
- The National Science Foundation, through grant numbers CTS-0105360, CTS-9708497, CTS-9311420, CTS-8957123, and CNS-0540147.
- The industrial members of the Texas-Wisconsin Modeling and Control Consortium (TWMCC (http://www.che.utexas.edu/twmcc)).
- The Paul A. Elfers Endowed Chair in Chemical Engineering at the University of Wisconsin-Madison.
- Digital Equipment Corporation, for an equipment grant as part of their External Research Program.
- Sun Microsystems, Inc., for an Academic Equipment grant.
- International Business Machines, Inc., for providing equipment as part of a grant to the University of Texas College of Engineering.
- Texaco Chemical Company, for providing funding to continue the development of this software.

- The University of Texas College of Engineering, for providing a Challenge for Excellence Research Supplement, and for providing an Academic Development Funds grant.
- The State of Texas, for providing funding through the Texas Advanced Technology Program under Grant No. 003658-078.
- Noel Bell, Senior Engineer, Texaco Chemical Company, Austin Texas.
- John A. Turner, Group Leader, Continuum Dynamics (CCS-2), Los Alamos National Laboratory, for registering the octave.org domain name.
- James B. Rawlings, Professor, University of Wisconsin-Madison, Department of Chemical and Biological Engineering.
- Richard Stallman, for writing GNU.

This project would not have been possible without the GNU software used in and used to produce Octave.

## How You Can Contribute to Octave

There are a number of ways that you can contribute to help make Octave a better system. Perhaps the most important way to contribute is to write high-quality code for solving new problems, and to make your code freely available for others to use.

If you find Octave useful, consider providing additional funding to continue its development. Even a modest amount of additional funding could make a significant difference in the amount of time that is available for development and support.

If you cannot provide funding or contribute code, you can still help make Octave better and more reliable by reporting any bugs you find and by offering suggestions for ways to improve Octave. See Appendix D [Trouble], page 505, for tips on how to write useful bug reports.

## Distribution

Octave is *free* software. This means that everyone is free to use it and free to redistribute it on certain conditions. Octave is not in the public domain. It is copyrighted and there are restrictions on its distribution, but the restrictions are designed to ensure that others will have the same freedom to use and redistribute Octave that you have. The precise conditions can be found in the GNU General Public License that comes with Octave and that also appears in Appendix G [Copying], page 527.

Octave is available on CD-ROM with various collections of other free software, and from the Free Software Foundation. Ordering a copy of Octave from the Free Software Foundation helps to fund the development of more free software. For more information, write to

Free Software Foundation
51 Franklin Street, Fifth Floor
Boston, MA 02110-1301–1307
USA

Octave can also be downloaded from http://www.octave.org, where additional information also is available.

# 1 A Brief Introduction to Octave

This manual documents how to install, run, and use GNU Octave, and how to report bugs.

GNU Octave is a high-level language, primarily intended for numerical computations. It provides a convenient command line interface for solving linear and nonlinear problems numerically, and for performing other numerical experiments. It may also be used as a batch-oriented language.

GNU Octave is also freely redistributable software. You may redistribute it and/or modify it under the terms of the GNU General Public License as published by the Free Software Foundation. The GPL is included in this manual in Appendix G [Copying], page 527.

This document corresponds to Octave version 3.0.2.

## 1.1 Running Octave

On most systems, the way to invoke Octave is with the shell command `octave`. Octave displays an initial message and then a prompt indicating it is ready to accept input. You can begin typing Octave commands immediately afterward.

If you get into trouble, you can usually interrupt Octave by typing `Control-C` (usually written `C-c` for short). `C-c` gets its name from the fact that you type it by holding down CTRL and then pressing c. Doing this will normally return you to Octave's prompt.

To exit Octave, type `quit`, or `exit` at the Octave prompt.

On systems that support job control, you can suspend Octave by sending it a SIGTSTP signal, usually by typing `C-z`.

## 1.2 Simple Examples

The following chapters describe all of Octave's features in detail, but before doing that, it might be helpful to give a sampling of some of its capabilities.

If you are new to Octave, I recommend that you try these examples to begin learning Octave by using it. Lines marked with `octave:13>` are lines you type, ending each with a carriage return. Octave will respond with an answer, or by displaying a graph.

### 1.2.1 Creating a Matrix

To create a new matrix and store it in a variable so that it you can refer to it later, type the command

    octave:1> A = [ 1, 1, 2; 3, 5, 8; 13, 21, 34 ]

Octave will respond by printing the matrix in neatly aligned columns. Ending a command with a semicolon tells Octave to not print the result of a command. For example

```
octave:2> B = rand (3, 2);
```
will create a 3 row, 2 column matrix with each element set to a random value between zero and one.

To display the value of any variable, simply type the name of the variable. For example, to display the value stored in the matrix B, type the command
```
octave:3> B
```

### 1.2.2 Matrix Arithmetic

Octave has a convenient operator notation for performing matrix arithmetic. For example, to multiply the matrix A by a scalar value, type the command
```
octave:4> 2 * A
```
To multiply the two matrices A and B, type the command
```
octave:5> A * B
```
and to form the matrix product $A^T A$, type the command
```
octave:6> A' * A
```

### 1.2.3 Solving Linear Equations

To solve the set of linear equations Ax = b, use the left division operator, \:
```
octave:7> A \ b
```
This is conceptually equivalent to $A^{-1}b$, but avoids computing the inverse of a matrix directly.

If the coefficient matrix is singular, Octave will print a warning message and compute a minimum norm solution.

### 1.2.4 Integrating Differential Equations

Octave has built-in functions for solving nonlinear differential equations of the form

$$\frac{dx}{dt} = f(x, t), \qquad x(t = t_0) = x_0$$

For Octave to integrate equations of this form, you must first provide a definition of the function $f(x, t)$. This is straightforward, and may be accomplished by entering the function body directly on the command line. For example, the following commands define the right hand side function for an interesting pair of nonlinear differential equations. Note that while you are entering a function, Octave responds with a different prompt, to indicate that it is waiting for you to complete your input.

# Chapter 1: A Brief Introduction to Octave

```
octave:8> function xdot = f (x, t)
>
>   r = 0.25;
>   k = 1.4;
>   a = 1.5;
>   b = 0.16;
>   c = 0.9;
>   d = 0.8;
>
>   xdot(1) = r*x(1)*(1 - x(1)/k) - a*x(1)*x(2)/(1 + b*x(1));
>   xdot(2) = c*a*x(1)*x(2)/(1 + b*x(1)) - d*x(2);
>
> endfunction
```

Given the initial condition

```
x0 = [1; 2];
```

and the set of output times as a column vector (note that the first output time corresponds to the initial condition given above)

```
t = linspace (0, 50, 200)';
```

it is easy to integrate the set of differential equations:

```
x = lsode ("f", x0, t);
```

The function `lsode` uses the Livermore Solver for Ordinary Differential Equations, described in A. C. Hindmarsh, *ODEPACK, a Systematized Collection of ODE Solvers*, in: Scientific Computing, R. S. Stepleman et al. (Eds.), North-Holland, Amsterdam, 1983, pages 55–64.

## 1.2.5 Producing Graphical Output

To display the solution of the previous example graphically, use the command

```
plot (t, x)
```

If you are using a graphical user interface, Octave will automatically create a separate window to display the plot.

To save a plot once it has been displayed on the screen, use the print command. For example,

```
print -deps foo.eps
```

will create a file called 'foo.eps' that contains a rendering of the current plot.

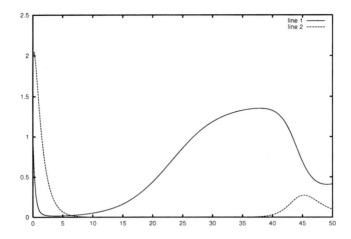

The command

    help print

explains more options for the print command and provides a list of additional output file formats.

### 1.2.6 Editing What You Have Typed

At the Octave prompt, you can recall, edit, and reissue previous commands using Emacs- or vi-style editing commands. The default keybindings use Emacs-style commands. For example, to recall the previous command, press Control-p (usually written C-p for short). Doing this will normally bring back the previous line of input. C-n will bring up the next line of input, C-b will move the cursor backward on the line, C-f will move the cursor forward on the line, etc.

A complete description of the command line editing capability is given in this manual in Section 2.4 [Command Line Editing], page 23.

### 1.2.7 Help and Documentation

Octave has an extensive help facility. The same documentation that is available in printed form is also available from the Octave prompt, because both forms of the documentation are created from the same input file.

In order to get good help you first need to know the name of the command that you want to use. This name of the function may not always be obvious, but a good place to start is to just type help. This will show you all the operators, reserved words, functions, built-in variables, and function files. An alternative is to search the documentation using the lookfor function. This function is described in Section 2.3 [Getting Help], page 22.

Once you know the name of the function you wish to use, you can get more help on the function by simply including the name as an argument to help. For example,

```
help plot
```
will display the help text for the `plot` function.

Octave sends output that is too long to fit on one screen through a pager like `less` or `more`. Type a ⟨RET⟩ to advance one line, a ⟨SPC⟩ to advance one page, and ⟨q⟩ to exit the pager.

The part of Octave's help facility that allows you to read the complete text of the printed manual from within Octave normally uses a separate program called Info. When you invoke Info you will be put into a menu driven program that contains the entire Octave manual. Help for using Info is provided in this manual in Section 2.3 [Getting Help], page 22.

## 1.3 Conventions

This section explains the notational conventions that are used in this manual. You may want to skip this section and refer back to it later.

### 1.3.1 Fonts

Examples of Octave code appear in this font or form: `svd (a)`. Names that represent arguments or metasyntactic variables appear in this font or form: *first-number*. Commands that you type at the shell prompt sometimes appear in this font or form: `octave --no-init-file`. Commands that you type at the Octave prompt sometimes appear in this font or form: *foo --bar --baz*. Specific keys on your keyboard appear in this font or form: ⟨ANY⟩.

### 1.3.2 Evaluation Notation

In the examples in this manual, results from expressions that you evaluate are indicated with ⇒. For example,

```
sqrt (2)
     ⇒ 1.4142
```

You can read this as "`sqrt (2)` evaluates to 1.4142".

In some cases, matrix values that are returned by expressions are displayed like this

```
[1, 2; 3, 4] == [1, 3; 2, 4]
     ⇒ [ 1, 0; 0, 1 ]
```

and in other cases, they are displayed like this

```
eye (3)
     ⇒  1  0  0
        0  1  0
        0  0  1
```

in order to clearly show the structure of the result.

Sometimes to help describe one expression, another expression is shown that produces identical results. The exact equivalence of expressions is indicated with ≡. For example,

```
rot90 ([1, 2; 3, 4], -1)
≡
rot90 ([1, 2; 3, 4], 3)
≡
rot90 ([1, 2; 3, 4], 7)
```

### 1.3.3 Printing Notation

Many of the examples in this manual print text when they are evaluated. Examples in this manual indicate printed text with ⊣. The value that is returned by evaluating the expression (here 1) is displayed with ⇒ and follows on a separate line.

```
printf ("foo %s\n", "bar")
    ⊣ foo bar
    ⇒ 1
```

### 1.3.4 Error Messages

Some examples signal errors. This normally displays an error message on your terminal. Error messages are shown on a line starting with `error:`.

```
struct_elements ([1, 2; 3, 4])
error: struct_elements: wrong type argument 'matrix'
```

### 1.3.5 Format of Descriptions

Functions, commands, and variables are described in this manual in a uniform format. The first line of a description contains the name of the item followed by its arguments, if any. The category—function, variable, or whatever—is printed next to the right margin. The description follows on succeeding lines, sometimes with examples.

#### 1.3.5.1 A Sample Function Description

In a function description, the name of the function being described appears first. It is followed on the same line by a list of parameters. The names used for the parameters are also used in the body of the description.

Here is a description of an imaginary function foo:

foo ($x$, $y$, ...)                                                          Function

The function foo subtracts $x$ from $y$, then adds the remaining arguments to the result. If $y$ is not supplied, then the number 19 is used by default.

```
foo (1, [3, 5], 3, 9)
    ⇒ [ 14, 16 ]
foo (5)
    ⇒ 14
```

More generally,

```
foo (w, x, y, ...)
≡
x - w + y + ...
```

Chapter 1: A Brief Introduction to Octave                                    15

Any parameter whose name contains the name of a type (e.g. *integer*, *integer1* or *matrix*) is expected to be of that type. Parameters named *object* may be of any type. Parameters with other sorts of names (e.g. *new_file*) are discussed specifically in the description of the function. In some sections, features common to parameters of several functions are described at the beginning.

Functions in Octave may be defined in several different ways. The category name for functions may include another name that indicates the way that the function is defined. These additional tags include

Function File
: The function described is defined using Octave commands stored in a text file. See Section 11.7 [Function Files], page 133.

Built-in Function
: The function described is written in a language like C++, C, or Fortran, and is part of the compiled Octave binary.

Loadable Function
: The function described is written in a language like C++, C, or Fortran. On systems that support dynamic linking of user-supplied functions, it may be automatically linked while Octave is running, but only if it is needed.

Mapping Function
: The function described works element-by-element for matrix and vector arguments.

### 1.3.5.2 A Sample Command Description

Command descriptions have a format similar to function descriptions, except that the word 'Function' is replaced by 'Command. Commands are functions that may be called without surrounding their arguments in parentheses. For example, here is the description for Octave's `cd` command:

`cd dir`                                                                        Command
`chdir dir`                                                                     Command
: Change the current working directory to *dir*. For example, `cd ~/octave` changes the current working directory to '`~/octave`'. If the directory does not exist, an error message is printed and the working directory is not changed.

### 1.3.5.3 A Sample Variable Description

A *variable* is a name that can hold a value. Although any variable can be set by the user, *built-in variables* typically exist specifically so that users can change them to alter the way Octave behaves (built-in variables are also sometimes called *user options*). Ordinary variables and built-in variables are described using a format like that for functions except that there are no arguments.

Here is a description of the imaginary variable `do_what_i_mean_not_what_i_say`.

`do_what_i_mean_not_what_i_say`                                    Built-in Variable
  If the value of this variable is nonzero, Octave will do what you actually wanted, even if you have typed a completely different and meaningless list of commands.

Other variable descriptions have the same format, but 'Built-in Variable' is replaced by 'Variable', for ordinary variables, or 'Constant' for symbolic constants whose values cannot be changed.

# 2 Getting Started

This chapter explains some of Octave's basic features, including how to start an Octave session, get help at the command prompt, edit the command line, and write Octave programs that can be executed as commands from your shell.

## 2.1 Invoking Octave from the Command Line

Normally, Octave is used interactively by running the program `octave` without any arguments. Once started, Octave reads commands from the terminal until you tell it to exit.

You can also specify the name of a file on the command line, and Octave will read and execute the commands from the named file and then exit when it is finished.

You can further control how Octave starts by using the command-line options described in the next section, and Octave itself can remind you of the options available. Type `octave --help` to display all available options and briefly describe their use (`octave -h` is a shorter equivalent).

### 2.1.1 Command Line Options

Here is a complete list of all the command line options that Octave accepts.

`--debug`
`-d`   Enter parser debugging mode. Using this option will cause Octave's parser to print a lot of information about the commands it reads, and is probably only useful if you are actually trying to debug the parser.

`--echo-commands`
`-x`   Echo commands as they are executed.

`--eval` *code*
   Evaluate *code* and exit when done unless `--persist` is also specified.

`--exec-path` *path*
   Specify the path to search for programs to run. The value of *path* specified on the command line will override any value of `OCTAVE_EXEC_PATH` found in the environment, but not any commands in the system or user startup files that set the built-in variable `EXEC_PATH`.

`--help`
`-h`
`-?`   Print short help message and exit.

`--image-path` *path*
   Specify the path to search for images. The value of *path* specified on the command line will set the value of `IMAGE_PATH` found in the environment.

`--info-file` *filename*
: Specify the name of the info file to use. The value of *filename* specified on the command line will override any value of `OCTAVE_INFO_FILE` found in the environment, but not any commands in the system or user startup files that use the `info_file` function.

`--info-program` *program*
: Specify the name of the info program to use. The value of *program* specified on the command line will override any value of `OCTAVE_INFO_PROGRAM` found in the environment, but not any commands in the system or user startup files that use the `info_program` function.

`--interactive`
`-i`
: Force interactive behavior. This can be useful for running Octave via a remote shell command or inside an Emacs shell buffer. For another way to run Octave within Emacs, see Appendix F [Emacs], page 519.

`--no-history`
`-H` Disable command-line history.

`--no-init-file`
: Don't read the '~/.octaverc' or '.octaverc' files.

`--no-line-editing`
: Disable command-line editing.

`--no-site-file`
: Don't read the site-wide 'octaverc' file.

`--norc`
`-f` Don't read any of the system or user initialization files at startup. This is equivalent to using both of the options `--no-init-file` and `--no-site-file`.

`--path` *path*
`-p` *path*
: Specify the path to search for function files. The value of *path* specified on the command line will override any value of `OCTAVE_PATH` found in the environment, but not any commands in the system or user startup files that set the internal load path through one of the path functions.

`--persist`
: Go to interactive mode after `--eval` or reading from a file named on the command line.

`--silent`
`--quiet`
`-q` Don't print the usual greeting and version message at startup.

`--traditional`
`--braindead`
: For compatibility with MATLAB, set initial values for user-preferences to the following values

## Chapter 2: Getting Started

```
        PS1                       = ">> "
        PS2                       = ""
        beep_on_error             = true
        crash_dumps_octave_core   = false
        default_save_options      = "-mat-binary"
        fixed_point_format        = true
        history_timestamp_format_string
                                  = "%%-- %D %I:%M %p --%%"
        page_screen_output        = false
        print_empty_dimensions    = false
```
and disable the following warnings

```
    Octave:fopen-file-in-path
    Octave:function-name-clash
    Octave:load-file-in-path
```

`--verbose`
`-V` Turn on verbose output.

`--version`
`-v` Print the program version number and exit.

`file`
Execute commands from *file*. Exit when done unless `--persist` is also specified.

Octave also includes several built-in variables that contain information about the command line, including the number of arguments and all of the options.

**argv ()**                                                                 *Built-in Function*
Return the command line arguments passed to Octave. For example, if you invoked Octave using the command

```
    octave --no-line-editing --silent
```

argv would return a cell array of strings with the elements `--no-line-editing` and `--silent`.

If you write an executable Octave script, `argv` will return the list of arguments passed to the script. See Section 2.6 [Executable Octave Programs], page 25, for an example of how to create an executable Octave script.

**program_name ()**                                                         *Built-in Function*
Return the last component of the value returned by `program_invocation_name`.

See also: program_invocation_name.

`program_invocation_name ()`                              Built-in Function
   Return the name that was typed at the shell prompt to run Octave.
   If executing a script from the command line (e.g. `octave foo.m`) or using an
   executable Octave script, the program name is set to the name of the script.
   See Section 2.6 [Executable Octave Programs], page 25, for an example of
   how to create an executable Octave script.

   **See also:** program_name.

   Here is an example of using these functions to reproduce Octave's command
line.
```
printf ("%s", program_name ());
arg_list = argv ();
for i = 1:nargin
  printf (" %s", arg_list{i});
endfor
printf ("\n");
```
See Section 8.1 [Index Expressions], page 93, for an explanation of how to properly index arrays of strings and substrings in Octave, and See Section 11.1 [Defining Functions], page 125, for information about the variable `nargin`.

### 2.1.2 Startup Files

When Octave starts, it looks for commands to execute from the files in the following list. These files may contain any valid Octave commands, including function definitions.

*octave-home*/share/octave/site/m/startup/octaverc
   Where *octave-home* is the directory in which all of Octave is installed (the default is '/usr/local'). This file is provided so that changes to the default Octave environment can be made globally for all users at your site for all versions of Octave you have installed. Some care should be taken when making changes to this file, since all users of Octave at your site will be affected.

*octave-home*/share/octave/*version*/m/startup/octaverc
   Where *octave-home* is the directory in which all of Octave is installed (the default is '/usr/local'), and *version* is the version number of Octave. This file is provided so that changes to the default Octave environment can be made globally for all users for a particular version of Octave. Some care should be taken when making changes to this file, since all users of Octave at your site will be affected.

~/.octaverc
   This file is normally used to make personal changes to the default Octave environment.

.octaverc
> This file can be used to make changes to the default Octave environment for a particular project. Octave searches for this file in the current directory after it reads '~/.octaverc'. Any use of the cd command in the '~/.octaverc' file will affect the directory that Octave searches for the file '.octaverc'.
>
> If you start Octave in your home directory, commands from the file '~/.octaverc' will only be executed once.

A message will be displayed as each of the startup files is read if you invoke Octave with the --verbose option but without the --silent option.

## 2.2 Quitting Octave

exit (*status*)  Built-in Function
quit (*status*)  Built-in Function
> Exit the current Octave session. If the optional integer value *status* is supplied, pass that value to the operating system as the Octave's exit status. The default value is zero.

atexit (*fcn*)  Built-in Function
> Register a function to be called when Octave exits. For example,
>
>     function bye_bye ()
>       disp ("Bye bye");
>     endfunction
>     atexit ("bye_bye");
>
> will print the message "Bye bye" when Octave exits.

atexit (*fcn, flag*)  Built-in Function
> Register or unregister a function to be called when Octave exits, depending on *flag*. If *flag* is true, the function is registered, if *flag* is false, it is unregistered. For example, after registering the function bye_bye as above,
>
>     atexit ("bye_bye", false);
>
> will remove the function from the list and Octave will not call the function bye_by when it exits.
>
> Note that atexit only removes the first occurrence of a function from the list, so if a function was placed in the list multiple times with atexit, it must also be removed from the list multiple times.

## 2.3 Commands for Getting Help

The entire text of this manual is available from the Octave prompt via the command `doc`. In addition, the documentation for individual user-written functions and variables is also available via the `help` command. This section describes the commands used for reading the manual and the documentation strings for user-supplied functions and variables. See Section 11.7 [Function Files], page 133, for more information about how to document the functions you write.

**help** *name*                                                                 Command

    Display the help text for *name*. If invoked without any arguments, `help` prints a list of all the available operators and functions.

    For example, the command `help help` prints a short message describing the help command.

    The help command can give you information about operators, but not the comma and semicolons that are used as command separators. To get help for those, you must type `help comma` or `help semicolon`.

    See also: doc, which, lookfor.

**doc** *function_name*                                                         Command

    Display documentation for the function *function_name* directly from an on-line version of the printed manual, using the GNU Info browser. If invoked without any arguments, the manual is shown from the beginning.

    For example, the command `doc rand` starts the GNU Info browser at this node in the on-line version of the manual.

    Once the GNU Info browser is running, help for using it is available using the command `C-h`.

    See also: help.

**lookfor** *str*                                                               Command
**lookfor -all** *str*                                                          Command
[*fun, helpstring*] = lookfor (*str*)                                           Function
[*fun, helpstring*] = lookfor ('-all', *str*)                                   Function

    Search for the string *str* in all of the functions found in the function search path. By default `lookfor` searches for *str* in the first sentence of the help string of each function found. The entire help string of each function found in the path can be searched if the '-all' argument is supplied. All searches are case insensitive.

    Called with no output arguments, `lookfor` prints the list of matching functions to the terminal. Otherwise the output arguments *fun* and *helpstring* define the matching functions and the first sentence of each of their help strings.

    Note that the ability of `lookfor` to correctly identify the first sentence of the help of the functions is dependent on the format of the functions help. All of the functions in Octave itself will correctly find the first sentence, but the same cannot be guaranteed for other functions. Therefore the use of the '-all' argument might be necessary to find related functions that are not part of Octave.

See also: help, which.

The following function can be used to change which programs are used for displaying the documentation, and where the documentation can be found.

`val = info_file ()`                                              Built-in Function
`old_val = info_file (new_val)`                  Built-in Function
    Query or set the internal variable that specifies the name of the Octave info file. The default value is `"octave-home/info/octave.info"`, in which *octave-home* is the directory where all of Octave is installed.

See also: info_program, doc, help, makeinfo_program.

`val = info_program ()`                                Built-in Function
`old_val = info_program (new_val)`             Built-in Function
    Query or set the internal variable that specifies the name of the info program to run. The default value is `"octave-home/libexec/octave/version/ exec/arch/info"` in which *octave-home* is the directory where all of Octave is installed, *version* is the Octave version number, and *arch* is the system type (for example, i686-pc-linux-gnu). The default initial value may be overridden by the environment variable `OCTAVE_INFO_PROGRAM`, or the command line argument `--info-program NAME`.

See also: info_file, doc, help, makeinfo_program.

`val = makeinfo_program ()`                        Built-in Function
`old_val = makeinfo_program (new_val)`       Built-in Function
    Query or set the internal variable that specifies the name of the makeinfo program that Octave runs to format help text containing Texinfo markup commands. The default initial value is `"makeinfo"`.

See also: info_file, info_program, doc, help.

`val = suppress_verbose_help_message ()`    Built-in Function
`old_val = suppress_verbose_help_message (new_val)`   Built-in Function
    Query or set the internal variable that controls whether Octave will add additional help information to the end of the output from the `help` command and usage messages for built-in commands.

## 2.4 Command Line Editing

Octave uses the standard GNU readline library to provide an extensive set of command-line editing and history features. These are describe in Appendix A [Command Line Editing and History], page 477.

## 2.5 How Octave Reports Errors

Octave reports two kinds of errors for invalid programs.

A *parse error* occurs if Octave cannot understand something you have typed. For example, if you misspell a keyword,

```
octave:13> functon y = f (x) y = x^2; endfunction
```

Octave will respond immediately with a message like this:

```
parse error:

  functon y = f (x) y = x^2; endfunction
          ^
```

For most parse errors, Octave uses a caret (^) to mark the point on the line where it was unable to make sense of your input. In this case, Octave generated an error message because the keyword `function` was misspelled. Instead of seeing `function f`, Octave saw two consecutive variable names, which is invalid in this context. It marked the error at y because the first name by itself was accepted as valid input.

Another class of error message occurs at evaluation time. These errors are called *run-time errors*, or sometimes *evaluation errors* because they occur when your program is being *run*, or *evaluated*. For example, if after correcting the mistake in the previous function definition, you type

```
octave:13> f ()
```

Octave will respond with

```
error: 'x' undefined near line 1 column 24
error: evaluating expression near line 1, column 24
error: evaluating assignment expression near line 1, column 22
error: called from 'f'
```

This error message has several parts, and gives you quite a bit of information to help you locate the source of the error. The messages are generated from the point of the innermost error, and provide a traceback of enclosing expressions and function calls.

In the example above, the first line indicates that a variable named x was found to be undefined near line 1 and column 24 of some function or expression. For errors occurring within functions, lines are counted from the beginning of the file containing the function definition. For errors occurring at the top level, the line number indicates the input line number, which is usually displayed in the prompt string.

The second and third lines in the example indicate that the error occurred within an assignment expression, and the last line of the error message indicates that the error occurred within the function f. If the function f had been called from another function, for example, g, the list of errors would have ended with one more line:

```
error: called from 'g'
```

These lists of function calls usually make it fairly easy to trace the path your program took before the error occurred, and to correct the error before trying again.

## 2.6 Executable Octave Programs

Once you have learned Octave, you may want to write self-contained Octave scripts, using the #! script mechanism. You can do this on GNU systems and on many Unix systems[1].

Self-contained Octave scripts are useful when you want to write a program which users can invoke without knowing that the program is written in the Octave language.

For example, you could create a text file named 'hello', containing the following lines:

    #! octave-interpreter-name -qf
    # a sample Octave program
    printf ("Hello, world!\n");

(where *octave-interpreter-name* should be replaced with the full file name for your Octave binary). Note that this will only work if #! appears at the very beginning of the file. After making this file executable (with the chmod command), you can simply type:

    hello

at the shell, and the system will arrange to run Octave as if you had typed:

    octave hello

The line beginning with #! lists the full file name of an interpreter to be run, and an optional initial command line argument to pass to that interpreter. The operating system then runs the interpreter with the given argument and the full argument list of the executed program. The first argument in the list is the full file name of the Octave program. The rest of the argument list will either be options to Octave, or data files, or both. The -qf option is usually specified in stand-alone Octave programs to prevent them from printing the normal startup message, and to keep them from behaving differently depending on the contents of a particular user's '~/.octaverc' file. See Section 2.1 [Invoking Octave from the Command Line], page 17.

Note that some operating systems may place a limit on the number of characters that are recognized after #!. Also, the various shells/systems parse differently the arguments appearing in a #! line. The majority of them group together all the arguments in a string and pass it to the interpreter as a single argument. In this case, the following script:

    #! octave-interpreter-name -q -f # comment

is equivalent to type at the command line:

    octave "-q -f # comment"

which would obviously produce an error message. Unfortunately, it is impossible for Octave to know whether it has been called from the command line or from a #! script, so some care is needed when using the #! mechanism.

Note that when Octave is started from an executable script, the built-in function argv returns a cell array containing the command line arguments passed to

---

[1] The #! mechanism works on Unix systems derived from Berkeley Unix, System V Release 4, and some System V Release 3 systems.

an executable Octave script, not the arguments passed to the Octave interpreter on the #! line of the script. For example, the following program will reproduce the command line that is used to execute script, not -qf.

```
#! /bin/octave -qf
printf ("%s", program_name ());
arg_list = argv ();
for i = 1:nargin
  printf (" %s", arg_list{i});
endfor
printf ("\n");
```

## 2.7 Comments in Octave Programs

A *comment* is some text that is included in a program for the sake of human readers, and that is not really part of the program. Comments can explain what the program does, and how it works. Nearly all programming languages have provisions for comments, because programs are typically hard to understand without them.

In the Octave language, a comment starts with either the sharp sign character, #, or the percent symbol % and continues to the end of the line. The Octave interpreter ignores the rest of a line following a sharp sign or percent symbol. For example, we could have put the following into the function f:

```
function xdot = f (x, t)

# usage: f (x, t)
#
# This function defines the right hand
# side functions for a set of nonlinear
# differential equations.

  r = 0.25;
  ...
endfunction
```

The help command (see Section 2.3 [Getting Help], page 22) is able to find the first block of comments in a function (even those that are composed directly on the command line). This means that users of Octave can use the same commands to get help for built-in functions, and for functions that you have defined. For example, after defining the function f above, the command *help f* produces the output

```
usage: f (x, t)

This function defines the right hand
side functions for a set of nonlinear
differential equations.
```

Although it is possible to put comment lines into keyboard-composed throwaway Octave programs, it usually isn't very useful, because the purpose of a

Chapter 2: Getting Started                                              27

comment is to help you or another person understand the program at a later time.

# 3 Data Types

All versions of Octave include a number of built-in data types, including real and complex scalars and matrices, character strings, a data structure type, and an array that can contain all data types.

It is also possible to define new specialized data types by writing a small amount of C++ code. On some systems, new data types can be loaded dynamically while Octave is running, so it is not necessary to recompile all of Octave just to add a new type.[1]

typeinfo (*expr*)                                                                               Built-in Function
      Return the type of the expression *expr*, as a string. If *expr* is omitted, return an array of strings containing all the currently installed data types.

## 3.1 Built-in Data Types

The standard built-in data types are real and complex scalars and matrices, ranges, character strings, a data structure type, and cell arrays. Additional built-in data types may be added in future versions.

The data type of a variable can be determined and changed through the use of the following functions.

class (*expr*)                                                                                           Built-in Function
      Return the class of the expression *expr*, as a string.

isa (*x, class*)                                                                                           Function File
      Return true if *x* is a value from the class *class*.

cast (*val, type*)                                                                                 Function File
      Convert *val* to data type *type*.

**See also:** int8, uint8, int16, uint16, int32, uint32, int64, uint64, double.

typecast (*x, type*)                                                                     Loadable Function
      Converts from one datatype to another without changing the underlying data. The argument *type* defines the type of the return argument and must be one of 'uint8', 'uint16', 'uint32', 'uint64', 'int8', 'int16', 'int32', 'int64', 'single' or 'double'.

An example of the use of typecast on a little-endian machine is

```
x = uint16 ([1, 65535]);
typecast (x, 'uint8')
⇒ [   0,   1, 255, 255]
```

**See also:** cast, swapbytes.

---

[1] See the additional document *Dynamically Linked Functions in GNU Octave* at http://www.network-theory.co.uk/octave/manual/

**swapbytes** (*x*)                                                                       Function File
Swaps the byte order on values, converting from little endian to big endian and visa-versa. For example

    swapbytes (uint16 (1:4))
    ⇒ [ 256   512   768   1024]

See also: typecast, cast.

### 3.1.1 Numeric Objects

Octave's built-in numeric objects include real, complex, and integer scalars and matrices. All built-in floating point numeric data is currently stored as double precision numbers. On systems that use the IEEE floating point format, values in the range of approximately $2.2251 \times 10^{-308}$ to $1.7977 \times 10^{308}$ can be stored, and the relative precision is approximately $2.2204 \times 10^{-16}$. The exact values are given by the variables realmin, realmax, and eps, respectively.

Matrix objects can be of any size, and can be dynamically reshaped and resized. It is easy to extract individual rows, columns, or submatrices using a variety of powerful indexing features. See Section 8.1 [Index Expressions], page 93.

See Chapter 4 [Numeric Data Types], page 33, for more information.

### 3.1.2 Missing Data

It is possible to represent missing data explicitly in Octave using NA (short for "Not Available"). Missing data can only be represented when data is represented as floating point numbers. In this case missing data is represented as a special case of the representation of NaN.

**NA** (*x*)                                                                        Built-in Function
**NA** (*n*, *m*)                                                                 Built-in Function
**NA** (*n*, *m*, *k*, ...)                                                     Built-in Function
**NA** (..., *class*)                                                       Built-in Function
Return a matrix or N-dimensional array whose elements are all equal to the special constant used to designate missing values.

**isna** (*x*)                                                                       Mapping Function
Return 1 for elements of *x* that are NA (missing) values and zero otherwise. For example,

    isna ([13, Inf, NA, NaN])
    ⇒ [ 0, 0, 1, 0 ]

### 3.1.3 String Objects

A character string in Octave consists of a sequence of characters enclosed in either double-quote or single-quote marks. Internally, Octave currently stores strings as matrices of characters. All the indexing operations that work for matrix objects also work for strings.

See Chapter 5 [Strings], page 47, for more information.

Chapter 3: Data Types                                                    31

### 3.1.4 Data Structure Objects

Octave's data structure type can help you to organize related objects of different types. The current implementation uses an associative array with indices limited to strings, but the syntax is more like C-style structures.

See Section 6.1 [Data Structures], page 63, for more information.

### 3.1.5 Cell Array Objects

A Cell Array in Octave is general array that can hold any number of different data types.

See Section 6.2 [Cell Arrays], page 71, for more information.

## 3.2 Object Sizes

The following functions allow you to determine the size of a variable or expression. These functions are defined for all objects. They return −1 when the operation doesn't make sense. For example, Octave's data structure type doesn't have rows or columns, so the rows and columns functions return −1 for structure arguments.

`ndims (a)`                                                    Built-in Function
    Returns the number of dimensions of array a. For any array, the result will always be larger than or equal to 2. Trailing singleton dimensions are not counted.

`columns (a)`                                                  Built-in Function
    Return the number of columns of a.

    **See also:** size, numel, rows, length, isscalar, isvector, and ismatrix.

`rows (a)`                                                     Built-in Function
    Return the number of rows of a.

    **See also:** size, numel, columns, length, isscalar, isvector, ismatrix.

`numel (a)`                                                    Built-in Function
    Returns the number of elements in the object a.

    **See also:** size.

`length (a)`                                                   Built-in Function
    Return the 'length' of the object a. For matrix objects, the length is the number of rows or columns, whichever is greater (this odd definition is used for compatibility with MATLAB).

**size (a, n)**  *Built-in Function*
Return the number rows and columns of a.
With one input argument and one output argument, the result is returned in a row vector. If there are multiple output arguments, the number of rows is assigned to the first, and the number of columns to the second, etc. For example,

```
size ([1, 2; 3, 4; 5, 6])
     ⇒ [ 3, 2 ]

[nr, nc] = size ([1, 2; 3, 4; 5, 6])
     ⇒ nr = 3
     ⇒ nc = 2
```

If given a second argument, size will return the size of the corresponding dimension. For example

```
size ([1, 2; 3, 4; 5, 6], 2)
     ⇒ 2
```

returns the number of columns in the given matrix.

See also: numel.

**isempty (a)**  *Built-in Function*
Return 1 if a is an empty matrix (either the number of rows, or the number of columns, or both are zero). Otherwise, return 0.

**sizeof (val)**  *Built-in Function*
Return the size of val in bytes

**size_equal (a, b, ...)**  *Built-in Function*
Return true if the dimensions of all arguments agree. Trailing singleton dimensions are ignored.

See also: size, numel.

**squeeze (x)**  *Built-in Function*
Remove singleton dimensions from x and return the result. Note that for compatibility with MATLAB, all objects have a minimum of two dimensions and row vectors are left unchanged.

# 4 Numeric Data Types

A *numeric constant* may be a scalar, a vector, or a matrix, and it may contain complex values.

The simplest form of a numeric constant, a scalar, is a single number that can be an integer, a decimal fraction, a number in scientific (exponential) notation, or a complex number. Note that by default numeric constants are represented within Octave in double-precision floating point format (complex constants are stored as pairs of double-precision floating point values). It is however possible to represent real integers as described in Section 4.3 [Integer Data Types], page 38. Here are some examples of real-valued numeric constants, which all have the same value:

    105
    1.05e+2
    1050e-1

To specify complex constants, you can write an expression of the form

    3 + 4i
    3.0 + 4.0i
    0.3e1 + 40e-1i

all of which are equivalent. The letter i in the previous example stands for the pure imaginary constant, defined as $\sqrt{-1}$.

For Octave to recognize a value as the imaginary part of a complex constant, a space must not appear between the number and the i. If it does, Octave will print an error message, like this:

    octave:13> 3 + 4 i

    parse error:

    3 + 4 i
          ^

You may also use j, I, or J in place of the i above. All four forms are equivalent.

**double** (*x*)                                                               Built-in Function

    Convert *x* to double precision type.

**single** (*val*)                                                            Function File

    Convert the numeric value *val* to single precision.

    **Note:** this function currently returns its argument in double precision. Support for a single-precision numeric data type will be added in future versions of Octave.

**complex** (*val*)                                               Built-in Function
**complex** (*re*, *im*)                                  Built-in Function

    Convert *x* to a complex value.

## 4.1 Matrices

It is easy to define a matrix of values in Octave. The size of the matrix is determined automatically, so it is not necessary to explicitly state the dimensions. The expression

```
a = [1, 2; 3, 4]
```

results in the matrix

$$a = \begin{bmatrix} 1 & 2 \\ 3 & 4 \end{bmatrix}$$

Elements of a matrix may be arbitrary expressions, provided that the dimensions all make sense when combining the various pieces. For example, given the above matrix, the expression

```
[ a, a ]
```

produces the matrix

```
ans =

  1 2 1 2
  3 4 3 4
```

but the expression

```
[ a, 1 ]
```

produces the error

```
error: number of rows must match near line 13, column 6
```

(assuming that this expression was entered as the first thing on line 13, of course).

Inside the square brackets that delimit a matrix expression, Octave looks at the surrounding context to determine whether spaces and newline characters should be converted into element and row separators, or simply ignored, so an expression like

```
a = [ 1 2
      3 4 ]
```

will work. However, some possible sources of confusion remain. For example, in the expression

```
[ 1 - 1 ]
```

the - is treated as a binary operator and the result is the scalar 0, but in the expression

```
[ 1 -1 ]
```

the - is treated as a unary operator and the result is the vector [ 1, -1 ]. Similarly, the expression

```
[ sin (pi) ]
```

will be parsed as

```
[ sin, (pi) ]
```

and will result in an error since the sin function will be called with no arguments. To get around this, you must omit the space between sin and the opening parenthesis, or enclose the expression in a set of parentheses:

# Chapter 4: Numeric Data Types

    [ (sin (pi)) ]

Whitespace surrounding the single quote character (', used as a transpose operator and for delimiting character strings) can also cause confusion. Given a = 1, the expression

    [ 1 a' ]

results in the single quote character being treated as a transpose operator and the result is the vector [ 1, 1 ], but the expression

    [ 1 a ' ]

produces the error message

    error: unterminated string constant

because not doing so would cause trouble when parsing the valid expression

    [ a 'foo' ]

For clarity, it is probably best to always use commas and semicolons to separate matrix elements and rows.

When you type a matrix or the name of a variable whose value is a matrix, Octave responds by printing the matrix in with neatly aligned rows and columns. If the rows of the matrix are too large to fit on the screen, Octave splits the matrix and displays a header before each section to indicate which columns are being displayed. You can use the following variables to control the format of the output.

`val = output_max_field_width ()`                       Built-in Function
`old_val = output_max_field_width (`*new_val*`)`        Built-in Function

    Query or set the internal variable that specifies the maximum width of a numeric output field.

    **See also:** format, output_precision.

`val = output_precision ()`                             Built-in Function
`old_val = output_precision (`*new_val*`)`              Built-in Function

    Query or set the internal variable that specifies the minimum number of significant figures to display for numeric output.

    **See also:** format, output_max_field_width.

It is possible to achieve a wide range of output styles by using different values of output_precision and output_max_field_width. Reasonable combinations can be set using the format function. See Section 14.1 [Basic Input and Output], page 161.

`val = split_long_rows ()`                              Built-in Function
`old_val = split_long_rows (`*new_val*`)`               Built-in Function

    Query or set the internal variable that controls whether rows of a matrix may be split when displayed to a terminal window. If the rows are split, Octave will display the matrix in a series of smaller pieces, each of which can fit within the limits of your terminal width and each set of rows is labeled so that you can easily see which columns are currently being displayed. For example:

```
octave:13> rand (2,10)
ans =

Columns 1 through 6:

  0.75883   0.93290   0.40064   0.43818   0.94958   0.16467
  0.75697   0.51942   0.40031   0.61784   0.92309   0.40201

Columns 7 through 10:

  0.90174   0.11854   0.72313   0.73326
  0.44672   0.94303   0.56564   0.82150
```

Octave automatically switches to scientific notation when values become very large or very small. This guarantees that you will see several significant figures for every value in a matrix. If you would prefer to see all values in a matrix printed in a fixed point format, you can set the built-in variable `fixed_point_format` to a nonzero value. But doing so is not recommended, because it can produce output that can easily be misinterpreted.

`val = fixed_point_format ()`                                  Built-in Function
`old_val = fixed_point_format (new_val)`            Built-in Function

Query or set the internal variable that controls whether Octave will use a scaled format to print matrix values such that the largest element may be written with a single leading digit with the scaling factor is printed on the first line of output. For example,

```
octave:1> logspace (1, 7, 5)'
ans =

  1.0e+07  *

  0.00000
  0.00003
  0.00100
  0.03162
  1.00000
```

Notice that first value appears to be zero when it is actually 1. For this reason, you should be careful when setting `fixed_point_format` to a nonzero value.

## 4.1.1 Empty Matrices

A matrix may have one or both dimensions zero, and operations on empty matrices are handled as described by Carl de Boor in *An Empty Exercise*, SIGNUM, Volume 25, pages 2–6, 1990 and C. N. Nett and W. M. Haddad, in *A System-Theoretic Appropriate Realization of the Empty Matrix Concept*, IEEE Transactions on Automatic Control, Volume 38, Number 5, May 1993. Briefly, given a scalar $s$, an $m \times n$ matrix $M_{m \times n}$, and an $m \times n$ empty matrix $[\,]_{m \times n}$ (with either one or both dimensions equal to zero), the following are true:

$$s \cdot [\,]_{m \times n} = [\,]_{m \times n} \cdot s = [\,]_{m \times n}$$
$$[\,]_{m \times n} + [\,]_{m \times n} = [\,]_{m \times n}$$
$$[\,]_{0 \times m} \cdot M_{m \times n} = [\,]_{0 \times n}$$
$$M_{m \times n} \cdot [\,]_{n \times 0} = [\,]_{m \times 0}$$
$$[\,]_{m \times 0} \cdot [\,]_{0 \times n} = 0_{m \times n}$$

By default, dimensions of the empty matrix are printed along with the empty matrix symbol, []. The built-in variable print_empty_dimensions controls this behavior.

val = print_empty_dimensions ()                                   Built-in Function
old_val = print_empty_dimensions (*new_val*)            Built-in Function

Query or set the internal variable that controls whether the dimensions of empty matrices are printed along with the empty matrix symbol, []. For example, the expression

    zeros (3, 0)

will print

    ans = [](3x0)

Empty matrices may also be used in assignment statements as a convenient way to delete rows or columns of matrices. See Section 8.6 [Assignment Expressions], page 103.

When Octave parses a matrix expression, it examines the elements of the list to determine whether they are all constants. If they are, it replaces the list with a single matrix constant.

## 4.2 Ranges

A *range* is a convenient way to write a row vector with evenly spaced elements. A range expression is defined by the value of the first element in the range, an optional value for the increment between elements, and a maximum value which the elements of the range will not exceed. The base, increment, and limit are separated by colons (the : character) and may contain any arithmetic expressions and function calls. If the increment is omitted, it is assumed to be 1. For example, the range

```
1 : 5
```

defines the set of values [ 1, 2, 3, 4, 5 ], and the range

```
1 : 3 : 5
```

defines the set of values [ 1, 4 ].

Although a range constant specifies a row vector, Octave does *not* convert range constants to vectors unless it is necessary to do so. This allows you to write a constant like 1 : 10000 without using 80,000 bytes of storage on a typical 32-bit workstation.

Note that the upper (or lower, if the increment is negative) bound on the range is not always included in the set of values, and that ranges defined by floating point values can produce surprising results because Octave uses floating point arithmetic to compute the values in the range. If it is important to include the endpoints of a range and the number of elements is known, you should use the linspace function instead (see Section 16.4 [Special Utility Matrices], page 244).

When Octave parses a range expression, it examines the elements of the expression to determine whether they are all constants. If they are, it replaces the range expression with a single range constant.

## 4.3 Integer Data Types

Octave supports integer matrices as an alternative to using double precision. It is possible to use both signed and unsigned integers represented by 8, 16, 32, or 64 bits. It should be noted that most computations require floating point data, meaning that integers will often change type when involved in numeric computations. For this reason integers are most often used to store data, and not for calculations.

In general most integer matrices are created by casting existing matrices to integers. The following example shows how to cast a matrix into 32 bit integers.

```
float = rand (2, 2)
    ⇒ float = 0.37569   0.92982
              0.11962   0.50876
integer = int32 (float)
    ⇒ integer = 0  1
                0  1
```

As can be seen, floating point values are rounded to the nearest integer when converted.

# Chapter 4: Numeric Data Types

**isinteger** (*x*)      *Built-in Function*
Return true if *x* is an integer object (int8, uint8, int16, etc.). Note that isinteger (14) is false because numeric constants in expressions are double precision floating point values.

**See also:** isreal, isnumeric, class, isa.

**int8** (*x*)      *Built-in Function*
Convert *x* to 8-bit integer type.

**uint8** (*x*)      *Built-in Function*
Convert *x* to unsigned 8-bit integer type.

**int16** (*x*)      *Built-in Function*
Convert *x* to 16-bit integer type.

**uint16** (*x*)      *Built-in Function*
Convert *x* to unsigned 16-bit integer type.

**int32** (*x*)      *Built-in Function*
Convert *x* to 32-bit integer type.

**uint32** (*x*)      *Built-in Function*
Convert *x* to unsigned 32-bit integer type.

**int64** (*x*)      *Built-in Function*
Convert *x* to 64-bit integer type.

**uint64** (*x*)      *Built-in Function*
Convert *x* to unsigned 64-bit integer type.

**intmax** (*type*)      *Built-in Function*
Return the largest integer that can be represented in an integer type. The variable *type* can be

int8
: signed 8-bit integer.

int16
: signed 16-bit integer.

int32
: signed 32-bit integer.

int64
: signed 64-bit integer.

uint8
: unsigned 8-bit integer.

`uint16`
> unsigned 16-bit integer.

`uint32`
> unsigned 32-bit integer.

`uint64`
> unsigned 64-bit integer.

The default for *type* is `uint32`.

**See also:** intmin, bitmax.

`intmin (type)`                                                                                     Built-in Function

Return the smallest integer that can be represented in an integer type. The variable *type* can be

`int8`
> signed 8-bit integer.

`int16`
> signed 16-bit integer.

`int32`
> signed 32-bit integer.

`int64`
> signed 64-bit integer.

`uint8`
> unsigned 8-bit integer.

`uint16`
> unsigned 16-bit integer.

`uint32`
> unsigned 32-bit integer.

`uint64`
> unsigned 64-bit integer.

The default for *type* is `uint32`.

**See also:** intmax, bitmax.

## 4.3.1 Integer Arithmetic

Octave supports some integer operations such as addition and multiplication. The operators +, -, .*, and ./ work on integers of the same type. So, it is possible to add two 32 bit integers, but not to add a 32 bit integer and a 16 bit integer.

The arithmetic operations on integers are performed by casting the integer values to double precision values, performing the operation, and then re-casting the values back to the original integer type. As the double precision type of Octave is only capable of representing integers with up to 53 bits of precision, it is not possible to perform arithmetic with 64 bit integer types.

When doing integer arithmetic one should consider the possibility of underflow and overflow. This happens when the result of the computation can't be represented using the chosen integer type. As an example it is not possible to represent the result of $10 - 20$ when using unsigned integers. Octave makes sure that the result of integer computations is the integer that is closest to the true result. So, the result of $10 - 20$ when using unsigned integers is zero.

When doing integer division Octave will round the result to the nearest integer. This is different from most programming languages, where the result is often floored to the nearest integer. So, the result of int32(5)./int32(8) is 1.

## 4.4 Bit Manipulations

Octave provides a number of functions for the manipulation of numeric values on a bit by bit basis. The basic functions to set and obtain the values of individual bits are bitset and bitget.

x = bitset (a, n)  *Function File*
x = bitset (a, n, v)  *Function File*

Set or reset bit(s) $n$ of unsigned integers in a. $v = 0$ resets and $v = 1$ sets the bits. The lowest significant bit is: $n = 1$

    dec2bin (bitset (10, 1))
    ⇒ 1011

See also: bitand, bitor, bitxor, bitget, bitcmp, bitshift, bitmax.

X = bitget (a,n)  *Function File*

Return the status of bit(s) $n$ of unsigned integers in a the lowest significant bit is $n = 1$.

    bitget (100, 8:-1:1)
    ⇒ 0 1 1 0 0 1 0 0

See also: bitand, bitor, bitxor, bitset, bitcmp, bitshift, bitmax.

The arguments to all of Octave's bitwise operations can be scalar or arrays, except for bitcmp, whose $k$ argument must a scalar. In the case where more than one argument is an array, then all arguments must have the same shape, and the bitwise operator is applied to each of the elements of the argument individually. If at least one argument is a scalar and one an array, then the scalar argument is duplicated. Therefore

        bitget (100, 8:-1:1)
is the same as
        bitget (100 * ones (1, 8), 8:-1:1)

It should be noted that all values passed to the bit manipulation functions of Octave are treated as integers. Therefore, even though the example for bitset above passes the floating point value 10, it is treated as the bits [1, 0, 1, 0] rather than the bits of the native floating point format representation of 10.

As the maximum value that can be represented by a number is important for bit manipulation, particularly when forming masks, Octave supplies the function bitmax.

bitmax ()                                                           Built-in Function
    Return the largest integer that can be represented as a floating point value. On IEEE-754 compatible systems, bitmax is 2^53 - 1.

This is the double precision version of the functions intmax, previously discussed.

Octave also includes the basic bitwise 'and', 'or' and 'exclusive or' operators.

bitand (x, y)                                                       Built-in Function
    Return the bitwise AND of nonnegative integers. x, y must be in the range [0,bitmax]

    See also: bitor, bitxor, bitset, bitget, bitcmp, bitshift, bitmax.

bitor (x, y)                                                        Built-in Function
    Return the bitwise OR of nonnegative integers. x, y must be in the range [0,bitmax]

    See also: bitor, bitxor, bitset, bitget, bitcmp, bitshift, bitmax.

bitxor (x, y)                                                       Built-in Function
    Return the bitwise XOR of nonnegative integers. x, y must be in the range [0,bitmax]

    See also: bitand, bitor, bitset, bitget, bitcmp, bitshift, bitmax.

The bitwise 'not' operator is a unary operator that performs a logical negation of each of the bits of the value. For this to make sense, the mask against which the value is negated must be defined. Octave's bitwise 'not' operator is bitcmp.

bitcmp (a, k)                                                         Function File
    Return the k-bit complement of integers in a. If k is omitted k = log2 (bitmax) + 1 is assumed.

        bitcmp(7,4)
        ⇒ 8
        dec2bin(11)
        ⇒ 1011
        dec2bin(bitcmp(11, 6))
        ⇒ 110100

    See also: bitand, bitor, bitxor, bitset, bitget, bitcmp, bitshift, bitmax.

## Chapter 4: Numeric Data Types

Octave also includes the ability to left-shift and right-shift values bitwise.

**bitshift (a, k)**                                                              *Built-in Function*
**bitshift (a, k, n)**                                                *Built-in Function*

Return a $k$ bit shift of $n$-digit unsigned integers in a. A positive $k$ leads to a left shift. A negative value to a right shift. If $n$ is omitted it defaults to log2(bitmax)+1. $n$ must be in the range [1,log2(bitmax)+1] usually [1,33]

```
bitshift (eye (3), 1)
⇒
2 0 0
0 2 0
0 0 2

bitshift (10, [-2, -1, 0, 1, 2])
⇒ 2   5   10  20  40
```

See also: bitand, bitor, bitxor, bitset, bitget, bitcmp, bitmax.

Bits that are shifted out of either end of the value are lost. Octave also uses arithmetic shifts, where the sign bit of the value is kept during a right shift. For example

```
bitshift (-10, -1)
⇒ -5
bitshift (int8 (-1), -1)
⇒ -1
```

Note that `bitshift (int8 (-1), -1)` is -1 since the bit representation of -1 in the int8 data type is [1, 1, 1, 1, 1, 1, 1, 1].

## 4.5 Logical Values

Octave has built-in support for logical values, i.e. variables that are either `true` or `false`. When comparing two variables, the result will be a logical value whose value depends on whether or not the comparison is true.

The basic logical operations are &, |, and !, which correspond to "Logical And", "Logical Or", and "Logical Negation". These operations all follow the usual rules of logic.

It is also possible to use logical values as part of standard numerical calculations. In this case `true` is converted to 1, and `false` to 0, both represented using double precision floating point numbers. So, the result of `true*22 - false/6` is 22.

Logical values can also be used to index matrices and cell arrays. When indexing with a logical array the result will be a vector containing the values corresponding to `true` parts of the logical array. The following example illustrates this.

```
data = [ 1, 2; 3, 4 ];
idx = (data <= 2);
data(idx)
    ⇒ ans = [ 1; 4 ]
```

Instead of creating the idx array it is possible to replace data(idx) with data( data <= 2 ) in the above code.

Logical values can also be constructed by casting numeric objects to logical values, or by using the true or false functions.

**logical** (*arg*)                                                                  Function File
Convert *arg* to a logical value. For example,

```
logical ([-1, 0, 1])
```

is equivalent to

```
[-1, 0, 1] != 0
```

**true** (*x*)                                                                         Built-in Function
**true** (*n*, *m*)                                                                  Built-in Function
**true** (*n*, *m*, *k*, ...)                                                      Built-in Function
Return a matrix or N-dimensional array whose elements are all logical 1. The arguments are handled the same as the arguments for eye.

**false** (*x*)                                                                   Built-in Function
**false** (*n*, *m*)                                                           Built-in Function
**false** (*n*, *m*, *k*, ...)                                          Built-in Function
Return a matrix or N-dimensional array whose elements are all logical 0. The arguments are handled the same as the arguments for eye.

## 4.6 Predicates for Numeric Objects

Since the type of a variable may change during the execution of a program, it can be necessary to do type checking at run-time. Doing this also allows you to change the behaviour of a function depending on the type of the input. As an example, this naive implementation of abs returns the absolute value of the input if it is a real number, and the magnitude of the input if it is a complex number.

```
function a = abs (x)
  if (isreal (x))
    a = sign (x) .* x;
  elseif (iscomplex (x))
    a = sqrt (real(x).^2 + imag(x).^2);
  endif
endfunction
```

The following functions are available for determining the type of a variable.

**isnumeric** (*x*)                                                        Built-in Function
Return nonzero if *x* is a numeric object.

Chapter 4: Numeric Data Types                                             45

**isreal (*x*)**                                                    Built-in Function
  Return true if *x* is a real-valued numeric object.

**iscomplex (*x*)**                                                 Built-in Function
  Return true if *x* is a complex-valued numeric object.

**ismatrix (a)**                                                    Built-in Function
  Return 1 if a is a matrix. Otherwise, return 0.

**isvector (a)**                                                     Function File
  Return 1 if a is a vector. Otherwise, return 0.

  **See also:** size, rows, columns, length, isscalar, ismatrix.

**isscalar (a)**                                                     Function File
  Return 1 if a is a scalar. Otherwise, return 0.

  **See also:** size, rows, columns, length, isscalar, ismatrix.

**issquare (*x*)**                                                   Function File
  If *x* is a square matrix, then return the dimension of *x*. Otherwise, return
  0.

  **See also:** size, rows, columns, length, ismatrix, isscalar, isvector.

**issymmetric (*x*, *tol*)**                                         Function File
  If *x* is symmetric within the tolerance specified by *tol*, then return the dimension of *x*. Otherwise, return 0. If *tol* is omitted, use a tolerance equal to
  the machine precision. Matrix *x* is considered symmetric if norm (*x* - *x*.',
  inf) / norm (*x*, inf) < *tol*.

  **See also:** size, rows, columns, length, ishermitian, ismatrix, isscalar, issquare, isvector.

**isdefinite (*x*, *tol*)**                                          Function File
  Return 1 if *x* is symmetric positive definite within the tolerance specified by
  *tol* or 0 if *x* is symmetric positive semidefinite. Otherwise, return -1. If *tol*
  is omitted, use a tolerance equal to 100 times the machine precision.

  **See also:** issymmetric.

**islogical (*x*)**                                                 Built-in Function
  Return true if *x* is a logical object.

isprime (n)  Function File
   Return true if *n* is a prime number, false otherwise.
   Something like the following is much faster if you need to test a lot of small numbers:
       t = ismember (n, primes (max (n (:))));
   If max(n) is very large, then you should be using special purpose factorization code.
   **See also:** primes, factor, gcd, lcm.

# 5 Strings

A *string constant* consists of a sequence of characters enclosed in either double-quote or single-quote marks. For example, both of the following expressions

```
"parrot"
'parrot'
```

represent the string whose contents are parrot. Strings in Octave can be of any length.

Since the single-quote mark is also used for the transpose operator (see Section 8.3 [Arithmetic Ops], page 98) but double-quote marks have no other purpose in Octave, it is best to use double-quote marks to denote strings.

In double-quoted strings, the backslash character is used to introduce *escape sequences* that represent other characters. For example, \n embeds a newline character in a double-quoted string and \" embeds a double quote character.

In single-quoted strings, backslash is not a special character.

Here is an example showing the difference

```
toascii ("\n")
    ⇒ 10
toascii ('\n')
    ⇒ [ 92 110 ]
```

You may also insert a single quote character in a single-quoted string by using two single quote characters in succession. For example,

```
'I can''t escape'
    ⇒ I can't escape
```

Here is a table of all the escape sequences used in Octave. They are the same as those used in the C programming language.

\\    Represents a literal backslash, \.

\"    Represents a literal double-quote character, ".

\'    Represents a literal single-quote character, '.

\0    Represents the "nul" character, control-@, ASCII code 0.

\a    Represents the "alert" character, control-g, ASCII code 7.

\b    Represents a backspace, control-h, ASCII code 8.

\f    Represents a formfeed, control-l, ASCII code 12.

\n    Represents a newline, control-j, ASCII code 10.

\r    Represents a carriage return, control-m, ASCII code 13.

\t    Represents a horizontal tab, control-i, ASCII code 9.

\v    Represents a vertical tab, control-k, ASCII code 11.

Strings may be concatenated using the notation for defining matrices. For example, the expression

    [ "foo" , "bar" , "baz" ]

produces the string whose contents are foobarbaz. See Chapter 4 [Numeric Data Types], page 33, for more information about creating matrices.

## 5.1 Creating Strings

The easiest way to create a string is, as illustrated in the introduction, to enclose a text in double-quotes or single-quotes. It is however possible to create a string without actually writing a text. The function blanks creates a string of a given length consisting only of blank characters (ASCII code 32).

blanks (*n*)                                                           Function File
    Return a string of *n* blanks.

    **See also:** repmat.

The string representation used by Octave is an array of characters, so the result of blanks(10) is actually a row vector of length 10 containing the value 32 in all places. This lends itself to the obvious generalisation to character matrices. Using a matrix of characters, it is possible to represent a collection of same-length strings in one variable. The convention used in Octave is that each row in a character matrix is a separate string, but letting each column represent a string is equally possible.

The easiest way to create a character matrix is to put several strings together into a matrix.

    collection = [ "String #1"; "String #2" ];

This creates a 2-by-9 character matrix.

One relevant question is, what happens when character matrix is created from strings of different length. The answer is that Octave puts blank characters at the end of strings shorter than the longest string. While it is possible to use a different character than the blank character using the string_fill_char function, it shows a problem with character matrices. It simply isn't possible to represent strings of different lengths. The solution is to use a cell array of strings, which is described in Section 6.2.3 [Cell Arrays of Strings], page 75.

char (*x*)                                                                   Built-in Function
char (*cell_array*)                                       Built-in Function
char (*s1, s2, ...*)                                     Built-in Function
    Create a string array from a numeric matrix, cell array, or list of

    If the argument is a numeric matrix, each element of the matrix is converted to the corresponding ASCII character. For example,

        char ([97, 98, 99])
            ⇒ "abc"

    If the argument is a cell array of strings, the result is a string array with each element corresponding to one element of the cell array.

## Chapter 5: Strings

For multiple string arguments, the result is a string array with each element corresponding to the arguments.

The returned values are padded with blanks as needed to make each row of the string array have the same length.

**strcat** (*s1*, *s2*, ...)   *Function File*
Return a string containing all the arguments concatenated. For example,
```
s = [ "ab"; "cde" ];
strcat (s, s, s)
⇒ "ab ab ab "
   "cdecdecde"
```

**strvcat** (*s_1*, ..., *s_n*)   *Function File*
Return a matrix containing the strings (and cell-strings) *s_1*, ..., *s_n* as its rows. Each string is padded with blanks in order to form a valid matrix. Unlike *str2mat*, empty strings are ignored.

**See also:** strcat, str2mat.

**strtrunc** (*s*, *n*)   *Function File*
Truncate the character string *s* to length *n*. If *s* is a char matrix, then the number of columns are adjusted.

If *s* is a cell array of strings, then the operation is performed on its members and the new cell array is returned.

*val* = **string_fill_char** ()   *Built-in Function*
*old_val* = **string_fill_char** (*new_val*)   *Built-in Function*
Query or set the internal variable used to pad all rows of a character matrix to the same length. It must be a single character. The default value is " " (a single space). For example,
```
string_fill_char ("X");
[ "these"; "are"; "strings" ]
    ⇒ "theseXX"
      "areXXXX"
      "strings"
```

**str2mat** (*s_1*, ..., *s_n*)   *Function File*
Return a matrix containing the strings *s_1*, ..., *s_n* as its rows. Each string is padded with blanks in order to form a valid matrix.

This function is modelled after MATLAB. In Octave, you can create a matrix of strings by [*s_1*; ...; *s_n*] even if the strings are not all the same length.

**ischar** (*a*)   *Built-in Function*
Return 1 if *a* is a string. Otherwise, return 0.

s = mat2str (x, n)                                        Function File
s = mat2str (..., 'class')                                Function File
Format real/complex numerical matrices as strings. This function returns values that are suitable for the use of the eval function.

The precision of the values is given by n. If n is a scalar then both real and imaginary parts of the matrix are printed to the same precision. Otherwise n (1) defines the precision of the real part and n (2) defines the precision of the imaginary part. The default for n is 17.

If the argument 'class' is given, then the class of x is included in the string in such a way that the eval will result in the construction of a matrix of the same class.

```
    mat2str( [ -1/3 + i/7; 1/3 - i/7 ], [4 2] )
⇒  '[-0.3333+0.14i;0.3333-0.14i]'
    mat2str( [ -1/3 +i/7; 1/3 -i/7 ], [4 2] )
⇒  '[-0.3333+0i,0+0.14i;0.3333+0i,-0-0.14i]'
    mat2str( int16([1 -1]), 'class')
⇒  'int16([1,-1])'
```

See also: sprintf, int2str.

num2str (n)                                               Function File
num2str (x, precision)                                    Function File
num2str (x, format)                                       Function File
Convert a number to a string. This function is not very flexible. For better control over the results, use sprintf (see Section 14.2.4 [Formatted Output], page 176).

See also: sprintf, int2str.

int2str (n)                                               Function File
Convert an integer to a string. This function is not very flexible. For better control over the results, use sprintf (see Section 14.2.4 [Formatted Output], page 176).

See also: sprintf, num2str.

## 5.2 Comparing Strings

Since a string is a character array, comparisons between strings work element by element as the following example shows:
```
GNU = "GNU's Not UNIX";
spaces = (GNU == " ")
⇒ spaces =
       0  0  0  0  0  1  0  0  0  1  0  0  0  0
```
To determine if two strings are identical it is therefore necessary to use the `strcmp` or `strncpm` functions. Similar functions exist for doing case-insensitive comparisons.

**strcmp** (*s1, s2*)                                                           Built-in Function

Return 1 if the character strings *s1* and *s2* are the same, and 0 otherwise.

If either *s1* or *s2* is a cell array of strings, then an array of the same size is returned, containing the values described above for every member of the cell array. The other argument may also be a cell array of strings (of the same size or with only one element), char matrix or character string.

**Caution:** For compatibility with MATLAB, Octave's strcmp function returns 1 if the character strings are equal, and 0 otherwise. This is just the opposite of the corresponding C library function.

**See also:** strcmpi, strncmp, strncmpi.

**strcmpi** (*s1, s2*)                                                          Function File

Ignoring case, return 1 if the character strings *s1* and *s2* are the same, and 0 otherwise.

If either *s1* or *s2* is a cell array of strings, then an array of the same size is returned, containing the values described above for every member of the cell array. The other argument may also be a cell array of strings (of the same size or with only one element), char matrix or character string.

**Caution:** For compatibility with MATLAB, Octave's strcmpi function returns 1 if the character strings are equal, and 0 otherwise. This is just the opposite of the corresponding C library function.

**See also:** strcmp, strncmp, strncmpi.

**strncmp** (*s1, s2, n*)                                              Built-in Function

Return 1 if the first *n* characters of strings *s1* and *s2* are the same, and 0 otherwise.
```
      strncmp ("abce", "abcd", 3)
         ⇒ 1
```
If either *s1* or *s2* is a cell array of strings, then an array of the same size is returned, containing the values described above for every member of the cell array. The other argument may also be a cell array of strings (of the same size or with only one element), char matrix or character string.

```
strncmp ("abce", {"abcd", "bca", "abc"}, 3)
    ⇒ [1, 0, 1]
```

**Caution:** For compatibility with MATLAB, Octave's strncmp function returns 1 if the character strings are equal, and 0 otherwise. This is just the opposite of the corresponding C library function.

See also: strncmpi, strcmp, strcmpi.

strncmpi (s1, s2, n)                                        Function File
    Ignoring case, return 1 if the first *n* characters of character strings *s1* and *s2* are the same, and 0 otherwise.

If either *s1* or *s2* is a cell array of strings, then an array of the same size is returned, containing the values described above for every member of the cell array. The other argument may also be a cell array of strings (of the same size or with only one element), char matrix or character string.

**Caution:** For compatibility with MATLAB, Octave's strncmpi function returns 1 if the character strings are equal, and 0 otherwise. This is just the opposite of the corresponding C library function.

See also: strcmp, strcmpi, strncmp.

## 5.3 Manipulating Strings

Octave supports a wide range of functions for manipulating strings. Since a string is just a matrix, simple manipulations can be accomplished using standard operators. The following example shows how to replace all blank characters with underscores.

```
quote = ...
    "First things first, but not necessarily in that order";
quote( quote == " " ) = "_"
    ⇒ quote =
        First_things_first,_but_not_necessarily_in_that_order
```

For more complex manipulations, such as searching, replacing, and general regular expressions, the following functions come with Octave.

deblank (s)                                                  Function File
    Remove trailing blanks and nulls from *s*. If *s* is a matrix, *deblank* trims each row to the length of longest string. If *s* is a cell array, operate recursively on each element of the cell array.

findstr (s, t, overlap)                                      Function File
    Return the vector of all positions in the longer of the two strings *s* and *t* where an occurrence of the shorter of the two starts. If the optional argument *overlap* is nonzero, the returned vector can include overlapping positions (this is the default). For example,

```
findstr ("ababab", "a")
    ⇒ [ 1, 3, 5 ]
findstr ("abababa", "aba", 0)
    ⇒ [ 1, 5 ]
```

Chapter 5: Strings                                                    53

**index** (*s, t*)                                         Function File
**index** (*s, t, direction*)                              Function File
    Return the position of the first occurrence of the string *t* in the string *s*, or 0 if no occurrence is found. For example,

        index ("Teststring", "t")
        ⇒ 4

    If *direction* is "first", return the first element found. If *direction* is "last", return the last element found. The rindex function is equivalent to index with *direction* set to "last".

    **Caution:** This function does not work for arrays of character strings.

    **See also:** find, rindex.

**rindex** (*s, t*)                                        Function File
    Return the position of the last occurrence of the character string *t* in the character string *s*, or 0 if no occurrence is found. For example,

        rindex ("Teststring", "t")
        ⇒ 6

    **Caution:** This function does not work for arrays of character strings.

    **See also:** find, index.

*idx* = **strfind** (*str, pattern*)                       Function File
*idx* = **strfind** (*cellstr, pattern*)                   Function File
    Search for *pattern* in the string *str* and return the starting index of every such occurrence in the vector *idx*. If there is no such occurrence, or if *pattern* is longer than *str*, then *idx* is the empty array [].

    If the cell array of strings *cellstr* is specified instead of the string *str*, then *idx* is a cell array of vectors, as specified above.

    **See also:** findstr, strmatch, strcmp, strncmp, strcmpi, strncmpi.

**strmatch** (*s, a,* "exact")                             Function File
    Return indices of entries of a that match the string *s*. The second argument a may be a string matrix or a cell array of strings. If the third argument "exact" is not given, then *s* only needs to match a up to the length of *s*. Nul characters match blanks. Results are returned as a column vector.

[*tok, rem*] = **strtok** (*str, delim*)                   Function File
    Find all characters up to but not including the first character which is in the string delim. If *rem* is requested, it contains the remainder of the string, starting at the first deliminator. Leading delimiters are ignored. If *delim* is not specified, space is assumed.

split (*s*, *t*, *n*)   Function File
    Divides the string *s* into pieces separated by *t*, returning the result in a string array (padded with blanks to form a valid matrix). If the optional input *n* is supplied, split *s* into at most *n* different pieces.

    For example,

```
        split ("Test string", "t")
        ⇒ "Tes "
           "  s  "
           "ring"
        split ("Test string", "t", 2)
        ⇒ "Tes    "
           " string"
```

strrep (*s*, *x*, *y*)   Function File
    Replaces all occurrences of the substring *x* of the string *s* with the string *y*.
    For example,

```
        strrep ("This is a test string", "is", "&%$")
        ⇒ "Th&%$ &%$ a test string"
```

substr (*s*, *offset*, *len*)   Function File
    Return the substring of *s* which starts at character number *offset* and is *len* characters long.

    If *offset* is negative, extraction starts that far from the end of the string. If *len* is omitted, the substring extends to the end of S.

    For example,

```
        substr ("This is a test string", 6, 9)
        ⇒ "is a test"
```

This function is patterned after AWK. You can get the same result by *s* (*offset* : (*offset* + *len* - 1)).

[*s*, *e*, *te*, *m*, *t*, *nm*] = regexp (*str*, *pat*)   Loadable Function
[...] = regexp (*str*, *pat*, *opts*, ...)   Loadable Function
    Regular expression string matching. Matches *pat* in *str* and returns the position and matching substrings or empty values if there are none.

    The matched pattern *pat* can include any of the standard regex operators, including:

.    Match any character

* + ? {}
    Repetition operators, representing

    *    Match zero or more times

    +    Match one or more times

    ?    Match zero or one times

Chapter 5: Strings                                                          55

{}
: Match range operator, which is of the form {n} to match exactly n times, {m,} to match m or more times, {m,n} to match between m and n times.

[...] [^...]
: List operators, where for example [ab]c matches ac and bc

()
: Grouping operator

|
: Alternation operator. Match one of a choice of regular expressions. The alternatives must be delimited by the grouping operator () above

^ $
: Anchoring operator. ^ matches the start of the string *str* and $ the end

In addition the following escaped characters have special meaning. Note that *pat* should be quoted with single quotes rather than double quotes, to avoid the escape sequences being interpreted by Octave before being passed to regexp.

\b
: Match a word boundary

\B
: Match within a word

\w
: Matches any word character

\W
: Matches any non word character

\<
: Matches the beginning of a word

\>
: Matches the end of a word

\s
: Matches any whitespace character

\S
: Matches any non whitespace character

\d
: Matches any digit

\D
: Matches any non-digit

The outputs of regexp by default are in the order as given below

s
: The start indices of each of the matching substrings

e
: The end indices of each matching substring

te
: The extents of each of the matched token surrounded by (...) in *pat*.

m
: A cell array of the text of each match.

t
: A cell array of the text of each token matched.

nm
: A structure containing the text of each matched named token, with the name being used as the fieldname. A named token is denoted as (?<name>...)

Particular output arguments or the order of the output arguments can be selected by additional *opts* arguments. These are strings and the correspondence between the output arguments and the optional argument are

| | |
|---|---|
| 'start' | s |
| 'end' | e |
| 'tokenExtents' | te |
| 'match' | m |
| 'tokens' | t |
| 'names' | nm |

A further optional argument is 'once', that limits the number of returned matches to the first match. Additional arguments are

matchcase
: Make the matching case sensitive.

ignorecase
: Make the matching case insensitive.

stringanchors
: Match the anchor characters at the beginning and end of the string.

lineanchors
: Match the anchor characters at the beginning and end of the line.

dotall
: The character . matches the newline character.

dotexceptnewline
: The character . matches all but the newline character.

freespacing
: The pattern can include arbitrary whitespace and comments starting with #.

literalspacing
: The pattern is taken literally.

[s, e, te, m, t, nm] = regexpi (*str*, *pat*)    Loadable Function
[...] = regexpi (*str*, *pat*, *opts*, ...)     Loadable Function
: Case insensitive regular expression string matching. Matches *pat* in *str* and returns the position and matching substrings or empty values if there are none. See regexp for more details

*string* = regexprep (*string*, *pat*, *repstr*, *options*)    Loadable Function
: Replace matches of *pat* in *string* with *repstr*.

 The replacement can contain $i, which substitutes for the i-th set of parentheses in the match string. E.g.,

        regexprep("Bill Dunn",'(\w+) (\w+)','$2, $1')

 returns "Dunn, Bill"

 *options* may be zero or more of

Chapter 5: Strings    57

once
: Replace only the first occurrence of *pat* in the result.

warnings
: This option is present for compatibility but is ignored.

ignorecase or matchcase
: Ignore case for the pattern matching (see `regexpi`). Alternatively, use (?i) or (?-i) in the pattern.

lineanchors and stringanchors
: Whether characters ^ and $ match the beginning and ending of lines. Alternatively, use (?m) or (?-m) in the pattern.

dotexceptnewline and dotall
: Whether . matches newlines in the string. Alternatively, use (?s) or (?-s) in the pattern.

freespacing or literalspacing
: Whether whitespace and # comments can be used to make the regular expression more readable. Alternatively, use (?x) or (?-x) in the pattern.

**See also:** regexp, regexpi.

## 5.4 String Conversions

Octave supports various kinds of conversions between strings and numbers. As an example, it is possible to convert a string containing a hexadecimal number to a floating point number.

```
hex2dec ("FF")
  ⇒ ans = 255
```

bin2dec (*s*)                                                                                        Function File
: Return the decimal number corresponding to the binary number stored in the string *s*. For example,

```
bin2dec ("1110")
  ⇒ 14
```

If *s* is a string matrix, returns a column vector of converted numbers, one per row of *s*. Invalid rows evaluate to NaN.

**See also:** dec2hex, base2dec, dec2base, hex2dec, dec2bin.

dec2bin (*n*, *len*)                                                                                 Function File
: Return a binary number corresponding to the nonnegative decimal number *n*, as a string of ones and zeros. For example,

```
dec2bin (14)
⇒ "1110"
```

If *n* is a vector, returns a string matrix, one row per value, padded with leading zeros to the width of the largest value.

The optional second argument, *len*, specifies the minimum number of digits in the result.

**See also:** bin2dec, dec2base, base2dec, hex2dec, dec2hex.

dec2hex (*n*, *len*)                                                                 Function File
Return the hexadecimal string corresponding to the nonnegative integer *n*. For example,

```
dec2hex (2748)
⇒ "ABC"
```

If *n* is a vector, returns a string matrix, one row per value, padded with leading zeros to the width of the largest value.

The optional second argument, *len*, specifies the minimum number of digits in the result.

**See also:** hex2dec, dec2base, base2dec, bin2dec, dec2bin.

hex2dec (*s*)                                                                        Function File
Return the integer corresponding to the hexadecimal number stored in the string *s*. For example,

```
hex2dec ("12B")
⇒ 299
hex2dec ("12b")
⇒ 299
```

If *s* is a string matrix, returns a column vector of converted numbers, one per row of *s*. Invalid rows evaluate to NaN.

**See also:** dec2hex, base2dec, dec2base, bin2dec, dec2bin.

dec2base (*n*, *b*, *len*)                                                           Function File
Return a string of symbols in base *b* corresponding to the nonnegative integer *n*.

```
dec2base (123, 3)
⇒ "11120"
```

If *n* is a vector, return a string matrix with one row per value, padded with leading zeros to the width of the largest value.

If *b* is a string then the characters of *b* are used as the symbols for the digits of *n*. Space (' ') may not be used as a symbol.

```
dec2base (123, "aei")
⇒ "eeeia"
```

The optional third argument, *len*, specifies the minimum number of digits in the result.

**See also:** base2dec, dec2bin, bin2dec, hex2dec, dec2hex.

Chapter 5: Strings                                                                 59

base2dec (*s*, *b*)                                                      Function File
    Convert *s* from a string of digits of base *b* into an integer.
        base2dec ("11120", 3)
        $\Rightarrow$ 123

    If *s* is a matrix, returns a column vector with one value per row of *s*. If a row contains invalid symbols then the corresponding value will be NaN. Rows are right-justified before converting so that trailing spaces are ignored.

    If *b* is a string, the characters of *b* are used as the symbols for the digits of *s*. Space (' ') may not be used as a symbol.
        base2dec ("yyyzx", "xyz")
        $\Rightarrow$ 123

    See also: dec2base, dec2bin, bin2dec, hex2dec, dec2hex.

[*num*, *status*, *strarray*] = str2double (*str*, *cdelim*, *rdelim*,     Function File
    *ddelim*)
    Convert strings into numeric values.

    str2double can replace str2num, but avoids the use of eval on unknown data.

    *str* can be the form [+-]d[.]dd[[eE][+-]ddd] in which d can be any of digit from 0 to 9, and [] indicate optional elements.

    *num* is the corresponding numeric value. If the conversion fails, status is -1 and *num* is NaN.

    *status* is 0 if the conversion was successful and -1 otherwise.

    *strarray* is a cell array of strings.

    Elements which are not defined or not valid return NaN and the *status* becomes -1.

    If *str* is a character array or a cell array of strings, then *num* and *status* return matrices of appropriate size.

    *str* can also contain multiple elements separated by row and column delimiters (*cdelim* and *rdelim*).

    The parameters *cdelim*, *rdelim*, and *ddelim* are optional column, row, and decimal delimiters.

    The default row-delimiters are newline, carriage return and semicolon (ASCII 10, 13 and 59). The default column-delimiters are tab, space and comma (ASCII 9, 32, and 44). The default decimal delimiter is . (ASCII 46).

    *cdelim*, *rdelim*, and *ddelim* must contain only nul, newline, carriage return, semicolon, colon, slash, tab, space, comma, or () [] {} (ASCII 0, 9, 10, 11, 12, 13, 14, 32, 33, 34, 40, 41, 44, 47, 58, 59, 91, 93, 123, 124, 125).

    Examples:

```
str2double ("-.1e-5")
⇒ -1.0000e-006

str2double (".314e1, 44.44e-1, .7; -1e+1")
⇒
    3.1400    4.4440    0.7000
  -10.0000       NaN       NaN

line = "200, 300, NaN, -inf, yes, no, 999, maybe, NaN";
[x, status] = str2double (line)
⇒ x =
    200   300   NaN   -Inf   NaN   NaN   999   NaN   NaN
⇒ status =
      0     0     0      0    -1    -1     0    -1     0
```

**strjust** (*s*, ["left"|"right"|"center"])                                  Function File
    Shift the non-blank text of *s* to the left, right or center of the string. If *s* is a string array, justify each string in the array. Null characters are replaced by blanks. If no justification is specified, then all rows are right-justified.

**str2num** (*s*)                                                            Function File
    Convert the string *s* to a number.

**toascii** (*s*)                                                           Mapping Function
    Return ASCII representation of *s* in a matrix. For example,
```
        toascii ("ASCII")
        ⇒ [ 65, 83, 67, 73, 73 ]
```

**tolower** (*s*)                                                           Mapping Function
    Return a copy of the string *s*, with each upper-case character replaced by the corresponding lower-case one; nonalphabetic characters are left unchanged. For example,
```
        tolower ("MiXeD cAsE 123")
        ⇒ "mixed case 123"
```

**toupper** (*s*)                                                            Built-in Function
    Return a copy of the string *s*, with each lower-case character replaced by the corresponding upper-case one; nonalphabetic characters are left unchanged. For example,
```
        toupper ("MiXeD cAsE 123")
        ⇒ "MIXED CASE 123"
```

**do_string_escapes** (*string*)                                             Built-in Function
    Convert special characters in *string* to their escaped forms.

Chapter 5: Strings                                                               61

**undo_string_escapes** (*s*)                                        Built-in Function
Converts special characters in strings back to their escaped forms. For example, the expression

    bell = "\a";

assigns the value of the alert character (control-g, ASCII code 7) to the string variable `bell`. If this string is printed, the system will ring the terminal bell (if it is possible). This is normally the desired outcome. However, sometimes it is useful to be able to print the original representation of the string, with the special characters replaced by their escape sequences. For example,

    octave:13> undo_string_escapes (bell)
    ans = \a

replaces the unprintable alert character with its printable representation.

## 5.5 Character Class Functions

Octave also provides the following character class test functions patterned after the functions in the standard C library. They all operate on string arrays and return matrices of zeros and ones. Elements that are nonzero indicate that the condition was true for the corresponding character in the string array. For example,

    isalpha ("!Q@WERT^Y&")
    ⇒ [ 0, 1, 0, 1, 1, 1, 1, 0, 1, 0 ]

**isalnum** (*s*)                                                    Mapping Function
Return 1 for characters that are letters or digits (`isalpha` (*s*) or `isdigit` (*s*) is true).

**isalpha** (*s*)                                                    Mapping Function
**isletter** (*s*)                                                   Mapping Function
Return true for characters that are letters (`isupper` (*s*) or `islower` (*s*) is true).

**isascii** (*s*)                                                    Mapping Function
Return 1 for characters that are ASCII (in the range 0 to 127 decimal).

**iscntrl** (*s*)                                                    Mapping Function
Return 1 for control characters.

**isdigit** (*s*)                                                    Mapping Function
Return 1 for characters that are decimal digits.

**isgraph** (*s*)                                                    Mapping Function
Return 1 for printable characters (but not the space character).

**isletter** (*s*)                                                                  Function File
    Returns true if *s* is a letter, false otherwise.

    **See also:** isalpha.

**islower** (*s*)                                                                 Mapping Function
    Return 1 for characters that are lower case letters.

**isprint** (*s*)                                                                 Mapping Function
    Return 1 for printable characters (including the space character).

**ispunct** (*s*)                                                                Mapping Function
    Return 1 for punctuation characters.

**isspace** (*s*)                                                                Mapping Function
    Return 1 for whitespace characters (space, formfeed, newline, carriage return, tab, and vertical tab).

**isupper** (*s*)                                                                Mapping Function
    Return 1 for upper case letters.

**isxdigit** (*s*)                                                              Mapping Function
    Return 1 for characters that are hexadecimal digits.

# 6 Data Containers

Octave includes support for two different mechanisms to contain arbitrary data types in the same variable. Structures, which are C-like, and are indexed with named fields, and cell arrays, where each element of the array can have a different data type and or shape.

## 6.1 Data Structures

Octave includes support for organizing data in structures. The current implementation uses an associative array with indices limited to strings, but the syntax is more like C-style structures. Here are some examples of using data structures in Octave.

Elements of structures can be of any value type. For example, the three expressions
```
x.a = 1
x.b = [1, 2; 3, 4]
x.c = "string"
```
create a structure with three elements. To print the value of the structure, you can type its name, just as for any other variable:
```
octave:2> x
x =
{
  a = 1
  b =

    1  2
    3  4

  c = string
}
```
Note that Octave may print the elements in any order.

Structures may be copied.
```
octave:1> y = x
y =
{
  a = 1
  b =

    1  2
    3  4

  c = string
}
```

Since structures are themselves values, structure elements may reference other structures. The following statements change the value of the element b of the structure x to be a data structure containing the single element d, which has a value of 3.

```
octave:1> x.b.d = 3
x.b.d = 3
octave:2> x.b
ans =
{
  d = 3
}
octave:3> x
x =
{
  a = 1
  b =
  {
    d = 3
  }

  c = string
}
```

Note that when Octave prints the value of a structure that contains other structures, only a few levels are displayed. For example,

```
octave:1> a.b.c.d.e = 1;
octave:2> a
a =
{
  b =
  {
    c =
    {
      d: 1x1 struct
    }
  }
}
```

This prevents long and confusing output from large deeply nested structures.

*val* = **struct_levels_to_print** ()  *Built-in Function*
*old_val* = **struct_levels_to_print** (*new_val*)  *Built-in Function*
  Query or set the internal variable that specifies the number of structure levels to display.

Functions can return structures. For example, the following function separates the real and complex parts of a matrix and stores them in two elements of the same structure variable.

## Chapter 6: Data Containers

```
octave:1> function y = f (x)
> y.re = real (x);
> y.im = imag (x);
> endfunction
```

When called with a complex-valued argument, f returns the data structure containing the real and imaginary parts of the original function argument.

```
octave:2> f (rand (2) + rand (2) * I)
ans =
{
  im =

    0.26475  0.14828
    0.18436  0.83669

  re =

    0.040239  0.242160
    0.238081  0.402523
}
```

Function return lists can include structure elements, and they may be indexed like any other variable. For example,

```
octave:1> [ x.u, x.s(2:3,2:3), x.v ] = svd ([1, 2; 3, 4])
x.u =

  -0.40455  -0.91451
  -0.91451   0.40455

x.s =

  0.00000  0.00000  0.00000
  0.00000  5.46499  0.00000
  0.00000  0.00000  0.36597

x.v =

  -0.57605   0.81742
  -0.81742  -0.57605
```

It is also possible to cycle through all the elements of a structure in a loop, using a special form of the for statement (see Section 10.5 [The for Statement], page 118)

### 6.1.1 Structure Arrays

A structure array is a particular instance of a structure, where each of the fields of the structure is represented by a cell array. Each of these cell arrays has the same dimensions. An example of the creation of a structure array is

```
x(1).a = "string1"
x(2).a = "string2"
x(1).b = 1
x(2).b = 2
```

which creates a 2-by-1 structure array with two fields. As previously, to print the value of the structure array, you can type its name:

```
octave:2> x
x =
{
  a =

  (,
    [1] = string1
    [2] = string2
  ,)

  b =

  (,
    [1] = 1
    [2] = 2
  ,)

}
```

Individual elements of the structure array can be returned by indexing the variable like x (1), which returns a structure with the two fields like

```
octave:2> x(1)
ans =
{
  a = string1
  b = 1
}
```

Furthermore, the structure array can return a comma separated list (see Section 6.3 [Comma Separated Lists], page 78), if indexed by one of its own field names. For example

```
octave:3> x.a
ans =

(,
  [1] = string1
  [2] = string2
,)
```

The function size with return the size of the structure. For the example above

```
octave:4> size(x)
ans =

        1   2
```

Elements can be deleted from a structure array in a similar manner to a numerical array, by assigning the elements to an empty matrix. For example

```
in = struct ("call1", {x, Inf, "last"},
             "call2", {x, Inf, "first"});
in (1, :) = []
⇒ in =
      {
        call1 =

          (,
            [1] = Inf
            [2] = last
          ,)

        call2 =

          (,
            [1] = Inf
            [2] = first
          ,)

      }
```

### 6.1.2 Creating Structures

As well as indexing a structure with ".", Octave can create a structure with the struct command. struct takes pairs of arguments, where the first argument in the pair is the fieldname to include in the structure and the second is a scalar or cell array, representing the values to include in the structure or structure array. For example

```
struct ("field1", 1, "field2", 2)
⇒ ans =
      {
        field1 = 1
        field2 = 2
      }
```

If the values passed to struct are a mix of scalar and cell arrays, then the scalar arguments are expanded to create a structure array with a consistent dimension. For example

```
    struct ("field1", {1, "one"}, "field2", {2, "two"},
           "field3", 3)
⇒ ans =
       {
         field1 =

         (,
           [1] =  1
           [2] = one
         ,)

         field2 =

         (,
           [1] =  2
           [2] = two
         ,)

         field3 =

         (,
           [1] =  3
           [2] =  3
         ,)

       }
```

**struct ("field", *value*, "field", *value*, ...)**       Built-in Function
    Create a structure and initialize its value.

    If the values are cell arrays, create a structure array and initialize its values. The dimensions of each cell array of values must match. Singleton cells and non-cell values are repeated so that they fill the entire array. If the cells are empty, create an empty structure array with the specified field names.

**isstruct (*expr*)**       Built-in Function
    Return 1 if the value of the expression *expr* is a structure.

    Additional functions that can manipulate the fields of a structure are listed below.

**rmfield (*s*, *f*)**       Built-in Function
    Remove field *f* from the structure *s*. If *f* is a cell array of character strings or a character array, remove the named fields.

    **See also:** cellstr, iscellstr, setfield.

Chapter 6: Data Containers                                                                  69

[k1, ..., v1] = setfield (s, k1, v1, ...)                                   Function File
   Set field members in a structure.
       oo(1,1).f0= 1;
       oo = setfield(oo,{1,2},'fd',{3},'b', 6);
       oo(1,2).fd(3).b == 6
      ⇒ ans = 1
   Note that this function could be written
       i1= {1,2}; i2= 'fd'; i3= {3}; i4= 'b';
       oo( i1{:} ).( i2 )( i3{:} ).( i4 ) == 6;
   **See also:** getfield, rmfield, isfield, isstruct, fieldnames, struct.

[t, p] = orderfields (s1, s2)                                               Function File
   Return a struct with fields arranged alphabetically or as specified by s2 and
   a corresponding permutation vector.

   Given one struct, arrange field names in s1 alphabetically.

   Given two structs, arrange field names in s1 as they appear in s2. The
   second argument may also specify the order in a permutation vector or a
   cell array of strings.

   **See also:** getfield, rmfield, isfield, isstruct, fieldnames, struct.

### 6.1.3 Manipulating Structures

   Other functions that can manipulate the fields of a structure are given below.

fieldnames (struct)                                                        Built-in Function
   Return a cell array of strings naming the elements of the structure *struct*.
   It is an error to call fieldnames with an argument that is not a structure.

isfield (expr, name)                                                       Built-in Function
   Return true if the expression *expr* is a structure and it includes an element
   named *name*. The first argument must be a structure and the second must
   be a string.

[v1, ...] = getfield (s, key, ...)                                          Function File
   Extract fields from a structure. For example
       ss(1,2).fd(3).b=5;
       getfield (ss, {1,2}, "fd", {3}, "b")
      ⇒ ans = 5
   Note that the function call in the previous example is equivalent to the
   expression
       i1= {1,2}; i2= "fd"; i3= {3}; i4= "b";
       ss(i1{:}).(i2)(i3{:}).(i4)
   **See also:** setfield, rmfield, isfield, isstruct, fieldnames, struct.

substruct (*type*, *subs*, ...)  Function File
    Create a subscript structure for use with subsref or subsasgn.

    See also: subsref, subsasgn.

### 6.1.4 Processing Data in Structures

The simplest way to process data in a structure is within a for loop or other means of iterating over the fields. A similar effect can be achieved with the structfun function, where a user defined function is applied to each field of the structure.

structfun (*func*, *s*)  Function File
[a, b] = structfun (...)  Function File
structfun (..., 'ErrorHandler', *errfunc*)  Function File
structfun (..., 'UniformOutput', *val*)  Function File
    Evaluate the function named *name* on the fields of the structure *s*. The fields of *s* are passed to the function *func* individually.

    structfun accepts an arbitrary function *func* in the form of an inline function, function handle, or the name of a function (in a character string). In the case of a character string argument, the function must accept a single argument named *x*, and it must return a string value. If the function returns more than one argument, they are returned as separate output variables.

    If the parameter 'UniformOutput' is set to true (the default), then the function must return a single element which will be concatenated into the return value. If 'UniformOutput' is false, the outputs are placed in a structure with the same fieldnames as the input structure.

```
s.name1 = "John Smith";
s.name2 = "Jill Jones";
structfun (@(x) regexp (x, '(\w+)$', 'matches'){1}, s,
          'UniformOutput', false)
```

    Given the parameter 'ErrorHandler', then *errfunc* defines a function to call in case *func* generates an error. The form of the function is

```
function [...] = errfunc (se, ...)
```

where there is an additional input argument to *errfunc* relative to *func*, given by *se*. This is a structure with the elements 'identifier', 'message' and 'index', giving respectively the error identifier, the error message, and the index into the input arguments of the element that caused the error.

    See also: cellfun, arrayfun.

Alternatively, to process the data in a structure, the structure might be converted to another type of container before being treated.

struct2cell (*S*)  Built-in Function
    Create a new cell array from the objects stored in the struct object. If *f* is the number of fields in the structure, the resulting cell array will have a dimension vector corresponding to [*F* size(*S*)].

    See also: cell2struct, fieldnames.

# Chapter 6: Data Containers

## 6.2 Cell Arrays

Like structures, a cell array is able to store variables of different size or type. A cell array uses multi-dimensional integer indices, just like an ordinary N-dimensional numerical array. To distinguish them from ordinary arrays, cell arrays use braces { and } as allocation and indexing operators.

As an example, the following code creates a cell array containing a string and a 2-by-2 random matrix

```
c = {"a string", rand(2, 2)};
```

And a cell array can be indexed with the { and } operators, so the variable created in the previous example can be indexed like this

```
c{1}
    ⇒ ans = a string
```

As with numerical arrays several elements of a cell array can be extracted by indexing with a vector of indexes

```
c{1:2}
    ⇒ ans =

        (,
          [1] = a string
          [2] =

            0.593993   0.627732
            0.377037   0.033643

        ,)
```

The indexing operators can also be used to insert or overwrite elements of a cell array. The following code inserts the scalar 3 on the third place of the previously created cell array

```
c{3} = 3
    ⇒ c =

        {
          [1,1] = a string
          [1,2] =

            0.593993   0.627732
            0.377037   0.033643

          [1,3] =   3
        }
```

In general nested cell arrays are displayed hierarchically as above. In some circumstances it makes sense to reference them by their index, and this can be performed by the `celldisp` function.

celldisp (*c, name*)                                                                         Function File
    Recursively display the contents of a cell array. By default the values are displayed with the name of the variable *c*. However, this name can be replaced with the variable *name*.

    See also: disp.

### 6.2.1 Creating Cell Array

    The introductory example showed how to create a cell array containing currently available variables. In many situations, however, it is useful to create a cell array and then fill it with data.

    The cell function returns a cell array of a given size, containing empty matrices. This function is similar to to the zeros function for creating new numerical arrays. The following example creates a 2-by-2 cell array containing empty matrices

```
c = cell(2,2)
  ⇒ c =
    {
      [1,1] = [](0x0)
      [2,1] = [](0x0)
      [1,2] = [](0x0)
      [2,2] = [](0x0)
    }
```

    Just like numerical arrays, cell arrays can be multidimensional. The cell function accepts any number of positive integers to describe the size of the returned cell array. It is also possible to set the size of the cell array through a vector of positive integers. In the following example two cell arrays of equal size are created, and the size of the first one is displayed

```
c1 = cell(3, 4, 5);
c2 = cell( [3, 4, 5] );
size(c1)
  ⇒ ans =
    3   4   5
```

As can be seen, the size function also works for cell arrays. As do the other functions describing the size of an object, such as length, numel, rows, and columns.

    As an alternative to creating empty cell arrays, and then filling them, it is possible to convert numerical arrays into cell arrays using the num2cell and mat2cell functions.

cell (*x*)                                                                                             Built-in Function
cell (*n, m*)                                                                                     Built-in Function
    Create a new cell array object. If invoked with a single scalar argument, cell returns a square cell array with the dimension specified. If you supply two scalar arguments, cell takes them to be the number of rows and columns. If given a vector with two elements, cell uses the values of the elements as the number of rows and columns, respectively.

# Chapter 6: Data Containers

iscell (x)                                       Built-in Function
    Return true if x is a cell array object. Otherwise, return false.

c = num2cell (m)                                Loadable Function
c = num2cell (m, d)                             Loadable Function
    Convert to matrix m into a cell array. If d is defined the value c is of dimension 1 in this dimension and the elements of m are placed in slices in c.

**See also:** mat2cell.

b = mat2cell (a, m, n)                            Loadable Function
b = mat2cell (a, d1, d2, ...)                     Loadable Function
b = mat2cell (a, r)                                Loadable Function
    Converts the matrix a to a cell array If a is 2-D, then it is required that sum (m) == size (a, 1) and sum (n) == size (a, 2). Similarly, if a is a multi-dimensional and the number of dimensional arguments is equal to the dimensions of a, then it is required that sum (di) == size (a, i).

Given a single dimensional argument r, the other dimensional arguments are assumed to equal size (a,i).

An example of the use of mat2cell is

```
mat2cell (reshape(1:16,4,4),[3,1],[3,1])
⇒ {
  [1,1] =

    1   5   9
    2   6  10
    3   7  11

  [2,1] =

    4   8  12

  [1,2] =

    13
    14
    15

  [2,2] = 16
}
```

**See also:** num2cell, cell2mat.

## 6.2.2 Indexing Cell Arrays

As shown in the introductory example elements can be inserted from cell arrays using the { and } operators. Besides the change of operators, indexing works for cell arrays like for multidimensional arrays. As an example, all the rows of the first and third column of a cell array can be set to 0 with the following code

```
c{:, [1, 3]} = 0;
```

Accessing values in a cell array is, however, different from the same operation for numerical arrays. Accessing a single element of a cell array is very similar to numerical arrays, for example

```
element = c{1, 2};
```

This will, however, *not* work when accessing multiple elements of a cell array, because it might not be possible to represent all elements with a single variable as is the case with numerical arrays.

Accessing multiple elements of a cell array with the { and } operators will result in a comma-separated list (see Section 6.3 [Comma Separated Lists], page 78) of all the requested elements as discussed later.

One distinction between { and ( to index cell arrays is in the deletion of elements from the cell array. In a similar manner to a numerical array the () operator can be used to delete elements from the cell array. The {} operator however will remove the elements of the cell array, but not delete the space for them. For example

```
x = {"1", "2"; "3", "4"};
x{1, :} = []
⇒ x =
    {
      [1,1] = [](0x0)
      [2,1] = 3
      [1,2] = [](0x0)
      [2,2] = 4
    }

x(1, :) = []
⇒ x =
    {
      [1,1] = 3
      [1,2] = 4
    }
```

## 6.2.3 Cell Arrays of Strings

One common use of cell arrays is to store multiple strings in the same variable. It is possible to store multiple strings in a character matrix by letting each row be a string. This, however, introduces the problem that all strings must be of equal length. Therefore it is recommended to use cell arrays to store multiple strings. If, however, the character matrix representation is required for an operation, it can be converted to a cell array of strings using the cellstr function

```
a = ["hello"; "world"];
c = cellstr (a)
    ⇒ c =
        {
          [1,1] = hello
          [2,1] = world
        }
```

One further advantage of using cell arrays to store multiple strings is that most functions for string manipulations included with Octave support this representation. As an example, it is possible to compare one string with many others using the strcmp function. If one of the arguments to this function is a string and the other is a cell array of strings, each element of the cell array will be compared the string argument,

```
c = {"hello", "world"};
strcmp ("hello", c)
    ⇒ ans =
        1   0
```

The following functions for string manipulation support cell arrays of strings, strcmp, strcmpi, strncmp, strncmpi, str2double, str2mat, strappend, strtrunc, strvcat, strfind, and strmatch.

cellstr (*string*)  Built-in Function
    Create a new cell array object from the elements of the string array *string*.

iscellstr (*cell*)  Built-in Function
    Return true if every element of the cell array *cell* is a character string

[idxvec, errmsg] = cellidx (*listvar*, *strlist*)  Function File
    Return indices of string entries in *listvar* that match strings in *strlist*.

    Both *listvar* and *strlist* may be passed as strings or string matrices. If they are passed as string matrices, each entry is processed by deblank prior to searching for the entries.

    The first output is the vector of indices in *listvar*.

    If *strlist* contains a string not in *listvar*, then an error message is returned in errmsg. If only one output argument is requested, then cellidx prints errmsg to the screen and exits with an error.

### 6.2.4 Processing Data in Cell Arrays

Data that is stored in a cell array can be processed in several ways depending on the actual data. The simplest way to process that data is to iterate through it using one or more for loops. The same idea can be implemented more easily through the use of the cellfun function that calls a user-specified function on all elements of a cell array.

cellfun (*name*, *c*)                                           Loadable Function
cellfun ("size", *c*, *k*)                                      Loadable Function
cellfun ("isclass", *c*, *class*)                               Loadable Function
cellfun (*func*, *c*)                                           Loadable Function
cellfun (*func*, *c*, *d*)                                      Loadable Function
[a, b] = cellfun (...)                                          Loadable Function
cellfun (..., 'ErrorHandler', *errfunc*)                        Loadable Function
cellfun (..., 'UniformOutput', *val*)                           Loadable Function

>Evaluate the function named *name* on the elements of the cell array *c*. Elements in *c* are passed on to the named function individually. The function *name* can be one of the functions

>>isempty
>>>Return 1 for empty elements.

>>islogical
>>>Return 1 for logical elements.

>>isreal
>>>Return 1 for real elements.

>>length
>>>Return a vector of the lengths of cell elements.

>>ndims
>>>Return the number of dimensions of each element.

>>prodofsize
>>>Return the product of dimensions of each element.

>>size
>>>Return the size along the *k*-th dimension.

>>isclass
>>>Return 1 for elements of *class*.

>Additionally, cellfun accepts an arbitrary function *func* in the form of an inline function, function handle, or the name of a function (in a character string). In the case of a character string argument, the function must accept a single argument named *x*, and it must return a string value. The function can take one or more arguments, with the inputs args given by *c*, *d*, etc. Equally the function can return one or more output arguments. For example

## Chapter 6: Data Containers

```
cellfun (@atan2, {1, 0}, {0, 1})
⇒ans = [1.57080    0.00000]
```
Note that the default output argument is an array of the same size as the input arguments.

If the parameter 'UniformOutput' is set to true (the default), then the function must return a single element which will be concatenated into the return value. If 'UniformOutput' is false, the outputs are concatenated in a cell array. For example

```
cellfun ("tolower(x)", {"Foo", "Bar", "FooBar"},
         "UniformOutput",false)
⇒ ans = {"foo", "bar", "foobar"}
```

Given the parameter 'ErrorHandler', then *errfunc* defines a function to call in case *func* generates an error. The form of the function is

```
function [...] = errfunc (s, ...)
```

where there is an additional input argument to *errfunc* relative to *func*, given by *s*. This is a structure with the elements 'identifier', 'message' and 'index', giving respectively the error identifier, the error message, and the index into the input arguments of the element that caused the error. For example

```
function y = foo (s, x), y = NaN; endfunction
cellfun (@factorial, {-1,2},'ErrorHandler',@foo)
⇒ ans = [NaN 2]
```

**See also:** isempty, islogical, isreal, length, ndims, numel, size, isclass.

An alternative is to convert the data to a different container, such as a matrix or a data structure. Depending on the data this is possible using the `cell2mat` and `cell2struct` functions.

*m* = cell2mat (*c*)                                                    Function File
    Convert the cell array *c* into a matrix by concatenating all elements of *c* into a hyperrectangle. Elements of *c* must be numeric, logical or char, and cat must be able to concatenate them together.

    **See also:** mat2cell, num2cell.

cell2struct (*cell, fields, dim*)                                  Built-in Function
    Convert *cell* to a structure. The number of fields in *fields* must match the number of elements in *cell* along dimension *dim*, that is numel (*fields*) == size (*cell, dim*).

```
A = cell2struct ({'Peter', 'Hannah', 'Robert';
                  185, 170, 168},
                 {'Name','Height'}, 1);
A(1)
⇒ ans =
     {
       Height = 185
       Name   = Peter
     }
```

## 6.3 Comma Separated Lists

Comma separated lists are the basic argument type to all Octave functions. In the example

```
max (a, b)
```

a, b is a comma separated list. Comma separated lists can appear on both the right and left hand side of an equation. For example

```
[i, j] = ceil (find (x, [], "last"));
```

where i, j is equally a comma separated list. Comma separated lists cannot be directly manipulated by the user. However, both structures and cell arrays can be converted into comma separated lists, which makes them useful to keep the input arguments and return values of functions organized. Another example of where a comma separated list can be used is in the creation of a new array. If all the accessed elements of a cell array are scalars or column vectors, they can be concatenated into a new column vector containing the elements, by surrounding the list with [ and ] as in the following example

```
a = {1, [2, 3], 4};
b = [a{:}]
    ⇒ b =
        1  2  3  4
```

It is also possible to pass the accessed elements directly to a function. The list of elements from the cell array will be passed as an argument list to a given function as if it is called with the elements as arguments. The two calls to printf in the following example are identical but the latter is simpler and handles more situations

```
c = {"GNU", "Octave", "is", "Free", "Software"};
printf ("%s ", c{1}, c{2}, c{3}, c{4}, c{5});
    ⊣ GNU Octave is Free Software
printf ("%s ", c{:});
    ⊣ GNU Octave is Free Software
```

Just like it is possible to create a numerical array from selected elements of a cell array, it is possible to create a new cell array containing the selected elements. By surrounding the list with { and } a new cell array will be created, as the following example illustrates

```
a = {1, rand(2, 2), "three"};
b = { a{ [1, 3] } }
    ⇒ b =
        {
          [1,1] = 1
          [1,2] = three
        }
```

This syntax is however a bit cumbersome, and since this is a common operation, it is possible to achieve the same using the ( and ) operators for indexing. When a cell array is indexed using the ( and ) operators a new cell array containing the selected elements is returned. Using this syntax, the previous example can be simplified into the following

## Chapter 6: Data Containers

```
a = {1, rand(2, 2), "three"};
b = a( [1, 3] )
    ⇒ b =
        {
            [1,1] =  1
            [1,2] = three
        }
```

A comma separated list can equally appear on the left-hand side of an assignment. An example is

```
in {1} = ceil (rand (10, 1));
in {2} = [];
in {3} = "last";
in {4} = "first";
out = cell (4, 1);
[out{1:2}] = find (in{1 : 3});
[out{3:4}] = find (in{[1, 2, 4]});
```

Structure arrays can equally be used to create comma separated lists. This is done by addressing one of the fields of a structure array. For example

```
x = ceil (randn (10, 1));
in = struct ("call1", {x, Inf, "last"},
             "call2", {x, Inf, "first"});
out = struct ("call1", cell (2, 1), "call2", cell (2, 1));
[out.call1] = find (in.call1);
[out.call2] = find (in.call2);
```

# 7 Variables

Variables let you give names to values and refer to them later. You have already seen variables in many of the examples. The name of a variable must be a sequence of letters, digits and underscores, but it may not begin with a digit. Octave does not enforce a limit on the length of variable names, but it is seldom useful to have variables with names longer than about 30 characters. The following are all valid variable names

```
x
x15
__foo_bar_baz__
fucnrdthsucngtagdjb
```

However, names like `__foo_bar_baz__` that begin and end with two underscores are understood to be reserved for internal use by Octave. You should not use them in code you write, except to access Octave's documented internal variables and built-in symbolic constants.

Case is significant in variable names. The symbols a and A are distinct variables.

A variable name is a valid expression by itself. It represents the variable's current value. Variables are given new values with *assignment operators* and *increment operators*. See Section 8.6 [Assignment Expressions], page 103.

A number of variables have special built-in meanings. For example, ans holds the current working directory, and pi names the ratio of the circumference of a circle to its diameter. See Section 7.4 [Summary of Built-in Variables], page 89, for a list of all the predefined variables. Some of these built-in symbols are constants and may not be changed. Others can be used and assigned just like all other variables, but their values are also used or changed automatically by Octave.

Variables in Octave do not have fixed types, so it is possible to first store a numeric value in a variable and then to later use the same name to hold a string value in the same program. Variables may not be used before they have been given a value. Doing so results in an error.

`isvarname (`*name*`)`     Built-in Function
    Return true if *name* is a valid variable name

## 7.1 Global Variables

A variable that has been declared *global* may be accessed from within a function body without having to pass it as a formal parameter.

A variable may be declared global using a `global` declaration statement. The following statements are all global declarations.

```
global a
global a b
global c = 2
global d = 3 e f = 5
```

A global variable may only be initialized once in a `global` statement. For example, after executing the following code

```
global gvar = 1
global gvar = 2
```

the value of the global variable gvar is 1, not 2. Issuing a `clear gvar` command does not change the above behavior, but `clear all` does.

It is necessary declare a variable as global within a function body in order to access it. For example,

```
global x
function f ()
  x = 1;
endfunction
f ()
```

does *not* set the value of the global variable x to 1. In order to change the value of the global variable x, you must also declare it to be global within the function body, like this

```
function f ()
  global x;
  x = 1;
endfunction
```

Passing a global variable in a function parameter list will make a local copy and not modify the global value. For example, given the function

```
function f (x)
  x = 0
endfunction
```

and the definition of x as a global variable at the top level,

```
global x = 13
```

the expression

```
f (x)
```

will display the value of x from inside the function as 0, but the value of x at the top level remains unchanged, because the function works with a *copy* of its argument.

isglobal (*name*)                                                          Built-in Function
    Return 1 if *name* is globally visible. Otherwise, return 0. For example,

```
global x
isglobal ("x")
     ⇒ 1
```

## 7.2 Persistent Variables

A variable that has been declared *persistent* within a function will retain its contents in memory between subsequent calls to the same function. The difference between persistent variables and global variables is that persistent variables are local in scope to a particular function and are not visible elsewhere.

The following example uses a persistent variable to create a function that prints the number of times it has been called.

```
function count_calls ()
  persistent calls = 0;
  printf ("'count_calls' has been called %d times\n",
          ++calls);
endfunction

for i = 1:3
  count_calls ();
endfor

⊣ 'count_calls' has been called 1 times
⊣ 'count_calls' has been called 2 times
⊣ 'count_calls' has been called 3 times
```

As the example shows, a variable may be declared persistent using a persistent declaration statement. The following statements are all persistent declarations.

```
persistent a
persistent a b
persistent c = 2
persistent d = 3 e f = 5
```

The behavior of persistent variables is equivalent to the behavior of static variables in C. The command `static` in Octave is also recognized and is equivalent to persistent.

Like global variables, a persistent variable may only be initialized once. For example, after executing the following code

```
persistent pvar = 1
persistent pvar = 2
```

the value of the persistent variable pvar is 1, not 2.

If a persistent variable is declared but not initialized to a specific value, it will contain an empty matrix. So, it is also possible to initialize a persistent variable by checking whether it is empty, as the following example illustrates.

```
function count_calls ()
  persistent calls;
  if (isempty (calls))
    calls = 0;
  endif
  printf ("'count_calls' has been called %d times\n",
          ++calls);
endfunction
```

This implementation behaves in exactly the same way as the previous implementation of count_calls.

The value of a persistent variable is kept in memory until it is explicitly cleared. Assuming that the implementation of count_calls is saved on disc, we get the following behaviour.

```
for i = 1:2
  count_calls ();
endfor
⊣ 'count_calls' has been called 1 times
⊣ 'count_calls' has been called 2 times

clear
for i = 1:2
  count_calls();
endfor
⊣ 'count_calls' has been called 3 times
⊣ 'count_calls' has been called 4 times

clear all
for i = 1:2
  count_calls();
endfor
⊣ 'count_calls' has been called 1 times
⊣ 'count_calls' has been called 2 times

clear count_calls
for i = 1:2
  count_calls();
endfor
⊣ 'count_calls' has been called 1 times
⊣ 'count_calls' has been called 2 times
```

That is, the persistent variable is only removed from memory when the function containing the variable is removed. Note that if the function definition is typed directly into the Octave prompt, the persistent variable will be cleared by a simple clear command as the entire function definition will be removed from memory. If you do not want a persistent variable to be removed from memory even if the function is cleared, you should use the mlock function as described in See Section 11.7.4 [Function Locking], page 137.

## 7.3 Status of Variables

When creating simple one-shot programs it can be very convenient to see which variables are available at the prompt. The function who and its siblings whos and whos_line_format will show different information about what is in memory, as the following shows.

```
str = "A random string";
who -variables
    ⊣ *** local user variables:
    ⊣
    ⊣ __nargin__   str
```

**who** options pattern ...                                                  Command
**whos** options pattern ...                                                Command

List currently defined symbols matching the given patterns. The following are valid options. They may be shortened to one character but may not be combined.

**-all**
    List all currently defined symbols.

**-builtins**
    List built-in functions. This includes all currently compiled function files, but does not include all function files that are in the search path.

**-functions**
    List user-defined functions.

**-long**
    Print a long listing including the type and dimensions of any symbols. The symbols in the first column of output indicate whether it is possible to redefine the symbol, and whether it is possible for it to be cleared.

**-variables**
    List user-defined variables.

Valid patterns are the same as described for the clear command above. If no patterns are supplied, all symbols from the given category are listed. By default, only user defined functions and variables visible in the local scope are displayed.

The command whos is equivalent to who -long.

**whos** options pattern ...                                               Command
    See who.

$val$ = **whos_line_format** ()                                  Built-in Function
$old\_val$ = **whos_line_format** ($new\_val$)           Built-in Function
    Query or set the format string used by the whos.

The following escape sequences may be used in the format:

    **%b**    Prints number of bytes occupied by variables.

%c  Prints class names of variables.

%e  Prints elements held by variables.

%n  Prints variable names.

%p  Prints protection attributes of variables.

%s  Prints dimensions of variables.

%t  Prints type names of variables.

Every command may also have a modifier:

l   Left alignment.

r   Right alignment (this is the default).

c   Centered (may only be applied to command %s).

A command is composed like this:

%[modifier]<command>[:size_of_parameter[:center-specific[
    :print_dims[:balance]]]];

Command and modifier is already explained. Size_of_parameter tells how many columns the parameter will need for printing. print_dims tells how many dimensions to print. If number of dimensions exceeds print_dims, dimensions will be printed like x-D. center-specific and print_dims may only be applied to command %s. A negative value for print_dims will cause Octave to print all dimensions whatsoever. balance specifies the offset for printing of the dimensions string.

The default format is " %p:4; %ln:6; %cs:16:6:8:1; %rb:12; %lc:-1;\n".

Instead of displaying which variables are in memory, it is possible to determine if a given variable is available. That way it is possible to alter the behaviour of a program depending on the existence of a variable. The following example illustrates this.

```
if (! exist ("meaning", "var"))
  disp ("The program has no 'meaning'");
endif
```

exist (*name, type*)                                     Built-in Function

Return 1 if the name exists as a variable, 2 if the name is an absolute file name, an ordinary file in Octave's path, or (after appending .m) a function file in Octave's path, 3 if the name is a .oct or .mex file in Octave's path, 5 if the name is a built-in function, 7 if the name is a directory, or 103 if the name is a function not associated with a file (entered on the command line). Otherwise, return 0.

This function also returns 2 if a regular file called *name* exists in Octave's search path. If you want information about other types of files, you should use some combination of the functions file_in_path and stat instead.

If the optional argument *type* is supplied, check only for symbols of the specified type. Valid types are

## Chapter 7: Variables

"var"
: Check only for variables.

"builtin"
: Check only for built-in functions.

"file"
: Check only for files.

"dir"
: Check only for directories.

Usually Octave will manage the memory, but sometimes it can be practical to remove variables from memory manually. This is usually needed when working with large variables that fill a substantial part of the memory. On a computer that uses the IEEE floating point format, the following program allocates a matrix that requires around 128 MB memory.

    large_matrix = zeros (4000, 4000);

Since having this variable in memory might slow down other computations, it can be necessary to remove it manually from memory. The clear function allows this.

clear [options] pattern ...                                        Command
: Delete the names matching the given patterns from the symbol table. The pattern may contain the following special characters:

?
: Match any single character.

*
: Match zero or more characters.

[ list ]
: Match the list of characters specified by *list*. If the first character is ! or ^, match all characters except those specified by *list*. For example, the pattern [a-zA-Z] will match all lower and upper case alphabetic characters.

For example, the command

    clear foo b*r

clears the name foo and all names that begin with the letter b and end with the letter r.

If clear is called without any arguments, all user-defined variables (local and global) are cleared from the symbol table. If clear is called with at least one argument, only the visible names matching the arguments are cleared. For example, suppose you have defined a function foo, and then hidden it by performing the assignment foo = 2. Executing the command *clear foo* once will clear the variable definition and restore the definition of foo as a function. Executing *clear foo* a second time will clear the function definition.

The following options are available in both long and short form

-all, -a
: Clears all local and global user-defined variables and all functions from the symbol table.

-exclusive, -x
: Clears the variables that don't match the following pattern.

-functions, -f
: Clears the function names and the built-in symbols names.

-global, -g
: Clears the global symbol names.

-variables, -v
: Clears the local variable names.

With the exception of `exclusive`, all long options can be used without the dash as well.

Information about a function or variable such as its location in the file system can also be acquired from within Octave. This is usually only useful during development of programs, and not within a program.

**document** (*symbol*, *text*)  *Built-in Function*
: Set the documentation string for *symbol* to *text*.

**type options name ...**  *Command*
: Display the definition of each *name* that refers to a function.

Normally also displays whether each *name* is user-defined or built-in; the -q option suppresses this behaviour.

**which name ...**  *Command*
: Display the type of each *name*. If *name* is defined from a function file, the full name of the file is also displayed.

**See also:** help, lookfor.

Chapter 7: Variables                                                                    89

## 7.4 Summary of Built-in Variables

Here is a summary of all of Octave's built-in variables along with cross references to additional information and their default values. In the following table *octave-home* stands for the root directory where all of Octave is installed (the default is '/usr/local', *version* stands for the Octave version number (for example, 3.0.2) and *arch* stands for the type of system for which Octave was compiled (for example, x86_64-unknown-linux-gnu).

EDITOR
: See Appendix A [Commands For History], page 480.

    Default value: "emacs".

EXEC_PATH
: See Section 32.5 [Controlling Subprocesses], page 457.

    Default value: ":$PATH".

OCTAVE_HOME
: Default value: "/usr/local".

PAGER
: See Chapter 14 [Input and Output], page 161.

    Default value: "less", or "more".

PS1
: See Appendix A [Customizing the Prompt], page 483.

    Default value: "\s:\#> ".

PS2
: See Appendix A [Customizing the Prompt], page 483.

    Default value: "> ".

PS4
: See Appendix A [Customizing the Prompt], page 483.

    Default value: "+ ".

beep_on_error
: See Chapter 12 [Errors and Warnings], page 147.

    Default value: 0.

completion_append_char
: See Appendix A [Commands For Completion], page 479.

    Default value: " ".

default_save_options
: See Section 14.1.3 [Simple File I/O], page 167.

    Default value: "ascii".

crash_dumps_octave_core
: See Section 14.1.3 [Simple File I/O], page 167.

    Default value: 1.

fixed_point_format
: See Section 4.1 [Matrices], page 34.

    Default value: 0.

gnuplot_binary
: See Section 15.1.2 [Three-Dimensional Plotting], page 207.

    Default value: "gnuplot".

history_file
: See Appendix A [Commands For History], page 480.

    Default value: "~/.octave_hist".

history_size
: See Appendix A [Commands For History], page 480.

    Default value: 1024.

ignore_function_time_stamp
: See Section 11.7 [Function Files], page 133.

    Default value: "system".

max_recursion_depth
: See Section 8.2.2 [Recursion], page 97.

    Default value: 256.

output_max_field_width
: See Section 4.1 [Matrices], page 34.

    Default value: 10.

output_precision
: See Section 4.1 [Matrices], page 34.

    Default value: 5.

page_screen_output
: See Chapter 14 [Input and Output], page 161.

    Default value: 1.

print_answer_id_name
: See Section 14.1.1 [Terminal Output], page 161.

    Default value: 1.

print_empty_dimensions
: See Section 4.1.1 [Empty Matrices], page 37.

    Default value: 1.

return_last_computed_value
: See Section 11.5 [Returning From a Function], page 132.

    Default value: 0.

save_precision
: See Section 14.1.3 [Simple File I/O], page 167.

    Default value: 17.

Chapter 7: Variables                                                            91

`saving_history`
>   See Appendix A [Commands For History], page 480.
>
>   Default value: 1.

`sighup_dumps_octave_core`
>   See Section 14.1.3 [Simple File I/O], page 167.
>
>   Default value: 1.

`sigterm_dumps_octave_core`
>   See Section 14.1.3 [Simple File I/O], page 167.
>
>   Default value: 1.

`silent_functions`
>   See Section 11.1 [Defining Functions], page 125.
>
>   Default value: 0.

`split_long_rows`
>   See Section 4.1 [Matrices], page 34.
>
>   Default value: 1.

`struct_levels_to_print`
>   See Section 6.1 [Data Structures], page 63.
>
>   Default value: 2.

`suppress_verbose_help_message`
>   See Section 2.3 [Getting Help], page 22.
>
>   Default value: 1.

## 7.5 Defaults from the Environment

Octave uses the values of the following environment variables to set the default values for the corresponding built-in or internal variables. In addition, the values from the environment may be overridden by command-line arguments. See Section 2.1.1 [Command Line Options], page 17.

`EDITOR`
>   See Appendix A [Commands For History], page 480.
>
>   Built-in variable: `EDITOR`.

`OCTAVE_EXEC_PATH`
>   See Section 32.5 [Controlling Subprocesses], page 457.
>
>   Built-in variable: `EXEC_PATH`. Command-line argument: `--exec-path`.

`OCTAVE_PATH`
>   See Section 11.7 [Function Files], page 133.
>
>   Internal variable changed by function `path`. Command-line argument: `--path`.

OCTAVE_INFO_FILE
: See Section 2.3 [Getting Help], page 22.

Internal variable changed by function `info_file`. Command-line argument: `--info-file`.

OCTAVE_INFO_PROGRAM
: See Section 2.3 [Getting Help], page 22.

Internal variable changed by function `info_program`. Command-line argument: `--info-program`.

OCTAVE_HISTSIZE
: See Appendix A [Commands For History], page 480.

Built-in variable: `history_size`.

OCTAVE_HISTFILE
: See Appendix A [Commands For History], page 480.

Built-in variable: `history_file`.

# 8 Expressions

Expressions are the basic building block of statements in Octave. An expression evaluates to a value, which you can print, test, store in a variable, pass to a function, or assign a new value to a variable with an assignment operator.

An expression can serve as a statement on its own. Most other kinds of statements contain one or more expressions which specify data to be operated on. As in other languages, expressions in Octave include variables, array references, constants, and function calls, as well as combinations of these with various operators.

## 8.1 Index Expressions

An *index expression* allows you to reference or extract selected elements of a matrix or vector.

Indices may be scalars, vectors, ranges, or the special operator :, which may be used to select entire rows or columns.

Vectors are indexed using a single index expression. Matrices may be indexed using one or two indices. When using a single index expression, the elements of the matrix are taken in column-first order; the dimensions of the output match those of the index expression. For example,

```
a (2)        # a scalar
a (1:2)      # a row vector
a ([1; 2])   # a column vector
```

As a special case, when a colon is used as a single index, the output is a column vector containing all the elements of the vector or matrix. For example

```
a (:)        # a column vector
```

Given the matrix

```
a = [1, 2; 3, 4]
```

all of the following expressions are equivalent

```
a (1, [1, 2])
a (1, 1:2)
a (1, :)
```

and select the first row of the matrix.

Indexing a scalar with a vector of ones can be used to create a vector the same size as the index vector, with each element equal to the value of the original scalar. For example, the following statements

```
a = 13;
a ([1, 1, 1, 1])
```

produce a vector whose four elements are all equal to 13.

Similarly, indexing a scalar with two vectors of ones can be used to create a matrix. For example the following statements

```
a = 13;
a ([1, 1], [1, 1, 1])
```
create a 2 by 3 matrix with all elements equal to 13.

This is an obscure notation and should be avoided. It is better to use the function ones to generate a matrix of the appropriate size whose elements are all one, and then to scale it to produce the desired result. See Section 16.4 [Special Utility Matrices], page 244.

It is also possible to create a matrix with different values. The following example creates a 10 dimensional row vector $a$ containing the values $a_i = \sqrt{i}$.

```
for i = 1:10
  a(i) = sqrt (i);
endfor
```

Note that it is quite inefficient to create a vector using a loop like the one shown in the example above. In this particular case, it would have been much more efficient to use the expression

```
a = sqrt (1:10);
```

thus avoiding the loop entirely. In cases where a loop is still required, or a number of values must be combined to form a larger matrix, it is generally much faster to set the size of the matrix first, and then insert elements using indexing commands. For example, given a matrix a,

```
[nr, nc] = size (a);
x = zeros (nr, n * nc);
for i = 1:n
  x(:,(i-1)*nc+1:i*nc) = a;
endfor
```

is considerably faster than

```
x = a;
for i = 1:n-1
  x = [x, a];
endfor
```

particularly for large matrices because Octave does not have to repeatedly resize the result.

**subsref** (*val, idx*)  Built-in Function

Perform the subscripted element selection operation according to the subscript specified by *idx*.

The subscript *idx* is expected to be a structure array with fields type and subs. Valid values for type are "()", "{}", and ".". The subs field may be either ":" or a cell array of index values.

The following example shows how to extract the two first columns of a matrix

```
val = magic(3)
   ⇒ val = [ 8   1   6
             3   5   7
             4   9   2 ]
idx.type = "()";
idx.subs = {":", 1:2};
```

```
          subsref(val, idx)
       ⇒ [ 8   1
           3   5
           4   9 ]
```
Note that this is the same as writing `val(:,1:2)`.

See also: subsasgn, substruct.

*ind* = sub2ind (*dims*, *i*, *j*)                          Function File
*ind* = sub2ind (*dims*, *s1*, *s2*, ..., *sN*)             Function File
    Convert subscripts into a linear index.

    The following example shows how to convert the two-dimensional index (2,3) of a 3-by-3 matrix to a linear index.
```
          linear_index = sub2ind ([3, 3], 2, 3)
       ⇒ 8
```
See also: ind2sub.

[*s1*, *s2*, ..., *sN*] = ind2sub (*dims*, *ind*)           Function File
    Convert a linear index into subscripts.

    The following example shows how to convert the linear index 8 in a 3-by-3 matrix into a subscript.
```
          [r, c] = ind2sub ([3, 3], 8)
       ⇒ r =   2
         c =   3
```
See also: sub2ind.

## 8.2 Calling Functions

A *function* is a name for a particular calculation. Because it has a name, you can ask for it by name at any point in the program. For example, the function sqrt computes the square root of a number.

A fixed set of functions are *built-in*, which means they are available in every Octave program. The sqrt function is one of these. In addition, you can define your own functions. See Chapter 11 [Functions and Scripts], page 125, for information about how to do this.

The way to use a function is with a *function call* expression, which consists of the function name followed by a list of *arguments* in parentheses. The arguments are expressions which give the raw materials for the calculation that the function will do. When there is more than one argument, they are separated by commas. If there are no arguments, you can omit the parentheses, but it is a good idea to include them anyway, to clearly indicate that a function call was intended. Here are some examples:

```
sqrt (x^2 + y^2)     # One argument
ones (n, m)          # Two arguments
rand ()              # No arguments
```

Each function expects a particular number of arguments. For example, the sqrt function must be called with a single argument, the number to take the square root of:

sqrt (*argument*)

Some of the built-in functions take a variable number of arguments, depending on the particular usage, and their behavior is different depending on the number of arguments supplied.

Like every other expression, the function call has a value, which is computed by the function based on the arguments you give it. In this example, the value of sqrt (*argument*) is the square root of the argument. A function can also have side effects, such as assigning the values of certain variables or doing input or output operations.

Unlike most languages, functions in Octave may return multiple values. For example, the following statement

```
[u, s, v] = svd (a)
```

computes the singular value decomposition of the matrix a and assigns the three result matrices to u, s, and v.

The left side of a multiple assignment expression is itself a list of expressions, and is allowed to be a list of variable names or index expressions. See also Section 8.1 [Index Expressions], page 93, and Section 8.6 [Assignment Ops], page 103.

## 8.2.1 Call by Value

In Octave, unlike Fortran, function arguments are passed by value, which means that each argument in a function call is evaluated and assigned to a temporary location in memory before being passed to the function. There is currently no way to specify that a function parameter should be passed by reference instead of by value. This means that it is impossible to directly alter the value of a function parameter in the calling function. It can only change the local copy within the function body. For example, the function

```
function f (x, n)
  while (n-- > 0)
    disp (x);
  endwhile
endfunction
```

displays the value of the first argument n times. In this function, the variable n is used as a temporary variable without having to worry that its value might also change in the calling function. Call by value is also useful because it is always possible to pass constants for any function parameter without first having to determine that the function will not attempt to modify the parameter.

The caller may use a variable as the expression for the argument, but the called function does not know this: it only knows what value the argument had. For example, given a function called as

```
foo = "bar";
fcn (foo)
```
you should not think of the argument as being "the variable foo." Instead, think of the argument as the string value, "bar".

Even though Octave uses pass-by-value semantics for function arguments, values are not copied unnecessarily. For example,
```
x = rand (1000);
f (x);
```
does not actually force two 1000 by 1000 element matrices to exist *unless* the function f modifies the value of its argument. Then Octave must create a copy to avoid changing the value outside the scope of the function f, or attempting (and probably failing!) to modify the value of a constant or the value of a temporary result.

### 8.2.2 Recursion

With some restrictions[1], recursive function calls are allowed. A *recursive function* is one which calls itself, either directly or indirectly. For example, here is an inefficient[2] way to compute the factorial of a given integer:
```
function retval = fact (n)
  if (n > 0)
    retval = n * fact (n-1);
  else
    retval = 1;
  endif
endfunction
```
This function is recursive because it calls itself directly. It eventually terminates because each time it calls itself, it uses an argument that is one less than was used for the previous call. Once the argument is no longer greater than zero, it does not call itself, and the recursion ends.

The built-in variable max_recursion_depth specifies a limit to the recursion depth and prevents Octave from recursing infinitely.

*val* = max_recursion_depth ()  *Built-in Function*
*old_val* = max_recursion_depth (*new_val*)  *Built-in Function*
  Query or set the internal limit on the number of times a function may be called recursively. If the limit is exceeded, an error message is printed and control returns to the top level.

---

[1] Some of Octave's functions are implemented in terms of functions that cannot be called recursively. For example, the ODE solver lsode is ultimately implemented in a Fortran subroutine that cannot be called recursively, so lsode should not be called either directly or indirectly from within the user-supplied function that lsode requires. Doing so will result in an error.

[2] It would be much better to use prod (1:n), or gamma (n+1) instead, after first checking to ensure that the value n is actually a positive integer.

## 8.3 Arithmetic Operators

The following arithmetic operators are available, and work on scalars and matrices.

*x* + *y*
: Addition. If both operands are matrices, the number of rows and columns must both agree. If one operand is a scalar, its value is added to all the elements of the other operand.

*x* .+ *y*
: Element by element addition. This operator is equivalent to +.

*x* - *y*
: Subtraction. If both operands are matrices, the number of rows and columns of both must agree.

*x* .- *y*
: Element by element subtraction. This operator is equivalent to -.

*x* * *y*
: Matrix multiplication. The number of columns of *x* must agree with the number of rows of *y*.

*x* .* *y*
: Element by element multiplication. If both operands are matrices, the number of rows and columns must both agree.

*x* / *y*
: Right division. This is conceptually equivalent to the expression

    (inverse (y') * x')'

    but it is computed without forming the inverse of *y'*.

    If the system is not square, or if the coefficient matrix is singular, a minimum norm solution is computed.

*x* ./ *y*
: Element by element right division.

*x* \ *y*
: Left division. This is conceptually equivalent to the expression

    inverse (x) * y

    but it is computed without forming the inverse of *x*.

    If the system is not square, or if the coefficient matrix is singular, a minimum norm solution is computed.

*x* .\ *y*
: Element by element left division. Each element of *y* is divided by each corresponding element of *x*.

## Chapter 8: Expressions

x ^ y
x ** y
: Power operator. If $x$ and $y$ are both scalars, this operator returns $x$ raised to the power $y$. If $x$ is a scalar and $y$ is a square matrix, the result is computed using an eigenvalue expansion. If $x$ is a square matrix, the result is computed by repeated multiplication if $y$ is an integer, and by an eigenvalue expansion if $y$ is not an integer. An error results if both $x$ and $y$ are matrices.

    The implementation of this operator needs to be improved.

x .^ y
x .** y
: Element by element power operator. If both operands are matrices, the number of rows and columns must both agree.

-x
: Negation.

+x
: Unary plus. This operator has no effect on the operand.

x '
: Complex conjugate transpose. For real arguments, this operator is the same as the transpose operator. For complex arguments, this operator is equivalent to the expression

        conj (x.')

x.'
: Transpose.

Note that because Octave's element by element operators begin with a ., there is a possible ambiguity for statements like

    1./m

because the period could be interpreted either as part of the constant or as part of the operator. To resolve this conflict, Octave treats the expression as if you had typed

    (1) ./ m

and not

    (1.) / m

Although this is inconsistent with the normal behavior of Octave's lexer, which usually prefers to break the input into tokens by preferring the longest possible match at any given point, it is more useful in this case.

## 8.4 Comparison Operators

*Comparison operators* compare numeric values for relationships such as equality. They are written using *relational operators*.

All of Octave's comparison operators return a value of 1 if the comparison is true, or 0 if it is false. For matrix values, they all work on an element-by-element basis. For example,

```
[1, 2; 3, 4] == [1, 3; 2, 4]
    ⇒   1   0
        0   1
```

If one operand is a scalar and the other is a matrix, the scalar is compared to each element of the matrix in turn, and the result is the same size as the matrix.

$x < y$
: True if $x$ is less than $y$.

$x <= y$
: True if $x$ is less than or equal to $y$.

$x == y$
: True if $x$ is equal to $y$.

$x >= y$
: True if $x$ is greater than or equal to $y$.

$x > y$
: True if $x$ is greater than $y$.

$x \mathrel{!}= y$
$x \mathrel{\sim}= y$
: True if $x$ is not equal to $y$.

String comparisons may also be performed with the strcmp function, not with the comparison operators listed above. See Chapter 5 [Strings], page 47.

isequal (*x1*, *x2*, ...)  *Function File*
: Return true if all of *x1*, *x2*, ... are equal.

**See also:** isequalwithequalnans.

isequalwithequalnans (*x1*, *x2*, ...)  *Function File*
: Assuming NaN == NaN, return true if all of *x1*, *x2*, ... are equal.

**See also:** isequal.

## 8.5 Boolean Expressions

### 8.5.1 Element-by-element Boolean Operators

An *element-by-element boolean expression* is a combination of comparison expressions using the boolean operators "or" (|), "and" (&), and "not" (!), along with parentheses to control nesting. The truth of the boolean expression is computed by combining the truth values of the corresponding elements of the component expressions. A value is considered to be false if it is zero, and true otherwise.

Element-by-element boolean expressions can be used wherever comparison expressions can be used. They can be used in if and while statements. However, a matrix value used as the condition in an if or while statement is only true if *all* of its elements are nonzero.

Like comparison operations, each element of an element-by-element boolean expression also has a numeric value (1 if true, 0 if false) that comes into play if the result of the boolean expression is stored in a variable, or used in arithmetic.

Here are descriptions of the three element-by-element boolean operators.

*boolean1* & *boolean2*
    Elements of the result are true if both corresponding elements of *boolean1* and *boolean2* are true.

*boolean1* | *boolean2*
    Elements of the result are true if either of the corresponding elements of *boolean1* or *boolean2* is true.

! *boolean*
~ *boolean*
    Each element of the result is true if the corresponding element of *boolean* is false.

For matrix operands, these operators work on an element-by-element basis. For example, the expression

    [1, 0; 0, 1] & [1, 0; 2, 3]

returns a two by two identity matrix.

For the binary operators, the dimensions of the operands must conform if both are matrices. If one of the operands is a scalar and the other a matrix, the operator is applied to the scalar and each element of the matrix.

For the binary element-by-element boolean operators, both subexpressions *boolean1* and *boolean2* are evaluated before computing the result. This can make a difference when the expressions have side effects. For example, in the expression

    a & b++

the value of the variable *b* is incremented even if the variable *a* is zero.

This behavior is necessary for the boolean operators to work as described for matrix-valued operands.

## 8.5.2 Short-circuit Boolean Operators

Combined with the implicit conversion to scalar values in if and while conditions, Octave's element-by-element boolean operators are often sufficient for performing most logical operations. However, it is sometimes desirable to stop evaluating a boolean expression as soon as the overall truth value can be determined. Octave's *short-circuit* boolean operators work this way.

boolean1 && boolean2

> The expression *boolean1* is evaluated and converted to a scalar using the equivalent of the operation all (*boolean1*(:)). If it is false, the result of the overall expression is 0. If it is true, the expression *boolean2* is evaluated and converted to a scalar using the equivalent of the operation all (*boolean1*(:)). If it is true, the result of the overall expression is 1. Otherwise, the result of the overall expression is 0.
>
> **Warning:** there is one exception to the rule of evaluating all (*boolean1*(:)), which is when boolean1 is the empty matrix. The truth value of an empty matrix is always false so [] && true evaluates to false even though all ([]) is true.

boolean1 || boolean2

> The expression *boolean1* is evaluated and converted to a scalar using the equivalent of the operation all (*boolean1*(:)). If it is true, the result of the overall expression is 1. If it is false, the expression *boolean2* is evaluated and converted to a scalar using the equivalent of the operation all (*boolean1*(:)). If it is true, the result of the overall expression is 1. Otherwise, the result of the overall expression is 0.
>
> **Warning:** the truth value of an empty matrix is always false, see the previous list item for details.

The fact that both operands may not be evaluated before determining the overall truth value of the expression can be important. For example, in the expression

```
a && b++
```

the value of the variable *b* is only incremented if the variable *a* is nonzero.

This can be used to write somewhat more concise code. For example, it is possible write

```
function f (a, b, c)
  if (nargin > 2 && isstr (c))
    ...
```

instead of having to use two if statements to avoid attempting to evaluate an argument that doesn't exist. For example, without the short-circuit feature, it would be necessary to write

```
function f (a, b, c)
  if (nargin > 2)
    if (isstr (c))
      ...
```

Writing

```
function f (a, b, c)
  if (nargin > 2 & isstr (c))
    ...
```

would result in an error if f were called with one or two arguments because Octave would be forced to try to evaluate both of the operands for the operator &.

## 8.6 Assignment Expressions

An *assignment* is an expression that stores a new value into a variable. For example, the following expression assigns the value 1 to the variable z:

```
z = 1
```

After this expression is executed, the variable z has the value 1. Whatever old value z had before the assignment is forgotten. The = sign is called an *assignment operator*.

Assignments can store string values also. For example, the following expression would store the value "this food is good" in the variable message:

```
thing = "food"
predicate = "good"
message = [ "this " , thing , " is " , predicate ]
```

(This also illustrates concatenation of strings.)

Most operators (addition, concatenation, and so on) have no effect except to compute a value. If you ignore the value, you might as well not use the operator. An assignment operator is different. It does produce a value, but even if you ignore the value, the assignment still makes itself felt through the alteration of the variable. We call this a *side effect*.

The left-hand operand of an assignment need not be a variable (see Chapter 7 [Variables], page 81). It can also be an element of a matrix (see Section 8.1 [Index Expressions], page 93) or a list of return values (see Section 8.2 [Calling Functions], page 95). These are all called *lvalues*, which means they can appear on the left-hand side of an assignment operator. The right-hand operand may be any expression. It produces the new value which the assignment stores in the specified variable, matrix element, or list of return values.

It is important to note that variables do *not* have permanent types. The type of a variable is simply the type of whatever value it happens to hold at the moment. In the following program fragment, the variable foo has a numeric value at first, and a string value later on:

```
octave:13> foo = 1
foo = 1
octave:13> foo = "bar"
foo = bar
```

When the second assignment gives foo a string value, the fact that it previously had a numeric value is forgotten.

Assignment of a scalar to an indexed matrix sets all of the elements that are referenced by the indices to the scalar value. For example, if a is a matrix with at least two columns,

```
a(:, 2) = 5
```
sets all the elements in the second column of a to 5.

Assigning an empty matrix [] works in most cases to allow you to delete rows or columns of matrices and vectors. See Section 4.1.1 [Empty Matrices], page 37. For example, given a 4 by 5 matrix A, the assignment
```
A (3, :) = []
```
deletes the third row of A, and the assignment
```
A (:, 1:2:5) = []
```
deletes the first, third, and fifth columns.

An assignment is an expression, so it has a value. Thus, z = 1 as an expression has the value 1. One consequence of this is that you can write multiple assignments together:
```
x = y = z = 0
```
stores the value 0 in all three variables. It does this because the value of z = 0, which is 0, is stored into y, and then the value of y = z = 0, which is 0, is stored into x.

This is also true of assignments to lists of values, so the following is a valid expression
```
[a, b, c] = [u, s, v] = svd (a)
```
that is exactly equivalent to
```
[u, s, v] = svd (a)
a = u
b = s
c = v
```

In expressions like this, the number of values in each part of the expression need not match. For example, the expression
```
[a, b] = [u, s, v] = svd (a)
```
is equivalent to
```
[u, s, v] = svd (a)
a = u
b = s
```
The number of values on the left side of the expression can, however, not exceed the number of values on the right side. For example, the following will produce an error.
```
[a, b, c, d] = [u, s, v] = svd (a)
    ⊣ error: element number 4 undefined in return list
    ⊣ error: evaluating assignment expression near line 8, column 15
```
A very common programming pattern is to increment an existing variable with a given value, like this
```
a = a + 2;
```
This can be written in a clearer and more condensed form using the += operator
```
a += 2;
```
Similar operators also exist for subtraction (-=), multiplication (*=), and division (/=). An expression of the form

## Chapter 8: Expressions

> *expr1 op= expr2*

is evaluated as

> *expr1 = (expr1) op (expr2)*

where *op* can be either +, -, *, or /. So, the expression

> a *= b+1

is evaluated as

> a = a * (b+1)

and *not*

> a = a * b + 1

You can use an assignment anywhere an expression is called for. For example, it is valid to write x != (y = 1) to set y to 1 and then test whether x equals 1. But this style tends to make programs hard to read. Except in a one-shot program, you should rewrite it to get rid of such nesting of assignments. This is never very hard.

**subsasgn** (*val, idx, rhs*)                                                        Built-in Function

Perform the subscripted assignment operation according to the subscript specified by *idx*.

The subscript *idx* is expected to be a structure array with fields type and subs. Valid values for type are "()", "{}", and ".". The subs field may be either ":" or a cell array of index values.

The following example shows how to set the two first columns of a 3-by-3 matrix to zero.

```
val = magic(3);
idx.type = "()";
idx.subs = {":", 1:2};
subsasgn (val, idx, 0)
    ⇒ [ 0   0   6
        0   0   7
        0   0   2 ]
```

Note that this is the same as writing val(:,1:2) = 0.

**See also:** subsref, substruct.

## 8.7 Increment Operators

*Increment operators* increase or decrease the value of a variable by 1. The operator to increment a variable is written as ++. It may be used to increment a variable either before or after taking its value.

For example, to pre-increment the variable $x$, you would write ++$x$. This would add one to $x$ and then return the new value of $x$ as the result of the expression. It is exactly the same as the expression $x = x + 1$.

To post-increment a variable $x$, you would write $x$++. This adds one to the variable $x$, but returns the value that $x$ had prior to incrementing it. For example, if $x$ is equal to 2, the result of the expression $x$++ is 2, and the new value of $x$ is 3.

For matrix and vector arguments, the increment and decrement operators work on each element of the operand.

Here is a list of all the increment and decrement expressions.

++$x$
> This expression increments the variable $x$. The value of the expression is the *new* value of $x$. It is equivalent to the expression $x = x + 1$.

--$x$
> This expression decrements the variable $x$. The value of the expression is the *new* value of $x$. It is equivalent to the expression $x = x - 1$.

$x$++
> This expression causes the variable $x$ to be incremented. The value of the expression is the *old* value of $x$.

$x$--
> This expression causes the variable $x$ to be decremented. The value of the expression is the *old* value of $x$.

## 8.8 Operator Precedence

*Operator precedence* determines how operators are grouped, when different operators appear close by in one expression. For example, * has higher precedence than +. Thus, the expression a + b * c means to multiply b and c, and then add a to the product (i.e., a + (b * c)).

You can overrule the precedence of the operators by using parentheses. You can think of the precedence rules as saying where the parentheses are assumed if you do not write parentheses yourself. In fact, it is wise to use parentheses whenever you have an unusual combination of operators, because other people who read the program may not remember what the precedence is in this case. You might forget as well, and then you too could make a mistake. Explicit parentheses will help prevent any such mistake.

When operators of equal precedence are used together, the leftmost operator groups first, except for the assignment and exponentiation operators, which group in the opposite order. Thus, the expression a - b + c groups as (a - b) + c, but the expression a = b = c groups as a = (b = c).

Chapter 8: Expressions                                                                                          107

The precedence of prefix unary operators is important when another operator follows the operand. For example, -x^2 means -(x^2), because - has lower precedence than ^.

Here is a table of the operators in Octave, in order of increasing precedence.

statement separators
    ;, ,.

assignment
    =, +=, -=, *=,/=. This operator groups right to left.

logical "or" and "and"
    ||, &&.

element-wise "or" and "and"
    |, &.

relational
    <, <=, ==, >=, >, !=, ~=.

colon
    :.

add, subtract
    +, -.

multiply, divide
    *, /, \, .\, .*, ./.

transpose
    ', .'

unary plus, minus, increment, decrement, and ''not''
    +, -, ++, --, !, ~.

exponentiation
    ^, **, .^, .**.

# 9 Evaluation

Normally, you evaluate expressions simply by typing them at the Octave prompt, or by asking Octave to interpret commands that you have saved in a file.

Sometimes, you may find it necessary to evaluate an expression that has been computed and stored in a string, which is exactly what the eval function lets you do.

eval (*try*, *catch*)                                                             Built-in Function

Parse the string *try* and evaluate it as if it were an Octave program. If that fails, evaluate the optional string *catch*. The string *try* is evaluated in the current context, so any results remain available after eval returns.

The following example makes the variable a with the approximate value 3.1416 available.

```
eval("a = acos(-1);");
```

If an error occurs during the evaluation of *try* the *catch* string is evaluated, as the following example shows:

```
eval ('error ("This is a bad example");',
      'printf ("This error occurred:\n%s", lasterr ());');
     ⊣ This error occurred:
        error: This is a bad example
```

## 9.1 Calling a Function by its Name

The feval function allows you to call a function from a string containing its name. This is useful when writing a function that needs to call user-supplied functions. The feval function takes the name of the function to call as its first argument, and the remaining arguments are given to the function.

The following example is a simple-minded function using feval that finds the root of a user-supplied function of one variable using Newton's method.

```
function result = newtroot (fname, x)

# usage: newtroot (fname, x)
#
#   fname : a string naming a function f(x).
#   x     : initial guess

  delta = tol = sqrt (eps);
  maxit = 200;
  fx = feval (fname, x);
  for i = 1:maxit
    if (abs (fx) < tol)
      result = x;
      return;
    else
      fx_new = feval (fname, x + delta);
      deriv = (fx_new - fx) / delta;
      x = x - fx / deriv;
      fx = fx_new;
    endif
  endfor

  result = x;

endfunction
```

Note that this is only meant to be an example of calling user-supplied functions and should not be taken too seriously. In addition to using a more robust algorithm, any serious code would check the number and type of all the arguments, ensure that the supplied function really was a function, etc. See Section 4.6 [Predicates for Numeric Objects], page 44, for example, for a list of predicates for numeric objects, and see Section 7.3 [Status of Variables], page 85, for a description of the `exist` function.

`feval (name, ...)`                                           Built-in Function
    Evaluate the function named *name*. Any arguments after the first are passed on to the named function. For example,

        `feval ("acos", -1)`
            ⇒ 3.1416

calls the function `acos` with the argument -1.

    The function `feval` is necessary in order to be able to write functions that call user-supplied functions, because Octave does not have a way to declare a pointer to a function (like C) or to declare a special kind of variable that can be used to hold the name of a function (like EXTERNAL in Fortran). Instead, you must refer to functions by name, and use `feval` to call them.

    A similar function `run` exists for calling user script files, that are not necessarily on the user path

Chapter 9: Evaluation                                                          111

run (*f*)                                                              Function File
run *f*                                                                    Command
    Run scripts in the current workspace that are not necessarily on the path.
    If *f* is the script to run, including its path, then run change the directory to
    the directory where *f* is found. run then executes the script, and returns to
    the original directory.

    **See also:** system.

## 9.2 Evaluation in a Different Context

Before you evaluate an expression you need to substitute the values of the variables used in the expression. These are stored in the symbol table. Whenever the interpreter starts a new function it saves the current symbol table and creates a new one, initializing it with the list of function parameters and a couple of predefined variables such as nargin. Expressions inside the function use the new symbol table.

Sometimes you want to write a function so that when you call it, it modifies variables in your own context. This allows you to use a pass-by-name style of function, which is similar to using a pointer in programming languages such as C. The following functions allow evaluation and assignment in a specified context:

evalin (*context, try, catch*)                                        Built-in Function
    Like eval, except that the expressions are evaluated in the context *context*,
    which may be either "caller" or "base".

assignin (*context, varname, value*)                                  Built-in Function
    Assign *value* to *varname* in context *context*, which may be either "base" or
    "caller".

    Consider how you might write save and load as m-files. For example,
```
function create_data
  x = linspace (0, 10, 10);
  y = sin (x);
  save mydata x y
endfunction
```
    With evalin, you could write save as follows:
```
function save (file, name1, name2)
  f = open_save_file (file);
  save_var(f, name1, evalin ("caller", name1));
  save_var(f, name2, evalin ("caller", name2));
endfunction
```
Here, caller is the create_data function and name1 is the string "x", which evaluates simply as the value of x.

You later want to load the values back from mydata in a different context:

```
function process_data
  load mydata
  ... do work ...
endfunction
```

With assignin, you could write load as follows:

```
function load (file)
  f = open_load_file (file);
  [name, val] = load_var (f);
  assignin ("caller", name, val);
  [name, val] = load_var (f);
  assignin ("caller", name, val);
endfunction
```

Here, caller is the process_data function.

You can set and use variables at the command prompt using the context base rather than caller.

These functions are rarely used in practice. One example is the fail (code, pattern) function which evaluates code in the caller's context and checks that the error message it produces matches the given pattern. Other examples such as save and load are written in C++ where all Octave variables are in the caller context and evalin is not needed.

# 10 Statements

Statements may be a simple constant expression or a complicated list of nested loops and conditional statements.

*Control statements* such as if, while, and so on control the flow of execution in Octave programs. All the control statements start with special keywords such as if and while, to distinguish them from simple expressions. Many control statements contain other statements; for example, the if statement contains another statement which may or may not be executed.

Each control statement has a corresponding *end* statement that marks the end of the control statement. For example, the keyword endif marks the end of an if statement, and endwhile marks the end of a while statement. You can use the keyword end anywhere a more specific end keyword is expected, but using the more specific keywords is preferred because if you use them, Octave is able to provide better diagnostics for mismatched or missing end tokens.

The list of statements contained between keywords like if or while and the corresponding end statement is called the *body* of a control statement.

## 10.1 The if Statement

The if statement is Octave's decision-making statement. There are three basic forms of an if statement. In its simplest form, it looks like this:

```
if (condition)
  then-body
endif
```

*condition* is an expression that controls what the rest of the statement will do. The *then-body* is executed only if *condition* is true.

The condition in an if statement is considered true if its value is non-zero, and false if its value is zero. If the value of the conditional expression in an if statement is a vector or a matrix, it is considered true only if it is non-empty and *all* of the elements are non-zero.

The second form of an if statement looks like this:

```
if (condition)
  then-body
else
  else-body
endif
```

If *condition* is true, *then-body* is executed; otherwise, *else-body* is executed.

Here is an example:

```
if (rem (x, 2) == 0)
  printf ("x is even\n");
else
  printf ("x is odd\n");
endif
```

In this example, if the expression rem (x, 2) == 0 is true (that is, the value of x is divisible by 2), then the first printf statement is evaluated, otherwise the second printf statement is evaluated.

The third and most general form of the if statement allows multiple decisions to be combined in a single statement. It looks like this:

```
if (condition)
  then-body
elseif (condition)
  elseif-body
else
  else-body
endif
```

Any number of elseif clauses may appear. Each condition is tested in turn, and if one is found to be true, its corresponding *body* is executed. If none of the conditions are true and the else clause is present, its body is executed. Only one else clause may appear, and it must be the last part of the statement.

In the following example, if the first condition is true (that is, the value of x is divisible by 2), then the first printf statement is executed. If it is false, then the second condition is tested, and if it is true (that is, the value of x is divisible by 3), then the second printf statement is executed. Otherwise, the third printf statement is performed.

```
if (rem (x, 2) == 0)
  printf ("x is even\n");
elseif (rem (x, 3) == 0)
  printf ("x is odd and divisible by 3\n");
else
  printf ("x is odd\n");
endif
```

Note that the elseif keyword must not be spelled else if, as is allowed in Fortran. If it is, the space between the else and if will tell Octave to treat this as a new if statement within another if statement's else clause. For example, if you write

```
if (c1)
  body-1
else if (c2)
  body-2
endif
```

Octave will expect additional input to complete the first if statement. If you are using Octave interactively, it will continue to prompt you for additional input. If Octave is reading this input from a file, it may complain about missing or mismatched end statements, or, if you have not used the more specific end statements (endif, endfor, etc.), it may simply produce incorrect results, without producing any warning messages.

It is much easier to see the error if we rewrite the statements above like this,

```
if (c1)
  body-1
else
  if (c2)
    body-2
  endif
```
using the indentation to show how Octave groups the statements. See Chapter 11 [Functions and Scripts], page 125.

## 10.2 The switch Statement

It is very common to take different actions depending on the value of one variable. This is possible using the if statement in the following way

```
if (X == 1)
  do_something ();
elseif (X == 2)
  do_something_else ();
else
  do_something_completely_different ();
endif
```

This kind of code can however be very cumbersome to both write and maintain. To overcome this problem Octave supports the switch statement. Using this statement, the above example becomes

```
switch (X)
  case 1
    do_something ();
  case 2
    do_something_else ();
  otherwise
    do_something_completely_different ();
endswitch
```

This code makes the repetitive structure of the problem more explicit, making the code easier to read, and hence maintain. Also, if the variable X should change its name, only one line would need changing compared to one line per case when if statements are used.

The general form of the switch statement is

```
switch expression
  case label
    command_list
  case label
    command_list
  ...

  otherwise
    command_list
endswitch
```

where *label* can be any expression. However, duplicate *label* values are not detected, and only the *command_list* corresponding to the first match will be executed. For the switch statement to be meaningful at least one case *label command_list* clause must be present, while the otherwise *command_list* clause is optional.

If *label* is a cell array the corresponding *command_list* is executed if *any* of the elements of the cell array match *expression*. As an example, the following program will print Variable is either 6 or 7.

```
A = 7;
switch A
  case { 6, 7 }
    printf ("variable is either 6 or 7\n");
  otherwise
    printf ("variable is neither 6 nor 7\n");
endswitch
```

As with all other specific end keywords, endswitch may be replaced by end, but you can get better diagnostics if you use the specific forms.

One advantage of using the switch statement compared to using if statements is that the *labels* can be strings. If an if statement is used it is *not* possible to write

```
if (X == "a string") # This is NOT valid
```

since a character-to-character comparison between X and the string will be made instead of evaluating if the strings are equal. This special-case is handled by the switch statement, and it is possible to write programs that look like this

```
switch (X)
  case "a string"
    do_something
  ...
endswitch
```

### 10.2.1 Notes for the C programmer

The switch statement is also available in the widely used C programming language. There are, however, some differences between the statement in Octave and C

- Cases are exclusive, so they don't 'fall through' as do the cases in the switch statement of the C language.
- The *command_list* elements are not optional. Making the list optional would have meant requiring a separator between the label and the command list. Otherwise, things like

```
switch (foo)
  case (1) -2
  ...
```

would produce surprising results, as would

```
switch (foo)
  case (1)
  case (2)
    doit ();
  ...
```

particularly for C programmers. If doit() should be executed if *foo* is either 1 or 2, the above code should be written with a cell array like this

```
switch (foo)
  case { 1, 2 }
    doit ();
  ...
```

## 10.3 The while Statement

In programming, a *loop* means a part of a program that is (or at least can be) executed two or more times in succession.

The while statement is the simplest looping statement in Octave. It repeatedly executes a statement as long as a condition is true. As with the condition in an if statement, the condition in a while statement is considered true if its value is non-zero, and false if its value is zero. If the value of the conditional expression in a while statement is a vector or a matrix, it is considered true only if it is non-empty and *all* of the elements are non-zero.

Octave's while statement looks like this:

```
while (condition)
  body
endwhile
```

Here *body* is a statement or list of statements that we call the *body* of the loop, and *condition* is an expression that controls how long the loop keeps running.

The first thing the while statement does is test *condition*. If *condition* is true, it executes the statement *body*. After *body* has been executed, *condition* is tested again, and if it is still true, *body* is executed again. This process repeats until *condition* is no longer true. If *condition* is initially false, the body of the loop is never executed.

This example creates a variable fib that contains the first ten elements of the Fibonacci sequence.

```
fib = ones (1, 10);
i = 3;
while (i <= 10)
  fib (i) = fib (i-1) + fib (i-2);
  i++;
endwhile
```

Here the body of the loop contains two statements.

The loop works like this: first, the value of i is set to 3. Then, the while tests whether i is less than or equal to 10. This is the case when i equals 3, so the value of the i-th element of fib is set to the sum of the previous two values in the sequence. Then the i++ increments the value of i and the loop repeats. The loop terminates when i reaches 11.

A newline is not required between the condition and the body; but using one makes the program clearer unless the body is very simple.

## 10.4 The do-until Statement

The do-until statement is similar to the while statement, except that it repeatedly executes a statement until a condition becomes true, and the test of the condition is at the end of the loop, so the body of the loop is always executed at least once. As with the condition in an if statement, the condition in a do-until statement is considered true if its value is non-zero, and false if its value is zero. If the value of the conditional expression in a do-until statement is a vector or a matrix, it is considered true only if it is non-empty and *all* of the elements are non-zero.

Octave's do-until statement looks like this:

```
do
  body
until (condition)
```

Here *body* is a statement or list of statements that we call the *body* of the loop, and *condition* is an expression that controls how long the loop keeps running.

This example creates a variable fib that contains the first ten elements of the Fibonacci sequence.

```
fib = ones (1, 10);
i = 2;
do
  i++;
  fib (i) = fib (i-1) + fib (i-2);
until (i == 10)
```

A newline is not required between the do keyword and the body; but using one makes the program clearer unless the body is very simple.

## 10.5 The for Statement

The for statement makes it more convenient to count iterations of a loop. The general form of the for statement looks like this:

```
for var = expression
  body
endfor
```

where *body* stands for any statement or list of statements, *expression* is any valid expression, and *var* may take several forms. Usually it is a simple variable name or an indexed variable. If the value of *expression* is a structure, *var* may also be a vector with two elements. See Section 10.5.1 [Looping Over Structure Elements], page 120, below.

The assignment expression in the for statement works a bit differently than Octave's normal assignment statement. Instead of assigning the complete result of the expression, it assigns each column of the expression to *var* in turn. If

## Chapter 10: Statements

*expression* is a range, a row vector, or a scalar, the value of *var* will be a scalar each time the loop body is executed. If *var* is a column vector or a matrix, *var* will be a column vector each time the loop body is executed.

The following example shows another way to create a vector containing the first ten elements of the Fibonacci sequence, this time using the for statement:

```
fib = ones (1, 10);
for i = 3:10
  fib (i) = fib (i-1) + fib (i-2);
endfor
```

This code works by first evaluating the expression 3:10, to produce a range of values from 3 to 10 inclusive. Then the variable i is assigned the first element of the range and the body of the loop is executed once. When the end of the loop body is reached, the next value in the range is assigned to the variable i, and the loop body is executed again. This process continues until there are no more elements to assign.

Within Octave is it also possible to iterate over matrices or cell arrays using the for statement. For example consider

```
disp("Loop over a matrix")
for i = [1,3;2,4]
  i
endfor
disp("Loop over a cell array")
for i = {1,"two";"three",4}
  i
endfor
```

In this case the variable i takes on the value of the columns of the matrix or cell matrix. So the first loop iterates twice, producing two column vectors [1;2], followed by [3;4], and likewise for the loop over the cell array. This can be extended to loops over multidimensional arrays. For example

```
a = [1,3;2,4]; b = cat(3, a, 2*a);
for i = c
  i
endfor
```

In the above case, the multidimensional matrix c is reshaped to a two dimensional matrix as reshape (c, rows(c), prod(size(c)(2:end))) and then the same behavior as a loop over a two dimensional matrix is produced.

Although it is possible to rewrite all for loops as while loops, the Octave language has both statements because often a for loop is both less work to type and more natural to think of. Counting the number of iterations is very common in loops and it can be easier to think of this counting as part of looping rather than as something to do inside the loop.

### 10.5.1 Looping Over Structure Elements

A special form of the `for` statement allows you to loop over all the elements of a structure:

```
for [ val, key ] = expression
  body
endfor
```

In this form of the `for` statement, the value of *expression* must be a structure. If it is, *key* and *val* are set to the name of the element and the corresponding value in turn, until there are no more elements. For example,

```
x.a = 1
x.b = [1, 2; 3, 4]
x.c = "string"
for [val, key] = x
  key
  val
endfor
```

⊣ key = a
⊣ val = 1
⊣ key = b
⊣ val =
⊣
⊣   1   2
⊣   3   4
⊣
⊣ key = c
⊣ val = string

The elements are not accessed in any particular order. If you need to cycle through the list in a particular way, you will have to use the function `fieldnames` and sort the list yourself.

The *key* variable may also be omitted. If it is, the brackets are also optional. This is useful for cycling through the values of all the structure elements when the names of the elements do not need to be known.

## 10.6 The break Statement

The `break` statement jumps out of the innermost `for` or `while` loop that encloses it. The `break` statement may only be used within the body of a loop. The following example finds the smallest divisor of a given integer, and also identifies prime numbers:

# Chapter 10: Statements

```
num = 103;
div = 2;
while (div*div <= num)
  if (rem (num, div) == 0)
    break;
  endif
  div++;
endwhile
if (rem (num, div) == 0)
  printf ("Smallest divisor of %d is %d\n", num, div)
else
  printf ("%d is prime\n", num);
endif
```

When the remainder is zero in the first while statement, Octave immediately *breaks out* of the loop. This means that Octave proceeds immediately to the statement following the loop and continues processing. (This is very different from the exit statement which stops the entire Octave program.)

Here is another program equivalent to the previous one. It illustrates how the *condition* of a while statement could just as well be replaced with a break inside an if:

```
num = 103;
div = 2;
while (1)
  if (rem (num, div) == 0)
    printf ("Smallest divisor of %d is %d\n", num, div);
    break;
  endif
  div++;
  if (div*div > num)
    printf ("%d is prime\n", num);
    break;
  endif
endwhile
```

## 10.7 The continue Statement

The continue statement, like break, is used only inside for or while loops. It skips over the rest of the loop body, causing the next cycle around the loop to begin immediately. Contrast this with break, which jumps out of the loop altogether. Here is an example:

```
# print elements of a vector of random
# integers that are even.

# first, create a row vector of 10 random
# integers with values between 0 and 100:

vec = round (rand (1, 10) * 100);

# print what we're interested in:

for x = vec
  if (rem (x, 2) != 0)
    continue;
  endif
  printf ("%d\n", x);
endfor
```

If one of the elements of *vec* is an odd number, this example skips the print statement for that element, and continues back to the first statement in the loop.

This is not a practical example of the `continue` statement, but it should give you a clear understanding of how it works. Normally, one would probably write the loop like this:

```
for x = vec
  if (rem (x, 2) == 0)
    printf ("%d\n", x);
  endif
endfor
```

## 10.8 The `unwind_protect` Statement

Octave supports a limited form of exception handling modelled after the unwind-protect form of Lisp.

The general form of an `unwind_protect` block looks like this:

```
unwind_protect
  body
unwind_protect_cleanup
  cleanup
end_unwind_protect
```

where *body* and *cleanup* are both optional and may contain any Octave expressions or commands. The statements in *cleanup* are guaranteed to be executed regardless of how control exits *body*.

This is useful to protect temporary changes to global variables from possible errors. For example, the following code will always restore the original value of the global variable `frobnosticate` even if an error occurs while performing the indexing operation.

```
save_frobnosticate = frobnosticate;
unwind_protect
  frobnosticate = true;
  ...
unwind_protect_cleanup
  frobnosticate = save_frobnosticate;
end_unwind_protect
```

Without `unwind_protect`, the value of *frobnosticate* would not be restored if an error occurs while performing the indexing operation because evaluation would stop at the point of the error and the statement to restore the value would not be executed.

## 10.9 The try Statement

In addition to unwind_protect, Octave supports another limited form of exception handling.

The general form of a try block looks like this:

```
try
  body
catch
  cleanup
end_try_catch
```

where *body* and *cleanup* are both optional and may contain any Octave expressions or commands. The statements in *cleanup* are only executed if an error occurs in *body*.

No warnings or error messages are printed while *body* is executing. If an error does occur during the execution of *body*, *cleanup* can use the function `lasterr` to access the text of the message that would have been printed. This is the same as eval (*try*, *catch*) but it is more efficient since the commands do not need to be parsed each time the *try* and *catch* statements are evaluated. See Chapter 12 [Errors and Warnings], page 147, for more information about the `lasterr` function.

## 10.10 Continuation Lines

In the Octave language, most statements end with a newline character and you must tell Octave to ignore the newline character in order to continue a statement from one line to the next. Lines that end with the characters ... or \ are joined with the following line before they are divided into tokens by Octave's parser. For example, the lines

```
x = long_variable_name ...
    + longer_variable_name \
    - 42
```

form a single statement. The backslash character on the second line above is interpreted as a continuation character, *not* as a division operator.

For continuation lines that do not occur inside string constants, whitespace and comments may appear between the continuation marker and the newline character. For example, the statement

```
x = long_variable_name ...        # comment one
  + longer_variable_name \        # comment two
  - 42                            # last comment
```

is equivalent to the one shown above. Inside string constants, the continuation marker must appear at the end of the line just before the newline character.

Input that occurs inside parentheses can be continued to the next line without having to use a continuation marker. For example, it is possible to write statements like

```
if (fine_dining_destination == on_a_boat
    || fine_dining_destination == on_a_train)
  seuss (i, will, not, eat, them, sam, i, am, i,
         will, not, eat, green, eggs, and, ham);
endif
```

without having to add to the clutter with continuation markers.

# 11 Functions and Script Files

Complicated Octave programs can often be simplified by defining functions. Functions can be defined directly on the command line during interactive Octave sessions, or in external files, and can be called just like built-in functions.

## 11.1 Defining Functions

In its simplest form, the definition of a function named *name* looks like this:

    function name
      body
    endfunction

A valid function name is like a valid variable name: a sequence of letters, digits and underscores, not starting with a digit. Functions share the same pool of names as variables.

The function *body* consists of Octave statements. It is the most important part of the definition, because it says what the function should actually *do*.

For example, here is a function that, when executed, will ring the bell on your terminal (assuming that it is possible to do so):

    function wakeup
      printf ("\a");
    endfunction

The `printf` statement (see Chapter 14 [Input and Output], page 161) simply tells Octave to print the string "\a". The special character \a stands for the alert character (ASCII 7). See Chapter 5 [Strings], page 47.

Once this function is defined, you can ask Octave to evaluate it by typing the name of the function.

Normally, you will want to pass some information to the functions you define. The syntax for passing parameters to a function in Octave is

    function name (arg-list)
      body
    endfunction

where *arg-list* is a comma-separated list of the function's arguments. When the function is called, the argument names are used to hold the argument values given in the call. The list of arguments may be empty, in which case this form is equivalent to the one shown above. Variables used in the body of a function are local to the function, as are variables named in *arg-list*.[1]

To print a message along with ringing the bell, you might modify the wakeup to look like this:

---

[1] See Section 7.1 [Global Variables], page 82, for information about how to access global variables inside a function.

```
function wakeup (message)
  printf ("\a%s\n", message);
endfunction
```

Calling this function using a statement like this

```
wakeup ("Rise and shine!");
```

will cause Octave to ring your terminal's bell and print the message Rise and shine!, followed by a newline character (the \n in the first argument to the printf statement).

In most cases, you will also want to get some information back from the functions you define. Here is the syntax for writing a function that returns a single value:

```
function ret-var = name (arg-list)
  body
endfunction
```

The symbol *ret-var* is the name of the variable that will hold the value to be returned by the function. This variable is local to the function and must be defined before the end of the function body in order for the function to return a value.

For example, here is a function that computes the average of the elements of a vector:

```
function retval = avg (v)
  retval = sum (v) / length (v);
endfunction
```

If we had written avg like this instead,

```
function retval = avg (v)
  if (isvector (v))
    retval = sum (v) / length (v);
  endif
endfunction
```

and then called the function with a matrix instead of a vector as the argument, Octave would have printed an error message like this:

```
error: 'retval' undefined near line 1 column 10
error: evaluating index expression near line 7, column 1
```

because the body of the if statement was never executed, and retval was never defined. To prevent obscure errors like this, it is a good idea to always make sure that the return variables will always have values, and to produce meaningful error messages when problems are encountered. For example, avg could have been written like this:

```
function retval = avg (v)
  retval = 0;
  if (isvector (v))
    retval = sum (v) / length (v);
  else
    error ("avg: expecting vector argument");
  endif
endfunction
```

There is still one additional problem with this function. What if it is called without an argument? Without additional error checking, Octave will probably print an error message that won't really help you track down the source of the error. To allow you to catch errors like this, Octave provides each function with an automatic variable called nargin. Each time a function is called, nargin is automatically initialized to the number of arguments that have actually been passed to the function. For example, we might rewrite the avg function like this:

```
function retval = avg (v)
  retval = 0;
  if (nargin != 1)
    usage ("avg (vector)");
  endif
  if (isvector (v))
    retval = sum (v) / length (v);
  else
    error ("avg: expecting vector argument");
  endif
endfunction
```

Although Octave does not automatically report an error if you call a function with more arguments than expected, doing so probably indicates that something is wrong. Octave also does not automatically report an error if a function is called with too few arguments, but any attempt to use a variable that has not been given a value will result in an error. To avoid such problems and to provide useful messages, we check for both possibilities and issue our own error message.

nargin () Built-in Function
nargin (*fcn_name*) Built-in Function
    Within a function, return the number of arguments passed to the function. At the top level, return the number of command line arguments passed to Octave. If called with the optional argument *fcn_name*, return the maximum number of arguments the named function can accept, or -1 if the function accepts a variable number of arguments.

    **See also:** nargout, varargin, varargout.

inputname (*n*) Function File
    Return the text defining *n*-th input to the function.

val = silent_functions () Built-in Function
old_val = silent_functions (*new_val*) Built-in Function
    Query or set the internal variable that controls whether internal output from a function is suppressed. If this option is disabled, Octave will display the results produced by evaluating expressions within a function body that are not terminated with a semicolon.

## 11.2 Multiple Return Values

Unlike many other computer languages, Octave allows you to define functions that return more than one value. The syntax for defining functions that return multiple values is

```
function [ret-list] = name (arg-list)
  body
endfunction
```

where *name*, *arg-list*, and *body* have the same meaning as before, and *ret-list* is a comma-separated list of variable names that will hold the values returned from the function. The list of return values must have at least one element. If *ret-list* has only one element, this form of the function statement is equivalent to the form described in the previous section.

Here is an example of a function that returns two values, the maximum element of a vector and the index of its first occurrence in the vector.

```
function [max, idx] = vmax (v)
  idx = 1;
  max = v (idx);
  for i = 2:length (v)
    if (v (i) > max)
      max = v (i);
      idx = i;
    endif
  endfor
endfunction
```

In this particular case, the two values could have been returned as elements of a single array, but that is not always possible or convenient. The values to be returned may not have compatible dimensions, and it is often desirable to give the individual return values distinct names.

In addition to setting nargin each time a function is called, Octave also automatically initializes nargout to the number of values that are expected to be returned. This allows you to write functions that behave differently depending on the number of values that the user of the function has requested. The implicit assignment to the built-in variable ans does not figure in the count of output arguments, so the value of nargout may be zero.

The svd and lu functions are examples of built-in functions that behave differently depending on the value of nargout.

It is possible to write functions that only set some return values. For example, calling the function

```
function [x, y, z] = f ()
  x = 1;
  z = 2;
endfunction
```

as

```
[a, b, c] = f ()
```

produces:

```
a = 1

b = [](0x0)

c = 2
```
along with a warning.

nargout ()  Built-in Function
nargout (*fcn_name*)  Built-in Function

    Within a function, return the number of values the caller expects to receive. If called with the optional argument *fcn_name*, return the maximum number of values the named function can produce, or -1 if the function can produce a variable number of values.

    For example,
```
f ()
```
will cause nargout to return 0 inside the function f and
```
[s, t] = f ()
```
will cause nargout to return 2 inside the function f.

    At the top level, nargout is undefined.

    **See also:** nargin, varargin, varargout.

nargchk (*nargin_min, nargin_max, n*)  Function File

    If *n* is in the range *nargin_min* through *nargin_max* inclusive, return the empty matrix. Otherwise, return a message indicating whether *n* is too large or too small.

    This is useful for checking to see that the number of arguments supplied to a function is within an acceptable range.

## 11.3 Variable-length Argument Lists

    Sometimes the number of input arguments is not known when the function is defined. As an example think of a function that returns the smallest of all its input arguments. For example,
```
a = smallest (1, 2, 3);
b = smallest (1, 2, 3, 4);
```
In this example both a and b would be 1. One way to write the smallest function is
```
function val = smallest (arg1, arg2, arg3, arg4, arg5)
  body
endfunction
```
and then use the value of nargin to determine which of the input arguments should be considered. The problem with this approach is that it can only handle a limited number of input arguments.

    Octave supports the varargin keyword for handling a variable number of input arguments. Using varargin the function looks like this

```
function val = smallest (varargin)
  body
endfunction
```

In the function body the input arguments can be accessed through the variable varargin. This variable is a cell array containing all the input arguments. See Section 6.2 [Cell Arrays], page 71, for details on working with cell arrays. The smallest function can now be defined like this

```
function val = smallest (varargin)
  val = min ([varargin{:}]);
endfunction
```

This implementation handles any number of input arguments, but it's also a very simple solution to the problem.

A slightly more complex example of varargin is a function print_arguments that prints all input arguments. Such a function can be defined like this

```
function print_arguments (varargin)
  for i = 1:length (varargin)
    printf ("Input argument %d: ", i);
    disp (varargin{i});
  endfor
endfunction
```

This function produces output like this

```
print_arguments (1, "two", 3);
    ⊣ Input argument 1:  1
    ⊣ Input argument 2: two
    ⊣ Input argument 3:  3
```

[*reg, prop*] = parseparams (*params*)  *Function File*
Return in *reg* the cell elements of *param* up to the first string element and in *prop* all remaining elements beginning with the first string element. For example

```
[reg, prop] = parseparams ({1, 2, "linewidth", 10})
reg =
{
  [1,1] = 1
  [1,2] = 2
}
prop =
{
  [1,1] = linewidth
  [1,2] = 10
}
```

The parseparams function may be used to separate 'regular' arguments and additional arguments given as property/value pairs of the *varargin* cell array.

See also: varargin.

## 11.4 Variable-length Return Lists

It is possible to return a variable number of output arguments from a function using a syntax that's similar to the one used with the varargin keyword. To let a function return a variable number of output arguments the varargout keyword is used. As with varargin, varargout is a cell array that will contain the requested output arguments.

As an example the following function sets the first output argument to 1, the second to 2, and so on.

```
function varargout = one_to_n ()
  for i = 1:nargout
    varargout{i} = i;
  endfor
endfunction
```

When called this function returns values like this

```
[a, b, c] = one_to_n ()
  ⇒ a = 1
  ⇒ b = 2
  ⇒ c = 3
```

[r1, r2, ..., rn] = deal (a)      *Function File*
[r1, r2, ..., rn] = deal (a1, a2, ..., an)      *Function File*

Copy the input parameters into the corresponding output parameters. If only one input parameter is supplied, its value is copied to each of the outputs.

For example,

    [a, b, c] = deal (x, y, z);

is equivalent to

    a = x;
    b = y;
    c = z;

and

    [a, b, c] = deal (x);

is equivalent to

    a = b = c = x;

## 11.5 Returning From a Function

The body of a user-defined function can contain a `return` statement. This statement returns control to the rest of the Octave program. It looks like this:

    return

Unlike the `return` statement in C, Octave's `return` statement cannot be used to return a value from a function. Instead, you must assign values to the list of return variables that are part of the `function` statement. The `return` statement simply makes it easier to exit a function from a deeply nested loop or conditional statement.

Here is an example of a function that checks to see if any elements of a vector are nonzero.

```
function retval = any_nonzero (v)
  retval = 0;
  for i = 1:length (v)
    if (v (i) != 0)
      retval = 1;
      return;
    endif
  endfor
  printf ("no nonzero elements found\n");
endfunction
```

Note that this function could not have been written using the `break` statement to exit the loop once a nonzero value is found without adding extra logic to avoid printing the message if the vector does contain a nonzero element.

**return**                                                                Keyword

When Octave encounters the keyword `return` inside a function or script, it returns control to the caller immediately. At the top level, the return statement is ignored. A `return` statement is assumed at the end of every function definition.

## 11.6 Default Arguments

Since Octave supports variable number of input arguments, it is very useful to assign default values to some input arguments. When an input argument is declared in the argument list it is possible to assign a default value to the argument like this

```
function name (arg1 = val1, ...)
  body
endfunction
```

If no value is assigned to *arg1* by the user, it will have the value *val1*.

As an example, the following function implements a variant of the classic "Hello, World" program.

Chapter 11: Functions and Script Files                                   133

```
function hello (who = "World")
  printf ("Hello, %s!\n", who);
endfunction
```
When called without an input argument the function prints the following
```
hello ();
    ⊣ Hello, World!
```
and when it's called with an input argument it prints the following
```
hello ("Beautiful World of Free Software");
    ⊣ Hello, Beautiful World of Free Software!
```
Sometimes it is useful to explicitly tell Octave to use the default value of an input argument. This can be done writing a : as the value of the input argument when calling the function.
```
hello (:);
    ⊣ Hello, World!
```

## 11.7 Function Files

Except for simple one-shot programs, it is not practical to have to define all the functions you need each time you need them. Instead, you will normally want to save them in a file so that you can easily edit them, and save them for use at a later time.

Octave does not require you to load function definitions from files before using them. You simply need to put the function definitions in a place where Octave can find them.

When Octave encounters an identifier that is undefined, it first looks for variables or functions that are already compiled and currently listed in its symbol table. If it fails to find a definition there, it searches a list of directories (the *path*) for files ending in '.m' that have the same base name as the undefined identifier.[2] Once Octave finds a file with a name that matches, the contents of the file are read. If it defines a *single* function, it is compiled and executed. See Section 11.8 [Script Files], page 139, for more information about how you can define more than one function in a single file.

When Octave defines a function from a function file, it saves the full name of the file it read and the time stamp on the file. If the time stamp on the file changes, Octave may reload the file. When Octave is running interactively, time stamp checking normally happens at most once each time Octave prints the prompt. Searching for new function definitions also occurs if the current working directory changes.

Checking the time stamp allows you to edit the definition of a function while Octave is running, and automatically use the new function definition without having to restart your Octave session.

To avoid degrading performance unnecessarily by checking the time stamps on functions that are not likely to change, Octave assumes that function files in the directory tree '*octave-home*/share/octave/*version*/m' will not change,

---

[2] The .m suffix was chosen for compatibility with MATLAB.

so it doesn't have to check their time stamps every time the functions defined in those files are used. This is normally a very good assumption and provides a significant improvement in performance for the function files that are distributed with Octave.

If you know that your own function files will not change while you are running Octave, you can improve performance by calling ignore_function_time_stamp ("all"), so that Octave will ignore the time stamps for all function files. Passing "system" to this function resets the default behavior.

mfilename ()  Built-in Function
mfilename ("fullpath")  Built-in Function
mfilename ("fullpathext")  Built-in Function
 Return the name of the currently executing file. At the top-level, return the empty string. Given the argument "fullpath", include the directory part of the file name, but not the extension. Given the argument "fullpathext", include the directory part of the file name and the extension.

val = ignore_function_time_stamp ()  Built-in Function
old_val = ignore_function_time_stamp (new_val)  Built-in Function
 Query or set the internal variable that controls whether Octave checks the time stamp on files each time it looks up functions defined in function files. If the internal variable is set to "system", Octave will not automatically recompile function files in subdirectories of '*octave-home*/lib/*version*' if they have changed since they were last compiled, but will recompile other function files in the search path if they change. If set to "all", Octave will not recompile any function files unless their definitions are removed with clear. If set to "none", Octave will always check time stamps on files to determine whether functions defined in function files need to be recompiled.

### 11.7.1 Manipulating the load path

When a function is called, Octave searches a list of directories for a file that contains the function declaration. This list of directories is known as the load path. By default the load path contains a list of directories distributed with Octave plus the current working directory. To see your current load path call the path function without any input or output arguments.

It is possible to add or remove directories to or from the load path using addpath and rmpath. As an example, the following code adds ~/Octave to the load path.

 addpath("~/Octave")

After this the directory ~/Octave will be searched for functions.

addpath (*dir1*, ...)  Built-in Function
addpath (*dir1*, ..., *option*)  Built-in Function
 Add *dir1*, ... to the current function search path. If *option* is "-begin" or 0 (the default), prepend the directory name to the current path. If *option* is "-end" or 1, append the directory name to the current path. Directories added to the path must exist.

 **See also:** path, rmpath, genpath, pathdef, savepath, pathsep.

# Chapter 11: Functions and Script Files                                                  135

**genpath** (*dir*)                                                               Built-in Function
    Return a path constructed from *dir* and all its subdirectories.

**rmpath** (*dir1*, ...)                                                          Built-in Function
    Remove *dir1*, ... from the current function search path.

    **See also:** path, addpath, genpath, pathdef, savepath, pathsep.

**savepath** (*file*)                                                             Function File
    Save the current function search path to *file*. If *file* is omitted, '~/.octaverc' is used. If successful, savepath returns 0.

    **See also:** path, addpath, rmpath, genpath, pathdef, pathsep.

**path** (...)                                                                    Built-in Function
    Modify or display Octave's load path.

    If *nargin* and *nargout* are zero, display the elements of Octave's load path in an easy to read format.

    If *nargin* is zero and nargout is greater than zero, return the current load path.

    If *nargin* is greater than zero, concatenate the arguments, separating them with pathsep(). Set the internal search path to the result and return it.

    No checks are made for duplicate elements.

    **See also:** addpath, rmpath, genpath, pathdef, savepath, pathsep.

**val = pathdef** ()                                                              Built-in Function
    Return the default list of directories in which to search for function files.

    **See also:** path, addpath, rmpath, genpath, savepath, pathsep.

**pathsep** ()                                                                    Built-in Function
    Return the system-dependent character used to separate directories in a path.

    **See also:** filesep, dir, ls.

**rehash** ()                                                                     Built-in Function
    Reinitialize Octave's load path directory cache.

**file_in_loadpath** (*file*)                                                     Built-in Function
**file_in_loadpath** (*file*, "all")                                              Built-in Function
    Return the absolute name of *file* if it can be found in the list of directories specified by path. If no file is found, return an empty matrix.

    If the first argument is a cell array of strings, search each directory of the loadpath for element of the cell array and return the first that matches.

    If the second optional argument "all" is supplied, return a cell array containing the list of all files that have the same name in the path. If no files are found, return an empty cell array.

    **See also:** file_in_path, path.

## 11.7.2 Subfunctions

A function file may contain secondary functions called *subfunctions*. These secondary functions are only visible to the other functions in the same function file. For example, a file 'f.m' containing

```
function f ()
  printf ("in f, calling g\n");
  g ()
endfunction
function g ()
  printf ("in g, calling h\n");
  h ()
endfunction
function h ()
  printf ("in h\n")
endfunction
```

defines a main function f and two subfunctions. The subfunctions g and h may only be called from the main function f or from the other subfunctions, but not from outside the file 'f.m'.

## 11.7.3 Overloading and Autoloading

The dispatch function can be used to alias one function name to another. It can be used to alias all calls to a particular function name to another function, or the alias can be limited to only a particular variable type. Consider the example

```
function y = spsin (x)
  printf ("Calling spsin\n");
  fflush(stdout);
  y = spfun ("sin", x);
endfunction

dispatch ("sin", "spsin", "sparse matrix");
y0 = sin(eye(3));
y1 = sin(speye(3));
```

which aliases the user-defined function spsin to sin, but only for real sparse matrices. Note that the builtin sin already correctly treats sparse matrices and so this example is only illustrative.

dispatch (*f*, *r*, *type*)  Loadable Function

Replace the function *f* with a dispatch so that function *r* is called when *f* is called with the first argument of the named *type*. If the type is any then call *r* if no other type matches. The original function *f* is accessible using builtin (*f*, ...).

If *r* is omitted, clear dispatch function associated with *type*.

If both *r* and *type* are omitted, list dispatch functions for *f*.

See also: builtin.

Chapter 11: Functions and Script Files                                          137

[...] builtin (*f*, ...)                                                Loadable Function
    Call the base function *f* even if *f* is overloaded to some other function for the given type signature.

    See also: dispatch.

    A single dynamically linked file might define several functions. However, as Octave searches for functions based on the functions filename, Octave needs a manner in which to find each of the functions in the dynamically linked file. On operating systems that support symbolic links, it is possible to create a symbolic link to the original file for each of the functions which it contains.

    However, there is at least one well known operating system that doesn't support symbolic links. Making copies of the original file for each of the functions is undesirable as it increases the amount of disk space used by Octave. Instead Octave supplies the `autoload` function, that permits the user to define in which file a certain function will be found.

autoload (*function*, *file*)                                            Built-in Function
    Define *function* to autoload from *file*.

    The second argument, *file*, should be an absolute file name or a file name in the same directory as the function or script from which the autoload command was run. *file* should not depend on the Octave load path.

    Normally, calls to `autoload` appear in PKG_ADD script files that are evaluated when a directory is added to the Octave's load path. To avoid having to hardcode directory names in *file*, if *file* is in the same directory as the PKG_ADD script then

        autoload ("foo", "bar.oct");

    will load the function foo from the file bar.oct. The above when bar.oct is not in the same directory or uses like

        autoload ("foo", file_in_loadpath ("bar.oct"))

    are strongly discouraged, as their behavior might be unpredictable.

    With no arguments, return a structure containing the current autoload map.

    See also: PKG_ADD.

### 11.7.4 Function Locking

    It is sometime desirable to lock a function into memory with the `mlock` function. This is typically used for dynamically linked functions in Oct-files or mex-files that contain some initialization, and it is desirable that calling `clear` does not remove this initialization.

    As an example,

      mlock ("my_function");

prevents `my_function` from being removed from memory, even if `clear` is called. It is possible to determine if a function is locked into memory with the `mislocked`, and to unlock a function with `munlock`, which the following illustrates.

```
mlock ("my_function");
mislocked ("my_function")
⇒ ans = 1
munlock ("my_function");
mislocked ("my_function")
⇒ ans = 0
```
A common use of mlock is to prevent persistent variables from being removed from memory, as the following example shows:
```
function count_calls()
  persistent calls = 0;
  printf ("'count_calls' has been called %d times\n",
          ++calls);
endfunction
mlock ("count_calls");

count_calls ();
⊣ 'count_calls' has been called 1 times

clear count_calls
count_calls ();
⊣ 'count_calls' has been called 2 times
```
It is, however, often inconvenient to lock a function from the prompt, so it is also possible to lock a function from within its body. This is simply done by calling mlock from within the function.
```
function count_calls ()
  mlock ();
  persistent calls = 0;
  printf ("'count_calls' has been called %d times\n",
          ++calls);
endfunction
```
mlock might equally be used to prevent changes to a function from having effect in Octave, though a similar effect can be had with the ignore_function_time_stamp function.

mlock (*name*)　　　　　　　　　　　　　　　　　　　　　　　Built-in Function
    Lock the named function into memory. If no function is named then lock in the current function.

    **See also:** munlock, mislocked, persistent.

munlock (*fcn*)　　　　　　　　　　　　　　　　　　　　　　　Built-in Function
    Unlock the named function. If no function is named then unlock the current function.

    **See also:** mlock, mislocked, persistent.

mislocked (*fcn*)                                         Built-in Function
   Return true if the named function is locked. If no function is named then return true if the current function is locked.

   See also: mlock, munlock, persistent.

## 11.8 Script Files

A script file is a file containing (almost) any sequence of Octave commands. It is read and evaluated just as if you had typed each command at the Octave prompt, and provides a convenient way to perform a sequence of commands that do not logically belong inside a function.

Unlike a function file, a script file must *not* begin with the keyword function. If it does, Octave will assume that it is a function file, and that it defines a single function that should be evaluated as soon as it is defined.

A script file also differs from a function file in that the variables named in a script file are not local variables, but are in the same scope as the other variables that are visible on the command line.

Even though a script file may not begin with the function keyword, it is possible to define more than one function in a single script file and load (but not execute) all of them at once. To do this, the first token in the file (ignoring comments and other white space) must be something other than function. If you have no other statements to evaluate, you can use a statement that has no effect, like this:

```
# Prevent Octave from thinking that this
# is a function file:

1;

# Define function one:

function one ()
  ...
```

To have Octave read and compile these functions into an internal form, you need to make sure that the file is in Octave's load path (accessible through the path function), then simply type the base name of the file that contains the commands. (Octave uses the same rules to search for script files as it does to search for function files.)

If the first token in a file (ignoring comments) is function, Octave will compile the function and try to execute it, printing a message warning about any non-whitespace characters that appear after the function definition.

Note that Octave does not try to look up the definition of any identifier until it needs to evaluate it. This means that Octave will compile the following statements if they appear in a script file, or are typed at the command line,

```
# not a function file:
1;
function foo ()
  do_something ();
endfunction
function do_something ()
  do_something_else ();
endfunction
```

even though the function do_something is not defined before it is referenced in the function foo. This is not an error because Octave does not need to resolve all symbols that are referenced by a function until the function is actually evaluated.

Since Octave doesn't look for definitions until they are needed, the following code will always print bar = 3 whether it is typed directly on the command line, read from a script file, or is part of a function body, even if there is a function or script file called 'bar.m' in Octave's path.

```
eval ("bar = 3");
bar
```

Code like this appearing within a function body could fool Octave if definitions were resolved as the function was being compiled. It would be virtually impossible to make Octave clever enough to evaluate this code in a consistent fashion. The parser would have to be able to perform the call to eval at compile time, and that would be impossible unless all the references in the string to be evaluated could also be resolved, and requiring that would be too restrictive (the string might come from user input, or depend on things that are not known until the function is evaluated).

Although Octave normally executes commands from script files that have the name 'file.m', you can use the function source to execute commands from any file.

source (*file*)                                                                 Built-in Function
    Parse and execute the contents of *file*. This is equivalent to executing commands from a script file, but without requiring the file to be named 'file.m'.

## 11.9 Function Handles, Inline Functions, and Anonymous Functions

It can be very convenient store a function in a variable so that it can be passed to a different function. For example, a function that performs numerical minimisation needs access to the function that should be minimised.

Chapter 11: Functions and Script Files                                    141

## 11.9.1 Function Handles

A function handle is a pointer to another function and is defined with the syntax

  @*function-name*

For example

  f = @sin;

Creates a function handle called f that refers to the function sin.

Function handles are used to call other functions indirectly, or to pass a function as an argument to another function like quad or fsolve. For example

  f = @sin;
  quad (f, 0, pi)
   ⇒ 2

You may use feval to call a function using function handle, or simply write the name of the function handle followed by an argument list. If there are no arguments, you must use an empty argument list (). For example

  f = @sin;
  feval (f, pi/4)
   ⇒ 0.70711
  f (pi/4)
   ⇒ 0.70711

functions (*fcn_handle*)             Built-in Function
  Return a struct containing information about the function handle *fcn_handle*.

func2str (*fcn_handle*)             Built-in Function
  Return a string containing the name of the function referenced by the function handle *fcn_handle*.

str2func (*fcn_name*)              Built-in Function
  Return a function handle constructed from the string *fcn_name*.

## 11.9.2 Anonymous Functions

Anonymous functions are defined using the syntax

  @(*argument-list*) *expression*

Any variables that are not found in the argument list are inherited from the enclosing scope. Anonymous functions are useful for creating simple unnamed functions from expressions or for wrapping calls to other functions to adapt them for use by functions like quad. For example,

  f = @(x) x.^2;
  quad (f, 0, 10)
   ⇒ 333.33

creates a simple unnamed function from the expression x.^2 and passes it to quad,

```
quad (@(x) sin (x), 0, pi)
    ⇒ 2
```
wraps another function, and
```
a = 1;
b = 2;
quad (@(x) betainc (x, a, b), 0, 0.4)
    ⇒ 0.13867
```
adapts a function with several parameters to the form required by quad. In this example, the values of a and b that are passed to betainc are inherited from the current environment.

### 11.9.3 Inline Functions

An inline function is created from a string containing the function body using the inline function. The following code defines the function $f(x) = x^2 + 2$.
```
f = inline("x^2 + 2");
```
After this it is possible to evaluate $f$ at any $x$ by writing f(x).

inline (*str*)                                          Built-in Function
inline (*str, arg1, ...*)                           Built-in Function
inline (*str, n*)                                     Built-in Function

    Create an inline function from the character string *str*. If called with a single argument, the arguments of the generated function are extracted from the function itself. The generated function arguments will then be in alphabetical order. It should be noted that i, and j are ignored as arguments due to the ambiguity between their use as a variable or their use as an inbuilt constant. All arguments followed by a parenthesis are considered to be functions.

    If the second and subsequent arguments are character strings, they are the names of the arguments of the function.

    If the second argument is an integer $n$, the arguments are "x", "P1", ..., "P$N$".

    See also: argnames, formula, vectorize.

argnames (*fun*)                                      Built-in Function
    Return a cell array of character strings containing the names of the arguments of the inline function *fun*.

    See also: inline, formula, vectorize.

formula (*fun*)                                         Built-in Function
    Return a character string representing the inline function *fun*. Note that char (*fun*) is equivalent to formula (*fun*).

    See also: argnames, inline, vectorize.

vectorize (*fun*)                                      Built-in Function
    Create a vectorized version of the inline function *fun* by replacing all occurrences of *, /, etc., with .*, ./, etc.

## 11.10 Commands

Commands are a special class of functions that only accept string input arguments. A command can be called as an ordinary function, but it can also be called without the parentheses like the following example shows

    `my_command hello world`

which is the same as

    `my_command("hello", "world")`

The general form of a command call is

    `name arg1 arg2 ...`

which translates directly to

    `name ("arg1", "arg2", ...)`

A function can be used as a command if it accepts string input arguments. To do this, the function must be marked as a command, which can be done with the `mark_as_command` command like this

    `mark_as_command name`

where name is the function to be marked as a command.

One difficulty of commands occurs when one of the string input arguments are stored in a variable. Since Octave can't tell the difference between a variable name, and an ordinary string, it is not possible to pass a variable as input to a command. In such a situation a command must be called as a function.

**mark_as_command** (*name*)        Built-in Function
    Enter *name* into the list of commands.

    **See also:** unmark_command, iscommand.

**unmark_command** (*name*)        Built-in Function
    Remove *name* from the list of commands.

    **See also:** mark_as_command, iscommand.

**iscommand** (*name*)        Built-in Function
    Return true if *name* is a command style function. If *name* is omitted, return a list of identifiers which are marked as commands with `mark_as_command`.

    **See also:** mark_as_command, unmark_command.

**mark_as_rawcommand** (*name*)        Built-in Function
    Enter *name* into the list of raw input commands and to the list of command style functions. Raw input commands are like normal command style functions, but they receive their input unprocessed (i.e. strings still contain the quotes and escapes they had when input). However, comments and continuations are handled as usual, you cannot pass a token starting with a comment character ('#' or '%') to your function, and the last token cannot be a continuation token ('\' or '...').

    **See also:** unmark_rawcommand, israwcommand, iscommand, mark_as_command.

unmark_rawcommand (*name*)  Built-in Function
    Remove *name* from the list of raw input commands. Note that this does not remove *name* from the list of command style functions.

    See also: mark_as_rawcommand, israwcommand, iscommand, unmark_command.

israwcommand (*name*)  Built-in Function
    Return true if *name* is a raw input command function. If *name* is omitted, return a list of identifiers which are marked as raw input commands with mark_as_rawcommand.

    See also: mark_as_rawcommand, unmark_rawcommand.

## 11.11 Organization of Functions Distributed with Octave

Many of the standard functions provided in Octave are distributed as function files. They are loosely organized by topic, in subdirectories of '*octave-home*/lib/octave/*version*/m', to make it easier to find them.

The following is a list of all the function file subdirectories, and the types of functions you will find there.

'audio'
    Functions for playing and recording sounds.

'control'
    Functions for design and simulation of automatic control systems.

'elfun'
    Elementary functions.

'finance'
    Functions for computing interest payments, investment values, and rates of return.

'general'
    Miscellaneous matrix manipulations, like flipud, rot90, and triu, as well as other basic functions, like ismatrix, nargchk, etc.

'image'
    Image processing tools. These functions require the X Window System.

'io'
    Input-output functions.

'linear-algebra'
    Functions for linear algebra.

'miscellaneous'
    Functions that don't really belong anywhere else.

'optimization'
    Minimization of functions.

Chapter 11: Functions and Script Files    145

'path'
: Functions to manage the directory path Octave uses to find functions.

'pkg'
: Install external packages of functions in Octave.

'plot'
: Functions for displaying and printing two- and three-dimensional graphs.

'polynomial'
: Functions for manipulating polynomials.

'set'
: Functions for creating and manipulating sets of unique values.

'signal'
: Functions for signal processing applications.

'sparse'
: Functions for handling sparse matrices.

'specfun'
: Special functions.

'special-matrix'
: Functions that create special matrix forms.

'startup'
: Octave's system-wide startup file.

'statistics'
: Statistical functions.

'strings'
: Miscellaneous string-handling functions.

'testfun'
: Perform unit tests on other functions.

'time'
: Functions related to time keeping.

# 12 Errors and Warnings

Octave includes several functions for printing error and warning messages. When you write functions that need to take special action when they encounter abnormal conditions, you should print the error messages using the functions described in this chapter.

Since many of Octave's functions use these functions, it is also useful to understand them, so that errors and warnings can be handled.

## 12.1 Handling Errors

An error is something that occurs when a program is in a state where it doesn't make sense to continue. An example is when a function is called with too few input arguments. In this situation the function should abort with an error message informing the user of the lacking input arguments.

Since an error can occur during the evaluation of a program, it is very convenient to be able to detect that an error occurred, so that the error can be fixed. This is possible with the try statement described in Section 10.9 [The try Statement], page 123.

### 12.1.1 Raising Errors

The most common use of errors is for checking input arguments to functions. The following example calls the error function if the function f is called without any input arguments.

```
function f (arg1)
  if (nargin == 0)
    error("not enough input arguments");
  endif
endfunction
```

When the error function is called, it prints the given message and returns to the Octave prompt. This means that no code following a call to error will be executed.

error (*template*, ...)  Built-in Function
error (*id*, *template*, ...)  Built-in Function

Format the optional arguments under the control of the template string *template* using the same rules as the printf family of functions (see Section 14.2.4 [Formatted Output], page 176) and print the resulting message on the stderr stream. The message is prefixed by the character string error:.

Calling error also sets Octave's internal error state such that control will return to the top level without evaluating any more commands. This is useful for aborting from functions or scripts.

If the error message does not end with a new line character, Octave will print a traceback of all the function calls leading to the error. For example, given the following function definitions:

```
function f () g (); end
function g () h (); end
function h () nargin == 1 || error ("nargin != 1"); end
```
calling the function f will result in a list of messages that can help you to quickly locate the exact location of the error:
```
f ()
error: nargin != 1
error: evaluating index expression near line 1, column 30
error: evaluating binary operator '||' near line 1, column 27
error: called from 'h'
error: called from 'g'
error: called from 'f'
```
If the error message ends in a new line character, Octave will print the message but will not display any traceback messages as it returns control to the top level. For example, modifying the error message in the previous example to end in a new line causes Octave to only print a single message:
```
function h () nargin == 1 || error ("nargin != 1\n"); end
f ()
error: nargin != 1
```
Since it is common to use errors when there is something wrong with the input to a function, Octave supports functions to simplify such code. When the print_usage function is called, it reads the help text of the function calling print_usage, and presents a useful error. If the help text is written in Texinfo it is possible to present an error message that only contains the function prototypes as described by the @deftypefn parts of the help text. When the help text isn't written in Texinfo, the error message contains the entire help message.

Consider the following function.
```
## -*- texinfo -*-
## @deftypefn {Function File} f (@var{arg1})
## Function help text goes here...
## @end deftypefn
function f (arg1)
  if (nargin == 0)
    print_usage ();
  endif
endfunction
```
When it is called with no input arguments it produces the following error.
```
f ()
⊣ Invalid call to f.  Correct usage is:
⊣
⊣  -- Function File: f (ARG1)
⊣
⊣
⊣
⊣ error: evaluating if command near line 6, column 3
⊣ error: called from 'f' in file '/home/jwe/octave/f.m'
```

# Chapter 12: Errors and Warnings

**print_usage ()**                                                                  *Loadable Function*

Print the usage message for the currently executing function. The print_usage function is only intended to work inside a user-defined function.

**See also:** help.

**usage (***msg***)**                                                                *Built-in Function*

Print the message *msg*, prefixed by the string usage: , and set Octave's internal error state such that control will return to the top level without evaluating any more commands. This is useful for aborting from functions.

After usage is evaluated, Octave will print a traceback of all the function calls leading to the usage message.

You should use this function for reporting problems errors that result from an improper call to a function, such as calling a function with an incorrect number of arguments, or with arguments of the wrong type. For example, most functions distributed with Octave begin with code like this

```
if (nargin != 2)
  usage ("foo (a, b)");
endif
```

to check for the proper number of arguments.

**beep ()**                                                                                                   *Function File*

Produce a beep from the speaker (or visual bell).

**See also:** puts, fputs, printf, fprintf.

**val = beep_on_error ()**                                              *Built-in Function*
**old_val = beep_on_error (***new_val***)**                      *Built-in Function*

Query or set the internal variable that controls whether Octave will try to ring the terminal bell before printing an error message.

## 12.1.2 Catching Errors

When an error occurs, it can be detected and handled using the try statement as described in Section 10.9 [The try Statement], page 123. As an example, the following piece of code counts the number of errors that occurs during a for loop.

```
number_of_errors = 0;
for n = 1:100
  try
    ...
  catch
    number_of_errors++;
  end_try_catch
endfor
```

The above example treats all errors the same. In many situations it can however be necessary to discriminate between errors, and take different actions depending on the error. The lasterror function returns a structure containing information about the last error that occurred. As an example, the code above could be changed to count the number of errors related to the * operator.

```
number_of_errors = 0;
for n = 1:100
  try
    ...
  catch
    msg = lasterror.message;
    if (strfind (msg, "operator *"))
      number_of_errors++;
    endif
  end_try_catch
endfor
```

err = lasterror (*err*)                                                        Built-in Function
lasterror ('reset')                                                            Built-in Function

Returns or sets the last error message. Called without any arguments returns a structure containing the last error message, as well as other information related to this error. The elements of this structure are:

'message'
: The text of the last error message

'identifier'
: The message identifier of this error message

'stack'
: A structure containing information on where the message occurred. This might be an empty structure if this in the case where this information cannot be obtained. The fields of this structure are:

  'file'
  : The name of the file where the error occurred

  'name'
  : The name of function in which the error occurred

  'line'
  : The line number at which the error occurred

  'column'
  : An optional field with the column number at which the error occurred

The *err* structure may also be passed to lasterror to set the information about the last error. The only constraint on *err* in that case is that it is a scalar structure. Any fields of *err* that match the above are set to the value passed in *err*, while other fields are set to their default values.

If lasterror is called with the argument 'reset', all values take their default values.

## Chapter 12: Errors and Warnings

[*msg*, *msgid*] = lasterr (*msg*, *msgid*)            Built-in Function

    Without any arguments, return the last error message. With one argument, set the last error message to *msg*. With two arguments, also set the last message identifier.

    When an error has been handled it is possible to raise it again. This can be useful when an error needs to be detected, but the program should still abort. This is possible using the `rethrow` function. The previous example can now be changed to count the number of errors related to the * operator, but still abort if another kind of error occurs.

```
number_of_errors = 0;
for n = 1:100
  try
    ...
  catch
    msg = lasterror.message;
    if (strfind (msg, "operator *"))
      number_of_errors++;
    else
      rethrow (lasterror);
    endif
  end_try_catch
endfor
```

rethrow (*err*)            Built-in Function

    Reissues a previous error as defined by *err*. *err* is a structure that must contain at least the 'message' and 'identifier' fields. *err* can also contain a field 'stack' that gives information on the assumed location of the error. Typically *err* is returned from `lasterror`.

    See also: lasterror, lasterr, error.

err = errno ()            Built-in Function
err = errno (*val*)            Built-in Function
err = errno (*name*)            Built-in Function

    Return the current value of the system-dependent variable errno, set its value to *val* and return the previous value, or return the named error code given *name* as a character string, or -1 if *name* is not found.

errno_list ()            Built-in Function

    Return a structure containing the system-dependent errno values.

## 12.2 Handling Warnings

Like an error, a warning is issued when something unexpected happens. Unlike an error, a warning doesn't abort the currently running program. A simple example of a warning is when a number is divided by zero. In this case Octave will issue a warning and assign the value Inf to the result.

```
a = 1/0
     ⊣ warning: division by zero
     ⇒ a = Inf
```

### 12.2.1 Issuing Warnings

It is possible to issue warnings from any code using the warning function. In its most simple form, the warning function takes a string describing the warning as its input argument. As an example, the following code controls if the variable a is non-negative, and if not issues a warning and sets a to zero.

```
a = -1;
if (a < 0)
  warning ("'a' must be non-negative. Setting 'a' to zero.");
  a = 0;
endif
     ⊣ 'a' must be non-negative. Setting 'a' to zero.
```

Since warnings aren't fatal to a running program, it is not possible to catch a warning using the try statement or something similar. It is however possible to access the last warning as a string using the lastwarn function.

It is also possible to assign an identification string to a warning. If a warning has such an ID the user can enable and disable this warning as will be described in the next section. To assign an ID to a warning, simply call warning with two string arguments, where the first is the identification string, and the second is the actual warning.

**warning** (*template*, ...)    Built-in Function
**warning** (*id, template*, ...)    Built-in Function

Format the optional arguments under the control of the template string *template* using the same rules as the printf family of functions (see Section 14.2.4 [Formatted Output], page 176) and print the resulting message on the stderr stream. The message is prefixed by the character string warning: . You should use this function when you want to notify the user of an unusual condition, but only when it makes sense for your program to go on.

The optional message identifier allows users to enable or disable warnings tagged by *id*. The special identifier "all" may be used to set the state of all warnings.

**warning** ("on", *id*)    Built-in Function
**warning** ("off", *id*)    Built-in Function
**warning** ("error", *id*)    Built-in Function
**warning** ("query", *id*)    Built-in Function

Set or query the state of a particular warning using the identifier *id*. If the identifier is omitted, a value of "all" is assumed. If you set the state of a warning to "error", the warning named by *id* is handled as if it were an error instead.

See also: warning_ids.

[*msg, msgid*] = lastwarn (*msg, msgid*)　　　　　　　　　　Built-in Function
Without any arguments, return the last warning message. With one argument, set the last warning message to *msg*. With two arguments, also set the last message identifier.

## 12.2.2 Enabling and Disabling Warnings

The warning function also allows you to control which warnings are actually printed to the screen. If the warning function is called with a string argument that is either "on" or "off" all warnings will be enabled or disabled.

It is also possible to enable and disable individual warnings through their string identifications. The following code will issue a warning

```
warning ("non-negative-variable",
         "'a' must be non-negative. Setting 'a' to zero.");
```

while the following won't issue a warning

```
warning ("off", "non-negative-variable");
warning ("non-negative-variable",
         "'a' must be non-negative. Setting 'a' to zero.");
```

The functions distributed with Octave can issue one of the following warnings.

Octave:array-to-scalar
　　If the Octave:array-to-scalar warning is enabled, Octave will warn when an implicit conversion from an array to a scalar value is attempted. By default, the Octave:array-to-scalar warning is disabled.

Octave:array-to-vector
　　If the Octave:array-to-vector warning is enabled, Octave will warn when an implicit conversion from an array to a vector value is attempted. By default, the Octave:array-to-vector warning is disabled.

Octave:assign-as-truth-value
　　If the Octave:assign-as-truth-value warning is enabled, a warning is issued for statements like

```
if (s = t)
  ...
```

since such statements are not common, and it is likely that the intent was to write

```
if (s == t)
  ...
```

instead.

There are times when it is useful to write code that contains assignments within the condition of a while or if statement. For example, statements like

```
while (c = getc())
 ...
```
are common in C programming.

It is possible to avoid all warnings about such statements by disabling the Octave:assign-as-truth-value warning, but that may also let real errors like
```
if (x = 1)   # intended to test (x == 1)!
 ...
```
slip by.

In such cases, it is possible suppress errors for specific statements by writing them with an extra set of parentheses. For example, writing the previous example as
```
while ((c = getc()))
 ...
```
will prevent the warning from being printed for this statement, while allowing Octave to warn about other assignments used in conditional contexts.

By default, the Octave:assign-as-truth-value warning is enabled.

Octave:associativity-change

If the Octave:associativity-change warning is enabled, Octave will warn about possible changes in the meaning of some code due to changes in associativity for some operators. Associativity changes have typically been made for MATLAB compatibility. By default, the Octave:associativity-change warning is enabled.

Octave:divide-by-zero

If the Octave:divide-by-zero warning is enabled, a warning is issued when Octave encounters a division by zero. By default, the Octave:divide-by-zero warning is enabled.

Octave:empty-list-elements

If the Octave:empty-list-elements warning is enabled, a warning is issued when an empty matrix is found in a matrix list. For example,
```
a = [1, [], 3, [], 5]
```
By default, the Octave:empty-list-elements warning is enabled.

Octave:fortran-indexing

If the Octave:fortran-indexing warning is enabled, a warning is printed for expressions which select elements of a two-dimensional matrix using a single index. By default, the Octave:fortran-indexing warning is disabled.

Octave:function-name-clash

If the Octave:function-name-clash warning is enabled, a warning is issued when Octave finds that the name of a function defined in a function file differs from the name of the file. (If the names disagree, the name declared inside the file is ignored.) By default, the Octave:function-name-clash warning is enabled.

## Chapter 12: Errors and Warnings

`Octave:future-time-stamp`
> If the `Octave:future-time-stamp` warning is enabled, Octave will print a warning if it finds a function file with a time stamp that is in the future. By default, the `Octave:future-time-stamp` warning is enabled.

`Octave:imag-to-real`
> If the `Octave:imag-to-real` warning is enabled, a warning is printed for implicit conversions of complex numbers to real numbers. By default, the `Octave:imag-to-real` warning is disabled.

`Octave:matlab-incompatible`
> Print warnings for Octave language features that may cause compatibility problems with MATLAB.

`Octave:missing-semicolon`
> If the `Octave:missing-semicolon` warning is enabled, Octave will warn when statements in function definitions don't end in semicolons. By default the `Octave:missing-semicolon` warning is disabled.

`Octave:neg-dim-as-zero`
> If the `Octave:neg-dim-as-zero` warning is enabled, print a warning for expressions like
>
>     eye (-1)
>
> By default, the `Octave:neg-dim-as-zero` warning is disabled.

`Octave:num-to-str`
> If the `Octave:num-to-str` warning is enable, a warning is printed for implicit conversions of numbers to their ASCII character equivalents when strings are constructed using a mixture of strings and numbers in matrix notation. For example,
>
>     [ "f", 111, 111 ]
>     ⇒ "foo"
>
> elicits a warning if the `Octave:num-to-str` warning is enabled. By default, the `Octave:num-to-str` warning is enabled.

`Octave:precedence-change`
> If the `Octave:precedence-change` warning is enabled, Octave will warn about possible changes in the meaning of some code due to changes in precedence for some operators. Precedence changes have typically been made for MATLAB compatibility. By default, the `Octave:precedence-change` warning is enabled.

`Octave:reload-forces-clear`
> If several functions have been loaded from the same file, Octave must clear all the functions before any one of them can be reloaded. If the `Octave:reload-forces-clear` warning is enabled, Octave will warn you when this happens, and print a list of the additional functions that it is forced to clear. By default, the `Octave:reload-forces-clear` warning is enabled.

`Octave:resize-on-range-error`
    If the `Octave:resize-on-range-error` warning is enabled, print a warning when a matrix is resized by an indexed assignment with indices outside the current bounds. By default, the `Octave:resize-on-range-error` warning is disabled.

`Octave:separator-insert`
    Print warning if commas or semicolons might be inserted automatically in literal matrices.

`Octave:single-quote-string`
    Print warning if a single quote character is used to introduce a string constant.

`Octave:str-to-num`
    If the `Octave:str-to-num` warning is enabled, a warning is printed for implicit conversions of strings to their numeric ASCII equivalents. For example,

    `"abc" + 0`
    $\Rightarrow$ 97 98 99

    elicits a warning if the `Octave:str-to-num` warning is enabled. By default, the `Octave:str-to-num` warning is disabled.

`Octave:string-concat`
    If the `Octave:string-concat` warning is enabled, print a warning when concatenating a mixture of double and single quoted strings. By default, the `Octave:string-concat` warning is disabled.

`Octave:undefined-return-values`
    If the `Octave:undefined-return-values` warning is disabled, print a warning if a function does not define all the values in the return list which are expected. By default, the `Octave:undefined-return-values` warning is enabled.

`Octave:variable-switch-label`
    If the `Octave:variable-switch-label` warning is enabled, Octave will print a warning if a switch label is not a constant or constant expression. By default, the `Octave:variable-switch-label` warning is disabled.

# 13 Debugging

Octave includes a built-in debugger to aid in the development of scripts. This can be used to interrupt the execution of an Octave script at a certain point, or when certain conditions are met. Once execution has stopped, and debug mode is entered, the symbol table at the point where execution has stopped can be examined and modified to check for errors.

The normal command-line editing and history functions are available in debug mode. However, one limitation on the debug mode is that commands entered at the debug prompt are evaluated as strings, rather than being handled by the Octave parser. This means that all commands in debug mode must be contained on a single line. To leave the debug mode, use either `quit`, `exit`, `return` or `dbcont`.

## 13.1 Entering Debug Mode

There are two basic means of interrupting the execution of an Octave script. These are breakpoints see Section 13.2 [Breakpoints], page 158, discussed in the next section and interruption based on some condition.

Octave supports three functions which control the conditions for entering debug mode, `debug_on_interrupt`, `debug_on_warning` and `debug_on_error`.

`val = debug_on_interrupt ()`                     *Built-in Function*
`old_val = debug_on_interrupt (new_val)`      *Built-in Function*

> Query or set the internal variable that controls whether Octave will try to enter debugging mode when it receives an interrupt signal (typically generated with `C-c`). If a second interrupt signal is received before reaching the debugging mode, a normal interrupt will occur.

`val = debug_on_warning ()`                       *Built-in Function*
`old_val = debug_on_warning (new_val)`        *Built-in Function*

> Query or set the internal variable that controls whether Octave will try to enter the debugger when a warning is encountered.

`val = debug_on_error ()`                         *Built-in Function*
`old_val = debug_on_error (new_val)`           *Built-in Function*

> Query or set the internal variable that controls whether Octave will try to enter the debugger when an error is encountered. This will also inhibit printing of the normal traceback message (you will only see the top-level error message).

## 13.2 Breakpoints

Breakpoints can be set in any Octave function, using the `dbstop` function.

*rline* = dbstop (*func*, *line*, ...)                         Loadable Function
    Set a breakpoint in a function

    func
        String representing the function name. When already in debug mode
        this should be left out and only the line should be given.

    line
        Line number you would like the breakpoint to be set on. Multiple lines
        might be given as separate arguments or as a vector.

    The rline returned is the real line that the breakpoint was set at.

    **See also:** dbclear, dbstatus, dbnext.

Note that breakpoints cannot be set in built-in functions (e.g. `sin`, etc) or dynamically loaded function (i.e. oct-files). To set a breakpoint immediately on entering a function, the breakpoint should be set to line 1. The leading comment block will be ignored and the breakpoint will be set to the first executable statement in the function. For example

```
dbstop ("asind", 1)
⇒ 27
```

Note that the return value of 27 means that the breakpoint was effectively set to line 27. The status of breakpoints in a function can be queried with the `dbstatus` function.

*lst* = dbstatus (*func*)                         Loadable Function
    Return a vector containing the lines on which a function has breakpoints set.

    func
        String representing the function name. When already in debug mode
        this should be left out.

    **See also:** dbclear, dbwhere.

Taking the above as an example, dbstatus ("asind") should return 27. The breakpoints can then be cleared with the `dbclear` function

dbclear (*func*, *line*, ...)                         Loadable Function
    Delete a breakpoint in a function

    func
        String representing the function name. When already in debug mode
        this should be left out and only the line should be given.

    line
        Line number where you would like to remove the breakpoint. Multiple
        lines might be given as separate arguments or as a vector.

# Chapter 13: Debugging

No checking is done to make sure that the line you requested is really a breakpoint. If you get the wrong line nothing will happen.

**See also:** dbstop, dbstatus, dbwhere.

These functions can be used to clear all the breakpoints in a function. For example,

```
dbclear ("asind", dbstatus ("asind"));
```

Another simple way of setting a breakpoint in an Octave script is the use of the keyboard function.

keyboard (*prompt*)                                                                                          Built-in Function

This function is normally used for simple debugging. When the keyboard function is executed, Octave prints a prompt and waits for user input. The input strings are then evaluated and the results are printed. This makes it possible to examine the values of variables within a function, and to assign new values to variables. No value is returned from the keyboard function, and it continues to prompt for input until the user types quit, or exit.

If keyboard is invoked without any arguments, a default prompt of debug> is used.

The keyboard function is typically placed in a script at the point where the user desires that the execution is stopped. It automatically sets the running script into the debug mode.

## 13.3 Debug Mode

There are two additional support functions that allow the user to interrogate where in the execution of a script Octave entered the debug mode and to print the lines in the script surrounding that point.

dbwhere ()                                                                                                   Loadable Function

Show where we are in the code

**See also:** dbclear, dbstatus, dbstop.

dbtype ()                                                                                                    Loadable Function

List script file with line numbers.

**See also:** dbclear, dbstatus, dbstop.

Debug mode also allows single line stepping through a function using the commands dbstep and dbnext. These differ slightly in the way they treat the next executable line if the next line itself is a function defined in an m-file. The dbnext command will execute the next line, while staying in the existing function being debugged. The dbstep command will step in to the new function.

# 14 Input and Output

Octave supports several ways of reading and writing data to or from the prompt or a file. The simplest functions for data Input and Output (I/O) are easy to use, but only provides limited control of how data is processed. For more control, a set of functions modelled after the C standard library are also provided by Octave.

## 14.1 Basic Input and Output

### 14.1.1 Terminal Output

Since Octave normally prints the value of an expression as soon as it has been evaluated, the simplest of all I/O functions is a simple expression. For example, the following expression will display the value of pi

    pi
        ⊣ pi = 3.1416

This works well as long as it is acceptable to have the name of the variable (or ans) printed along with the value. To print the value of a variable without printing its name, use the function disp.

The format command offers some control over the way Octave prints values with disp and through the normal echoing mechanism.

ans                                                              Automatic Variable
    The most recently computed result that was not explicitly assigned to a variable. For example, after the expression

    3^2 + 4^2

is evaluated, the value returned by ans is 25.

disp (x)                                                              Built-in Function
    Display the value of x. For example,

    disp ("The value of pi is:"), disp (pi)

        ⊣ the value of pi is:
        ⊣ 3.1416

Note that the output from disp always ends with a newline.

If an output value is requested, disp prints nothing and returns the formatted output in a string.

See also: fdisp.

format options                                                    Command
    Control the format of the output produced by disp and Octave's normal
    echoing mechanism. Valid options are listed in the following table.

    short
        Octave will try to print numbers with at least 5 significant figures within
        a field that is a maximum of 10 characters wide (not counting additional
        spacing that is added between columns of a matrix).

        If Octave is unable to format a matrix so that columns line up on the
        decimal point and all the numbers fit within the maximum field width,
        it switches to an e format.

    long
        Octave will try to print numbers with at least 15 significant figures
        within a field that is a maximum of 20 characters wide (not counting
        additional spacing that is added between columns of a matrix).

        As will the short format, Octave will switch to an e format if it is
        unable to format a matrix so that columns line up on the decimal point
        and all the numbers fit within the maximum field width.

    long e
    short e
        The same as format long or format short but always display output
        with an e format. For example, with the short e format, pi is displayed
        as 3.14e+00.

    long E
    short E
        The same as format long e or format short e but always display output with an uppercase E format. For example, with the long E format,
        pi is displayed as 3.14159265358979E+00.

    long g
    short g
        Choose between normal long (or short) and long e (or short e) formats based on the magnitude of the number. For example, with the
        short g format, pi .^ [2; 4; 8; 16; 32] is displayed as

            ans =

                9.8696
                97.409
                9488.5
              9.0032e+07
              8.1058e+15

    long G
    short G
        The same as format long g or format short g but use an uppercase
        E format. For example, with the short G format, pi .^ [2; 4; 8; 16;
        32] is displayed as

Chapter 14: Input and Output                                                163

                    ans =

                          9.8696
                         97.409
                       9488.5
                          9.0032E+07
                          8.1058E+15

free
none
    Print output in free format, without trying to line up columns of matrices on the decimal point. This also causes complex numbers to be formatted like this (0.604194, 0.607088) instead of like this 0.60419 + 0.60709i.

bank
    Print in a fixed format with two places to the right of the decimal point.

+
+ chars
plus
plus chars
    Print a + symbol for nonzero matrix elements and a space for zero matrix elements. This format can be very useful for examining the structure of a large matrix.

    The optional argument chars specifies a list of 3 characters to use for printing values greater than zero, less than zero and equal to zero. For example, with the + "+-." format, [1, 0, -1; -1, 0, 1] is displayed as

              ans =

                +.-
                -.+

native-hex
    Print the hexadecimal representation numbers as they are stored in memory. For example, on a workstation which stores 8 byte real values in IEEE format with the least significant byte first, the value of pi when printed in hex format is 400921fb54442d18. This format only works for numeric values.

hex
    The same as native-hex, but always print the most significant byte first.

native-bit
    Print the bit representation of numbers as stored in memory. For example, the value of pi is

```
01000000000010010010000111111011
01010100010001000010110100011000
```
(shown here in two 32 bit sections for typesetting purposes) when printed in bit format on a workstation which stores 8 byte real values in IEEE format with the least significant byte first. This format only works for numeric types.

bit
: The same as native-bit, but always print the most significant bits first.

compact
: Remove extra blank space around column number labels.

loose
: Insert blank lines above and below column number labels (this is the default).

rat
: Print a rational approximation. That is the values are approximated by one small integer divided by another.

By default, Octave will try to print numbers with at least 5 significant figures within a field that is a maximum of 10 characters wide.

If Octave is unable to format a matrix so that columns line up on the decimal point and all the numbers fit within the maximum field width, it switches to an e format.

If format is invoked without any options, the default format state is restored.

val = print_answer_id_name ()    Built-in Function
old_val = print_answer_id_name (new_val)    Built-in Function

Query or set the internal variable that controls whether variable names are printed along with results produced by evaluating an expression.

### 14.1.1.1 Paging Screen Output

When running interactively, Octave normally sends any output intended for your terminal that is more than one screen long to a paging program, such as less or more. This avoids the problem of having a large volume of output stream by before you can read it. With less (and some versions of more) you can also scan forward and backward, and search for specific items.

Normally, no output is displayed by the pager until just before Octave is ready to print the top level prompt, or read from the standard input (for example, by using the fscanf or scanf functions). This means that there may be some delay before any output appears on your screen if you have asked Octave to perform a significant amount of work with a single command statement. The function fflush may be used to force output to be sent to the pager (or any other stream) immediately.

You can select the program to run as the pager using the PAGER function, and you can turn paging off by using the function more.

## Chapter 14: Input and Output

more     *Command*
more on     *Command*
more off     *Command*

    Turn output pagination on or off. Without an argument, more toggles the current state. The current state can be determined via page_screen_output.

val = PAGER ()     *Built-in Function*
old_val = PAGER (new_val)     *Built-in Function*

    Query or set the internal variable that specifies the program to use to display terminal output on your system. The default value is normally "less", "more", or "pg", depending on what programs are installed on your system. See Appendix E [Installation], page 511.

    **See also:** more, page_screen_output, page_output_immediately, PAGER_FLAGS.

val = PAGER_FLAGS ()     *Built-in Function*
old_val = PAGER_FLAGS (new_val)     *Built-in Function*

    Query or set the internal variable that specifies the options to pass to the pager.

    **See also:** PAGER.

val = page_screen_output ()     *Built-in Function*
old_val = page_screen_output (new_val)     *Built-in Function*

    Query or set the internal variable that controls whether output intended for the terminal window that is longer than one page is sent through a pager. This allows you to view one screenful at a time. Some pagers (such as less—see Appendix E [Installation], page 511) are also capable of moving backward on the output.

val = page_output_immediately ()     *Built-in Function*
val = page_output_immediately (new_val)     *Built-in Function*

    Query or set the internal variable that controls whether Octave sends output to the pager as soon as it is available. Otherwise, Octave buffers its output and waits until just before the prompt is printed to flush it to the pager.

fflush (fid)     *Built-in Function*

    Flush output to fid. This is useful for ensuring that all pending output makes it to the screen before some other event occurs. For example, it is always a good idea to flush the standard output stream before calling input.

    fflush returns 0 on success and an OS dependent error value ($-1$ on unix) on error.

    **See also:** fopen, fclose.

## 14.1.2 Terminal Input

Octave has three functions that make it easy to prompt users for input. The input and menu functions are normally used for managing an interactive dialog with a user, and the keyboard function is normally used for doing simple debugging.

**input** (*prompt*)                                                                Built-in Function
**input** (*prompt*, "s")                                         Built-in Function

Print a prompt and wait for user input. For example,

    input ("Pick a number, any number! ")

prints the prompt

    Pick a number, any number!

and waits for the user to enter a value. The string entered by the user is evaluated as an expression, so it may be a literal constant, a variable name, or any other valid expression.

Currently, input only returns one value, regardless of the number of values produced by the evaluation of the expression.

If you are only interested in getting a literal string value, you can call input with the character string "s" as the second argument. This tells Octave to return the string entered by the user directly, without evaluating it first.

Because there may be output waiting to be displayed by the pager, it is a good idea to always call fflush (stdout) before calling input. This will ensure that all pending output is written to the screen before your prompt. See Chapter 14 [Input and Output], page 161.

**menu** (*title*, *opt1*, ...)                                           Function File

Print a title string followed by a series of options. Each option will be printed along with a number. The return value is the number of the option selected by the user. This function is useful for interactive programs. There is no limit to the number of options that may be passed in, but it may be confusing to present more than will fit easily on one screen.

**See also:** disp, printf, input.

For input, the normal command line history and editing functions are available at the prompt.

Octave also has a function that makes it possible to get a single character from the keyboard without requiring the user to type a carriage return.

**kbhit** ()                                                                                               Built-in Function

Read a single keystroke from the keyboard. If called with one argument, don't wait for a keypress. For example,

    x = kbhit ();

will set x to the next character typed at the keyboard as soon as it is typed.

    x = kbhit (1);

identical to the above example, but don't wait for a keypress, returning the empty string if no key is available.

### 14.1.3 Simple File I/O

The save and load commands allow data to be written to and read from disk files in various formats. The default format of files written by the save command can be controlled using the functions default_save_options and save_precision.

As an example the following code creates a 3-by-3 matrix and saves it to the file myfile.mat.

```
A = [ 1:3; 4:6; 7:9 ];
save myfile.mat A
```

Once one or more variables have been saved to a file, they can be read into memory using the load command.

```
load myfile.mat
A
⊣ A =
⊣
⊣    1    2    3
⊣    4    5    6
⊣    7    8    9
```

**save options file *v1 v2* ...**                                          *Command*

Save the named variables *v1*, *v2*, ..., in the file *file*. The special filename - can be used to write the output to your terminal. If no variable names are listed, Octave saves all the variables in the current scope. Valid options for the save command are listed in the following table. Options that modify the output format override the format specified by default_save_options.

If save is invoked using the functional form

```
save ("-option1", ..., "file", "v1", ...)
```

then the *options*, *file*, and variable name arguments (*v1*, ...) must be specified as character strings.

-ascii
    Save a single matrix in a text file.

-binary
    Save the data in Octave's binary data format.

-float-binary
    Save the data in Octave's binary data format but only using single precision. You should use this format only if you know that all the values to be saved can be represented in single precision.

-V7
-v7
-7
-mat7-binary
    Save the data in MATLAB's v7 binary data format.

`-V6`
`-v6`
`-6`
`-mat`
`-mat-binary`
> Save the data in MATLAB's v6 binary data format.

`-V4`
`-v4`
`-4`
`-mat4-binary`
> Save the data in the binary format written by MATLAB version 4.

`-hdf5`
> Save the data in HDF5 format. (HDF5 is a free, portable binary format developed by the National Center for Supercomputing Applications at the University of Illinois.)

`-float-hdf5`
> Save the data in HDF5 format but only using single precision. You should use this format only if you know that all the values to be saved can be represented in single precision.

`-zip`
`-z`  Use the gzip algorithm to compress the file. This is equivalent to compressing the file with `gzip` outside Octave.

The list of variables to save may include wildcard patterns containing the following special characters:

?  Match any single character.

*  Match zero or more characters.

[ *list* ]
> Match the list of characters specified by *list*. If the first character is ! or ^, match all characters except those specified by *list*. For example, the pattern [a-zA-Z] will match all lower and upper case alphabetic characters.

`-text`
> Save the data in Octave's text data format.

Except when using the MATLAB binary data file format, saving global variables also saves the global status of the variable, so that if it is restored at a later time using `load`, it will be restored as a global variable.

The command

```
save -binary data a b*
```

saves the variable a and all variables beginning with b to the file 'data' in Octave's binary format.

## Chapter 14: Input and Output

`load options file v1 v2 ...`  Command

Load the named variables *v1*, *v2*, ..., from the file *file*. As with save, you may specify a list of variables and `load` will only extract those variables with names that match. For example, to restore the variables saved in the file 'data', use the command

    `load data`

If load is invoked using the functional form

    `load ("-option1", ..., "file", "v1", ...)`

then the *options*, *file*, and variable name arguments (*v1*, ...) must be specified as character strings.

If a variable that is not marked as global is loaded from a file when a global symbol with the same name already exists, it is loaded in the global symbol table. Also, if a variable is marked as global in a file and a local symbol exists, the local symbol is moved to the global symbol table and given the value from the file. Since it seems that both of these cases are likely to be the result of some sort of error, they will generate warnings.

If invoked with a single output argument, Octave returns data instead of inserting variables in the symbol table. If the data file contains only numbers (TAB- or space-delimited columns), a matrix of values is returned. Otherwise, load returns a structure with members corresponding to the names of the variables in the file.

The `load` command can read data stored in Octave's text and binary formats, and MATLAB's binary format. It will automatically detect the type of file and do conversion from different floating point formats (currently only IEEE big and little endian, though other formats may added in the future).

Valid options for `load` are listed in the following table.

`-force`
> The `-force` option is accepted but ignored for backward compatibility. Octave now overwrites variables currently in memory with the same name as those found in the file.

`-ascii`
> Force Octave to assume the file contains columns of numbers in text format without any header or other information. Data in the file will be loaded as a single numeric matrix with the name of the variable derived from the name of the file.

`-binary`
> Force Octave to assume the file is in Octave's binary format.

`-mat`
`-mat-binary`
`-6`
`-v6`
`-7`
`-v7`
> Force Octave to assume the file is in MATLAB's version 6 or 7 binary format.

`-V4`
`-v4`
`-4`
`-mat4-binary`
> Force Octave to assume the file is in the binary format written by MATLAB version 4.

`-hdf5`
> Force Octave to assume the file is in HDF5 format. (HDF5 is a free, portable binary format developed by the National Center for Supercomputing Applications at the University of Illinois.) Note that Octave can read HDF5 files not created by itself, but may skip some datasets in formats that it cannot support.

`-import`
> The -import is accepted but ignored for backward compatibility. Octave can now support multi-dimensional HDF data and automatically modifies variable names if they are invalid Octave identifiers.

`-text`
> Force Octave to assume the file is in Octave's text format.

There are three functions that modify the behavior of save.

`val = default_save_options ()`      Built-in Function
`old_val = default_save_options (new_val)`      Built-in Function
> Query or set the internal variable that specifies the default options for the save command, and defines the default format. Typical values include "-ascii", "-ascii -zip". The default value is -ascii.

See also: save.

`val = save_precision ()`      Built-in Function
`old_val = save_precision (new_val)`      Built-in Function
> Query or set the internal variable that specifies the number of digits to keep when saving data in text format.

`val = save_header_format_string ()`      Built-in Function
`old_val = save_header_format_string (new_val)`      Built-in Function
> Query or set the internal variable that specifies the format string used for the comment line written at the beginning of text-format data files saved by Octave. The format string is passed to strftime and should begin with the character # and contain no newline characters. If the value of save_header_format_string is the empty string, the header comment is omitted from text-format data files. The default value is
>
> `"# Created by Octave VERSION, %a %b %d %H:%M:%S %Y %Z <USER@HOST>"`

See also: strftime.

# Chapter 14: Input and Output

**native_float_format ()**  *Built-in Function*
    Return the native floating point format as a string

It is possible to write data to a file in a similar way to the `disp` function for writing data to the screen. The `fdisp` works just like `disp` except its first argument is a file pointer as created by `fopen`. As an example, the following code writes to data `myfile.txt`.

```
fid = fopen ("myfile.txt", "w");
fdisp (fid, "3/8 is ");
fdisp (fid, 3/8);
fclose (fid);
```

See Section 14.2.1 [Opening and Closing Files], page 174, for details on how to use `fopen` and `fclose`.

**fdisp** (*fid*, *x*)  *Built-in Function*
    Display the value of *x* on the stream *fid*. For example,

        `fdisp (stdout, "The value of pi is:"), fdisp (stdout, pi)`

        ⊣ the value of pi is:
        ⊣ 3.1416

Note that the output from `fdisp` always ends with a newline.

**See also:** disp.

### 14.1.3.1 Saving Data on Unexpected Exits

If Octave for some reason exits unexpectedly it will by default save the variables available in the workspace to a file in the current directory. By default this file is named `octave-core` and can be loaded into memory with the `load` command. While the default behaviour most often is reasonable it can be changed through the following functions.

*val* = **crash_dumps_octave_core** ()  *Built-in Function*
*old_val* = **crash_dumps_octave_core** (*new_val*)  *Built-in Function*
    Query or set the internal variable that controls whether Octave tries to save all current variables to the file 'octave-core' if it crashes or receives a hangup, terminate or similar signal.

    **See also:** octave_core_file_limit, octave_core_file_name, octave_core_file_options.

*val* = **sighup_dumps_octave_core** ()  *Built-in Function*
*old_val* = **sighup_dumps_octave_core** (*new_val*)  *Built-in Function*
    Query or set the internal variable that controls whether Octave tries to save all current variables to the file 'octave-core' if it receives a hangup signal.

*val* = **sigterm_dumps_octave_core** ()  *Built-in Function*
*old_val* = **sigterm_dumps_octave_core** (*new_val*)  *Built-in Function*
    Query or set the internal variable that controls whether Octave tries to save all current variables to the file 'octave-core' if it receives a terminate signal.

val = octave_core_file_options ()                               Built-in Function
old_val = octave_core_file_options (new_val)                    Built-in Function
  Query or set the internal variable that specifies the options used for saving the workspace data if Octave aborts. The value of octave_core_file_options should follow the same format as the options for the save function. The default value is Octave's binary format.

  See also: crash_dumps_octave_core, octave_core_file_name, octave_core_file_limit.

val = octave_core_file_limit ()                                 Built-in Function
old_val = octave_core_file_limit (new_val)                      Built-in Function
  Query or set the internal variable that specifies the maximum amount of memory (in kilobytes) of the top-level workspace that Octave will attempt to save when writing data to the crash dump file (the name of the file is specified by octave_core_file_name). If octave_core_file_options flags specify a binary format, then octave_core_file_limit will be approximately the maximum size of the file. If a text file format is used, then the file could be much larger than the limit. The default value is -1 (unlimited)

  See also: crash_dumps_octave_core, octave_core_file_name, octave_core_file_options.

val = octave_core_file_name ()                                  Built-in Function
old_val = octave_core_file_name (new_val)                       Built-in Function
  Query or set the internal variable that specifies the name of the file used for saving data from the top-level workspace if Octave aborts. The default value is 'octave-core'

  See also: crash_dumps_octave_core, octave_core_file_name, octave_core_file_options.

### 14.1.4 Rational Approximations

s = rat (x, tol)                                                Function File
[n, d] = rat (x, tol)                                           Function File
  Find a rational approximation to x within tolerance defined by *tol* using a continued fraction expansion. E.g,

         rat(pi) = 3 + 1/(7 + 1/16) = 355/113
         rat(e)  = 3 + 1/(-4 + 1/(2 + 1/(5 + 1/(-2 + 1/(-7))))) 
                 = 1457/536

  Called with two arguments returns the numerator and denominator separately as two matrices.

  See also: rats.

rats (*x*, *len*)                                    Built-in Function
Convert *x* into a rational approximation represented as a string. You can convert the string back into a matrix as follows:

    r = rats(hilb(4));
    x = str2num(r)

The optional second argument defines the maximum length of the string representing the elements of *x*. By default *len* is 9.

See also: format, rat.

## 14.2 C-Style I/O Functions

Octave's C-style input and output functions provide most of the functionality of the C programming language's standard I/O library. The argument lists for some of the input functions are slightly different, however, because Octave has no way of passing arguments by reference.

In the following, *file* refers to a file name and fid refers to an integer file number, as returned by fopen.

There are three files that are always available. Although these files can be accessed using their corresponding numeric file ids, you should always use the symbolic names given in the table below, since it will make your programs easier to understand.

stdin ()                                             Built-in Function
Return the numeric value corresponding to the standard input stream. When Octave is used interactively, this is filtered through the command line editing functions.

See also: stdout, stderr.

stdout ()                                            Built-in Function
Return the numeric value corresponding to the standard output stream. Data written to the standard output is normally filtered through the pager.

See also: stdin, stderr.

stderr ()                                            Built-in Function
Return the numeric value corresponding to the standard error stream. Even if paging is turned on, the standard error is not sent to the pager. It is useful for error messages and prompts.

See also: stdin, stdout.

## 14.2.1 Opening and Closing Files

When reading data from a file it must be opened for reading first, and likewise when writing to a file. The fopen function returns a pointer to an open file that is ready to be read or written. Once all data has been read from or written to the opened file it should be closed. The fclose function does this. The following code illustrates the basic pattern for writing to a file, but a very similar pattern is used when reading a file.

```
filename = "myfile.txt";
fid = fopen (filename, "w");
# Do the actual I/O here...
fclose (fid);
```

[*fid, msg*] = fopen (*name, mode, arch*)  Built-in Function
*fid_list* = fopen ("all")  Built-in Function
[*file, mode, arch*] = fopen (*fid*)  Built-in Function

The first form of the fopen function opens the named file with the specified mode (read-write, read-only, etc.) and architecture interpretation (IEEE big endian, IEEE little endian, etc.), and returns an integer value that may be used to refer to the file later. If an error occurs, *fid* is set to −1 and *msg* contains the corresponding system error message. The *mode* is a one or two character string that specifies whether the file is to be opened for reading, writing, or both.

The second form of the fopen function returns a vector of file ids corresponding to all the currently open files, excluding the stdin, stdout, and stderr streams.

The third form of the fopen function returns information about the open file given its file id.

For example,

    myfile = fopen ("splat.dat", "r", "ieee-le");

opens the file 'splat.dat' for reading. If necessary, binary numeric values will be read assuming they are stored in IEEE format with the least significant bit first, and then converted to the native representation.

Opening a file that is already open simply opens it again and returns a separate file id. It is not an error to open a file several times, though writing to the same file through several different file ids may produce unexpected results.

The possible values mode may have are

r   Open a file for reading.

w   Open a file for writing. The previous contents are discarded.

a   Open or create a file for writing at the end of the file.

r+  Open an existing file for reading and writing.

w+  Open a file for reading or writing. The previous contents are discarded.

a+  Open or create a file for reading or writing at the end of the file.

Chapter 14: Input and Output                                                 175

Append a "t" to the mode string to open the file in text mode or a "b" to open in binary mode. On Windows and Macintosh systems, text mode reading and writing automatically converts linefeeds to the appropriate line end character for the system (carriage-return linefeed on Windows, carriage-return on Macintosh). The default if no mode is specified is binary mode.

Additionally, you may append a "z" to the mode string to open a gzipped file for reading or writing. For this to be successful, you must also open the file in binary mode.

The parameter *arch* is a string specifying the default data format for the file. Valid values for *arch* are:

> native The format of the current machine (this is the default).
>
> ieee-be IEEE big endian format.
>
> ieee-le IEEE little endian format.
>
> vaxd VAX D floating format.
>
> vaxg VAX G floating format.
>
> cray Cray floating format.

however, conversions are currently only supported for native ieee-be, and ieee-le formats.

See also: fclose, fread, fseek.

fclose (*fid*)                                                    Built-in Function
Closes the specified file. If successful, fclose returns 0, otherwise, it returns -1.

See also: fopen, fseek, ftell.

## 14.2.2 Simple Output

Once a file has been opened for writing a string can be written to the file using the fputs function. The following example shows how to write the string Free Software is needed for Free Science to the file free.txt.

```
filename = "free.txt";
fid = fopen (filename, "w");
fputs (fid, "Free Software is needed for Free Science");
fclose (fid);
```

fputs (*fid, string*)                                              Built-in Function
Write a string to a file with no formatting.

Return a non-negative number on success and EOF on error.

A function much similar to fputs is available for writing data to the screen. The puts function works just like fputs except it doesn't take a file pointer as its input.

puts (*string*)                                                    Built-in Function
Write a string to the standard output with no formatting.

Return a non-negative number on success and EOF on error.

### 14.2.3 Line-Oriented Input

To read from a file it must be opened for reading using `fopen`. Then a line can be read from the file using `fgetl` as the following code illustrates

```
fid = fopen ("free.txt");
txt = fgetl (fid)
    ⊣ Free Software is needed for Free Science
fclose (fid);
```

This of course assumes that the file `free.txt` exists and contains the line Free Software is needed for Free Science.

**fgetl** (*fid, len*)   Built-in Function
> Read characters from a file, stopping after a newline, or EOF, or *len* characters have been read. The characters read, excluding the possible trailing newline, are returned as a string.
>
> If *len* is omitted, `fgetl` reads until the next newline character.
>
> If there are no more characters to read, `fgetl` returns $-1$.
>
> **See also:** fread, fscanf.

**fgets** (*fid, len*)   Built-in Function
> Read characters from a file, stopping after a newline, or EOF, or *len* characters have been read. The characters read, including the possible trailing newline, are returned as a string.
>
> If *len* is omitted, `fgets` reads until the next newline character.
>
> If there are no more characters to read, `fgets` returns $-1$.
>
> **See also:** fread, fscanf.

### 14.2.4 Formatted Output

This section describes how to call `printf` and related functions.

The following functions are available for formatted output. They are modelled after the C language functions of the same name, but they interpret the format template differently in order to improve the performance of printing vector and matrix values.

**printf** (*template, ...*)   Built-in Function
> Print optional arguments under the control of the template string *template* to the stream stdout and return the number of characters printed.
>
> **See also:** fprintf, sprintf, scanf.

**fprintf** (*fid, template, ...*)   Built-in Function
> This function is just like `printf`, except that the output is written to the stream *fid* instead of stdout.
>
> **See also:** printf, sprintf, fread, fscanf, fopen, fclose.

Chapter 14: Input and Output    177

sprintf (*template*, ...)    Built-in Function
   This is like printf, except that the output is returned as a string. Unlike
   the C library function, which requires you to provide a suitably sized string
   as an argument, Octave's sprintf function returns the string, automatically
   sized to hold all of the items converted.
   **See also:** printf, fprintf, sscanf.

   The printf function can be used to print any number of arguments. The template string argument you supply in a call provides information not only about the number of additional arguments, but also about their types and what style should be used for printing them.

   Ordinary characters in the template string are simply written to the output stream as-is, while *conversion specifications* introduced by a % character in the template cause subsequent arguments to be formatted and written to the output stream. For example,

```
pct = 37;
filename = "foo.txt";
printf ("Processed %d%% of '%s'.\nPlease be patient.\n",
        pct, filename);
```

produces output like

```
Processed 37% of 'foo.txt'.
Please be patient.
```

This example shows the use of the %d conversion to specify that a scalar argument should be printed in decimal notation, the %s conversion to specify printing of a string argument, and the %% conversion to print a literal % character.

   There are also conversions for printing an integer argument as an unsigned value in octal, decimal, or hexadecimal radix (%o, %u, or %x, respectively); or as a character value (%c).

   Floating-point numbers can be printed in normal, fixed-point notation using the %f conversion or in exponential notation using the %e conversion. The %g conversion uses either %e or %f format, depending on what is more appropriate for the magnitude of the particular number.

   You can control formatting more precisely by writing *modifiers* between the % and the character that indicates which conversion to apply. These slightly alter the ordinary behavior of the conversion. For example, most conversion specifications permit you to specify a minimum field width and a flag indicating whether you want the result left- or right-justified within the field.

   The specific flags and modifiers that are permitted and their interpretation vary depending on the particular conversion. They're all described in more detail in the following sections.

### 14.2.5 Output Conversion for Matrices

When given a matrix value, Octave's formatted output functions cycle through the format template until all the values in the matrix have been printed. For example,

        printf ("%4.2f %10.2e %8.4g\n", hilb (3));

⊣ 1.00     5.00e-01     0.3333
⊣ 0.50     3.33e-01     0.25
⊣ 0.33     2.50e-01     0.2

If more than one value is to be printed in a single call, the output functions do not return to the beginning of the format template when moving on from one value to the next. This can lead to confusing output if the number of elements in the matrices are not exact multiples of the number of conversions in the format template. For example,

        printf ("%4.2f %10.2e %8.4g\n", [1, 2], [3, 4]);

⊣ 1.00     2.00e+00     3
⊣ 4.00

If this is not what you want, use a series of calls instead of just one.

### 14.2.6 Output Conversion Syntax

This section provides details about the precise syntax of conversion specifications that can appear in a `printf` template string.

Characters in the template string that are not part of a conversion specification are printed as-is to the output stream.

The conversion specifications in a `printf` template string have the general form:

% flags width [ . precision ] type conversion

For example, in the conversion specifier %-10.8ld, the - is a flag, 10 specifies the field width, the precision is 8, the letter l is a type modifier, and d specifies the conversion style. (This particular type specifier says to print a numeric argument in decimal notation, with a minimum of 8 digits left-justified in a field at least 10 characters wide.)

In more detail, output conversion specifications consist of an initial % character followed in sequence by:

- Zero or more *flag characters* that modify the normal behavior of the conversion specification.

- An optional decimal integer specifying the *minimum field width*. If the normal conversion produces fewer characters than this, the field is padded with spaces to the specified width. This is a *minimum* value; if the normal conversion produces more characters than this, the field is *not* truncated. Normally, the output is right-justified within the field.

  You can also specify a field width of *. This means that the next argument in the argument list (before the actual value to be printed) is used as the field width. The value is rounded to the nearest integer. If the value is

Chapter 14: Input and Output         179

negative, this means to set the - flag (see below) and to use the absolute value as the field width.

- An optional *precision* to specify the number of digits to be written for the numeric conversions. If the precision is specified, it consists of a period (.) followed optionally by a decimal integer (which defaults to zero if omitted).

You can also specify a precision of *. This means that the next argument in the argument list (before the actual value to be printed) is used as the precision. The value must be an integer, and is ignored if it is negative.

- An optional *type modifier character*. This character is ignored by Octave's printf function, but is recognized to provide compatibility with the C language printf.

- A character that specifies the conversion to be applied.

The exact options that are permitted and how they are interpreted vary between the different conversion specifiers. See the descriptions of the individual conversions for information about the particular options that they use.

### 14.2.7 Table of Output Conversions

Here is a table summarizing what all the different conversions do:

%d, %i
> Print an integer as a signed decimal number. See Section 14.2.8 [Integer Conversions], page 180, for details. %d and %i are synonymous for output, but are different when used with scanf for input (see Section 14.2.13 [Table of Input Conversions], page 184).

%o    Print an integer as an unsigned octal number. See Section 14.2.8 [Integer Conversions], page 180, for details.

%u    Print an integer as an unsigned decimal number. See Section 14.2.8 [Integer Conversions], page 180, for details.

%x, %X
> Print an integer as an unsigned hexadecimal number. %x uses lower-case letters and %X uses upper-case. See Section 14.2.8 [Integer Conversions], page 180, for details.

%f    Print a floating-point number in normal (fixed-point) notation. See Section 14.2.9 [Floating-Point Conversions], page 181, for details.

%e, %E
> Print a floating-point number in exponential notation. %e uses lower-case letters and %E uses upper-case. See Section 14.2.9 [Floating-Point Conversions], page 181, for details.

%g, %G
> Print a floating-point number in either normal (fixed-point) or exponential notation, whichever is more appropriate for its magnitude. %g uses lower-case letters and %G uses upper-case. See Section 14.2.9 [Floating-Point Conversions], page 181, for details.

%c  Print a single character. See Section 14.2.10 [Other Output Conversions], page 182.

%s  Print a string. See Section 14.2.10 [Other Output Conversions], page 182.

%%  Print a literal % character. See Section 14.2.10 [Other Output Conversions], page 182.

If the syntax of a conversion specification is invalid, unpredictable things will happen, so don't do this. If there aren't enough function arguments provided to supply values for all the conversion specifications in the template string, or if the arguments are not of the correct types, the results are unpredictable. If you supply more arguments than conversion specifications, the extra argument values are simply ignored; this is sometimes useful.

### 14.2.8 Integer Conversions

This section describes the options for the %d, %i, %o, %u, %x, and %X conversion specifications. These conversions print integers in various formats.

The %d and %i conversion specifications both print an numeric argument as a signed decimal number; while %o, %u, and %x print the argument as an unsigned octal, decimal, or hexadecimal number (respectively). The %X conversion specification is just like %x except that it uses the characters ABCDEF as digits instead of abcdef.

The following flags are meaningful:

-  Left-justify the result in the field (instead of the normal right-justification).

+  For the signed %d and %i conversions, print a plus sign if the value is positive.

   For the signed %d and %i conversions, if the result doesn't start with a plus or minus sign, prefix it with a space character instead. Since the + flag ensures that the result includes a sign, this flag is ignored if you supply both of them.

#  For the %o conversion, this forces the leading digit to be 0, as if by increasing the precision. For %x or %X, this prefixes a leading 0x or 0X (respectively) to the result. This doesn't do anything useful for the %d, %i, or %u conversions.

0  Pad the field with zeros instead of spaces. The zeros are placed after any indication of sign or base. This flag is ignored if the - flag is also specified, or if a precision is specified.

If a precision is supplied, it specifies the minimum number of digits to appear; leading zeros are produced if necessary. If you don't specify a precision, the number is printed with as many digits as it needs. If you convert a value of zero with an explicit precision of zero, then no characters at all are produced.

Chapter 14: Input and Output                                                  181

## 14.2.9 Floating-Point Conversions

This section discusses the conversion specifications for floating-point numbers: the %f, %e, %E, %g, and %G conversions.

The %f conversion prints its argument in fixed-point notation, producing output of the form [-]ddd.ddd, where the number of digits following the decimal point is controlled by the precision you specify.

The %e conversion prints its argument in exponential notation, producing output of the form [-]d.ddde[+|-]dd. Again, the number of digits following the decimal point is controlled by the precision. The exponent always contains at least two digits. The %E conversion is similar but the exponent is marked with the letter E instead of e.

The %g and %G conversions print the argument in the style of %e or %E (respectively) if the exponent would be less than -4 or greater than or equal to the precision; otherwise they use the %f style. Trailing zeros are removed from the fractional portion of the result and a decimal-point character appears only if it is followed by a digit.

The following flags can be used to modify the behavior:

-   Left-justify the result in the field. Normally the result is right-justified.

\+   Always include a plus or minus sign in the result.

    If the result doesn't start with a plus or minus sign, prefix it with a space instead. Since the + flag ensures that the result includes a sign, this flag is ignored if you supply both of them.

\#   Specifies that the result should always include a decimal point, even if no digits follow it. For the %g and %G conversions, this also forces trailing zeros after the decimal point to be left in place where they would otherwise be removed.

0   Pad the field with zeros instead of spaces; the zeros are placed after any sign. This flag is ignored if the - flag is also specified.

The precision specifies how many digits follow the decimal-point character for the %f, %e, and %E conversions. For these conversions, the default precision is 6. If the precision is explicitly 0, this suppresses the decimal point character entirely. For the %g and %G conversions, the precision specifies how many significant digits to print. Significant digits are the first digit before the decimal point, and all the digits after it. If the precision is 0 or not specified for %g or %G, it is treated like a value of 1. If the value being printed cannot be expressed precisely in the specified number of digits, the value is rounded to the nearest number that fits.

## 14.2.10 Other Output Conversions

This section describes miscellaneous conversions for `printf`.

The %c conversion prints a single character. The - flag can be used to specify left-justification in the field, but no other flags are defined, and no precision or type modifier can be given. For example:

    printf ("%c%c%c%c%c", "h", "e", "l", "l", "o");

prints hello.

The %s conversion prints a string. The corresponding argument must be a string. A precision can be specified to indicate the maximum number of characters to write; otherwise characters in the string up to but not including the terminating null character are written to the output stream. The - flag can be used to specify left-justification in the field, but no other flags or type modifiers are defined for this conversion. For example:

    printf ("%3s%-6s", "no", "where");

prints  nowhere (note the leading and trailing spaces).

## 14.2.11 Formatted Input

Octave provides the `scanf`, `fscanf`, and `sscanf` functions to read formatted input. There are two forms of each of these functions. One can be used to extract vectors of data from a file, and the other is more 'C-like'.

[val, count] = fscanf (fid, template, size)  Built-in Function
[v1, v2, ..., count] = fscanf (fid, template, "C")  Built-in Function

In the first form, read from *fid* according to *template*, returning the result in the matrix *val*.

The optional argument *size* specifies the amount of data to read and may be one of

Inf
: Read as much as possible, returning a column vector.

nr
: Read up to *nr* elements, returning a column vector.

[nr, Inf]
: Read as much as possible, returning a matrix with *nr* rows. If the number of elements read is not an exact multiple of *nr*, the last column is padded with zeros.

[nr, nc]
: Read up to *nr* * *nc* elements, returning a matrix with *nr* rows. If the number of elements read is not an exact multiple of *nr*, the last column is padded with zeros.

If *size* is omitted, a value of Inf is assumed.

A string is returned if *template* specifies only character conversions.

The number of items successfully read is returned in *count*.

In the second form, read from *fid* according to *template*, with each conversion specifier in *template* corresponding to a single scalar return value. This form

Chapter 14: Input and Output       183

is more 'C-like', and also compatible with previous versions of Octave. The number of successful conversions is returned in *count*

See also: scanf, sscanf, fread, fprintf.

[*val*, *count*] = sscanf (*string*, *template*, *size*)       Built-in Function
[*v1*, *v2*, ..., *count*] = sscanf (*string*, *template*, "C")       Built-in Function
This is like fscanf, except that the characters are taken from the string *string* instead of from a stream. Reaching the end of the string is treated as an end-of-file condition.

See also: fscanf, scanf, sprintf.

Calls to scanf are superficially similar to calls to printf in that arbitrary arguments are read under the control of a template string. While the syntax of the conversion specifications in the template is very similar to that for printf, the interpretation of the template is oriented more towards free-format input and simple pattern matching, rather than fixed-field formatting. For example, most scanf conversions skip over any amount of "white space" (including spaces, tabs, and newlines) in the input file, and there is no concept of precision for the numeric input conversions as there is for the corresponding output conversions. Ordinarily, non-whitespace characters in the template are expected to match characters in the input stream exactly.

When a *matching failure* occurs, scanf returns immediately, leaving the first non-matching character as the next character to be read from the stream, and scanf returns all the items that were successfully converted.

The formatted input functions are not used as frequently as the formatted output functions. Partly, this is because it takes some care to use them properly. Another reason is that it is difficult to recover from a matching error.

### 14.2.12 Input Conversion Syntax

A scanf template string is a string that contains ordinary multibyte characters interspersed with conversion specifications that start with %.

Any whitespace character in the template causes any number of whitespace characters in the input stream to be read and discarded. The whitespace characters that are matched need not be exactly the same whitespace characters that appear in the template string. For example, write , in the template to recognize a comma with optional whitespace before and after.

Other characters in the template string that are not part of conversion specifications must match characters in the input stream exactly; if this is not the case, a matching failure occurs.

The conversion specifications in a scanf template string have the general form:

*% flags width type conversion*

In more detail, an input conversion specification consists of an initial % character followed in sequence by:

- An optional *flag character* \*, which says to ignore the text read for this specification. When scanf finds a conversion specification that uses this flag, it reads input as directed by the rest of the conversion specification, but it discards this input, does not return any value, and does not increment the count of successful assignments.

- An optional decimal integer that specifies the *maximum field width*. Reading of characters from the input stream stops either when this maximum is reached or when a non-matching character is found, whichever happens first. Most conversions discard initial whitespace characters, and these discarded characters don't count towards the maximum field width. Conversions that do not discard initial whitespace are explicitly documented.

- An optional type modifier character. This character is ignored by Octave's scanf function, but is recognized to provide compatibility with the C language scanf.

- A character that specifies the conversion to be applied.

The exact options that are permitted and how they are interpreted vary between the different conversion specifiers. See the descriptions of the individual conversions for information about the particular options that they allow.

### 14.2.13 Table of Input Conversions

Here is a table that summarizes the various conversion specifications:

%d   Matches an optionally signed integer written in decimal. See Section 14.2.14 [Numeric Input Conversions], page 185.

%i   Matches an optionally signed integer in any of the formats that the C language defines for specifying an integer constant. See Section 14.2.14 [Numeric Input Conversions], page 185.

%o   Matches an unsigned integer written in octal radix. See Section 14.2.14 [Numeric Input Conversions], page 185.

%u   Matches an unsigned integer written in decimal radix. See Section 14.2.14 [Numeric Input Conversions], page 185.

%x, %X
   Matches an unsigned integer written in hexadecimal radix. See Section 14.2.14 [Numeric Input Conversions], page 185.

%e, %f, %g, %E, %G
   Matches an optionally signed floating-point number. See Section 14.2.14 [Numeric Input Conversions], page 185.

%s   Matches a string containing only non-whitespace characters. See Section 14.2.15 [String Input Conversions], page 185.

%c   Matches a string of one or more characters; the number of characters read is controlled by the maximum field width given for the conversion. See Section 14.2.15 [String Input Conversions], page 185.

Chapter 14: Input and Output                                        185

%%  This matches a literal % character in the input stream. No corresponding argument is used.

If the syntax of a conversion specification is invalid, the behavior is undefined. If there aren't enough function arguments provided to supply addresses for all the conversion specifications in the template strings that perform assignments, or if the arguments are not of the correct types, the behavior is also undefined. On the other hand, extra arguments are simply ignored.

### 14.2.14 Numeric Input Conversions

This section describes the scanf conversions for reading numeric values.

The %d conversion matches an optionally signed integer in decimal radix.

The %i conversion matches an optionally signed integer in any of the formats that the C language defines for specifying an integer constant.

For example, any of the strings 10, 0xa, or 012 could be read in as integers under the %i conversion. Each of these specifies a number with decimal value 10.

The %o, %u, and %x conversions match unsigned integers in octal, decimal, and hexadecimal radices, respectively.

The %X conversion is identical to the %x conversion. They both permit either uppercase or lowercase letters to be used as digits.

Unlike the C language scanf, Octave ignores the h, l, and L modifiers.

### 14.2.15 String Input Conversions

This section describes the scanf input conversions for reading string and character values: %s and %c.

The %c conversion is the simplest: it matches a fixed number of characters, always. The maximum field with says how many characters to read; if you don't specify the maximum, the default is 1. This conversion does not skip over initial whitespace characters. It reads precisely the next n characters, and fails if it cannot get that many.

The %s conversion matches a string of non-whitespace characters. It skips and discards initial whitespace, but stops when it encounters more whitespace after having read something.

For example, reading the input:

    hello, world

with the conversion %10c produces " hello, wo", but reading the same input with the conversion %10s produces "hello,".

## 14.2.16 Binary I/O

Octave can read and write binary data using the functions fread and fwrite, which are patterned after the standard C functions with the same names. They are able to automatically swap the byte order of integer data and convert among the supported floating point formats as the data are read.

[val, count] = fread (fid, size, precision, skip, arch)    Built-in Function
    Read binary data of type *precision* from the specified file ID *fid*.

    The optional argument *size* specifies the amount of data to read and may be one of

Inf
    Read as much as possible, returning a column vector.

nr  Read up to *nr* elements, returning a column vector.

[nr, Inf]
    Read as much as possible, returning a matrix with *nr* rows. If the number of elements read is not an exact multiple of *nr*, the last column is padded with zeros.

[nr, nc]
    Read up to *nr * nc* elements, returning a matrix with *nr* rows. If the number of elements read is not an exact multiple of *nr*, the last column is padded with zeros.

If *size* is omitted, a value of Inf is assumed.

The optional argument *precision* is a string specifying the type of data to read and may be one of

"schar"
"signed char"
    Signed character.

"uchar"
"unsigned char"
    Unsigned character.

"int8"
"integer*1"
    8-bit signed integer.

"int16"
"integer*2"
    16-bit signed integer.

"int32"
"integer*4"
    32-bit signed integer.

"int64"
"integer*8"
    64-bit signed integer.

Chapter 14: Input and Output          187

"uint8"
: 8-bit unsigned integer.

"uint16"
: 16-bit unsigned integer.

"uint32"
: 32-bit unsigned integer.

"uint64"
: 64-bit unsigned integer.

"single"
"float32"
"real*4"
: 32-bit floating point number.

"double"
"float64"
"real*8"
: 64-bit floating point number.

"char"
"char*1"
: Single character.

"short"
: Short integer (size is platform dependent).

"int"
: Integer (size is platform dependent).

"long"
: Long integer (size is platform dependent).

"ushort"
"unsigned short"
: Unsigned short integer (size is platform dependent).

"uint"
"unsigned int"
: Unsigned integer (size is platform dependent).

"ulong"
"unsigned long"
: Unsigned long integer (size is platform dependent).

"float"
: Single precision floating point number (size is platform dependent).

The default precision is "uchar".

The *precision* argument may also specify an optional repeat count. For example, 32*single causes `fread` to read a block of 32 single precision floating point numbers. Reading in blocks is useful in combination with the *skip* argument.

The *precision* argument may also specify a type conversion. For example, int16=>int32 causes `fread` to read 16-bit integer values and return an array of 32-bit integer values. By default, `fread` returns a double precision array. The special form *TYPE is shorthand for TYPE=>TYPE.

The conversion and repeat counts may be combined. For example, the specification 32*single=>single causes `fread` to read blocks of single precision floating point values and return an array of single precision values instead of the default array of double precision values.

The optional argument *skip* specifies the number of bytes to skip after each element (or block of elements) is read. If it is not specified, a value of 0 is assumed. If the final block read is not complete, the final skip is omitted. For example,

        fread (f, 10, "3*single=>single", 8)

will omit the final 8-byte skip because the last read will not be a complete block of 3 values.

The optional argument *arch* is a string specifying the data format for the file. Valid values are

"native"
: The format of the current machine.

"ieee-be"
: IEEE big endian.

"ieee-le"
: IEEE little endian.

"vaxd"
: VAX D floating format.

"vaxg"
: VAX G floating format.

"cray"
: Cray floating format.

Conversions are currently only supported for "ieee-be" and "ieee-le" formats.

The data read from the file is returned in *val*, and the number of values read is returned in count

**See also:** fwrite, fopen, fclose.

Chapter 14: Input and Output          189

*count* = fwrite (*fid, data, precision, skip, arch*)          Built-in Function
    Write data in binary form of type *precision* to the specified file ID *fid*, returning the number of values successfully written to the file.

    The argument *data* is a matrix of values that are to be written to the file. The values are extracted in column-major order.

    The remaining arguments *precision*, *skip*, and *arch* are optional, and are interpreted as described for fread.

    The behavior of fwrite is undefined if the values in *data* are too large to fit in the specified precision.

    See also: fread, fopen, fclose.

## 14.2.17 Temporary Files

Sometimes one needs to write data to a file that is only temporary. This is most commonly used when an external program launched from within Octave needs to access data. When Octave exits all temporary files will be deleted, so this step need not be executed manually.

[*fid, name, msg*] = mkstemp (*template, delete*)          Built-in Function
    Return the file ID corresponding to a new temporary file with a unique name created from *template*. The last six characters of *template* must be XXXXXX and these are replaced with a string that makes the filename unique. The file is then created with mode read/write and permissions that are system dependent (on GNU/Linux systems, the permissions will be 0600 for versions of glibc 2.0.7 and later). The file is opened with the O_EXCL flag.

    If the optional argument *delete* is supplied and is true, the file will be deleted automatically when Octave exits, or when the function purge_tmp_files is called.

    If successful, *fid* is a valid file ID, *name* is the name of the file, and *msg* is an empty string. Otherwise, *fid* is -1, *name* is empty, and *msg* contains a system-dependent error message.

    See also: tmpfile, tmpnam, P_tmpdir.

[*fid, msg*] = tmpfile ()          Built-in Function
    Return the file ID corresponding to a new temporary file with a unique name. The file is opened in binary read/write ("w+b") mode. The file will be deleted automatically when it is closed or when Octave exits.

    If successful, *fid* is a valid file ID and *msg* is an empty string. Otherwise, *fid* is -1 and *msg* contains a system-dependent error message.

    See also: tmpnam, mkstemp, P_tmpdir.

tmpnam (*dir, prefix*)  Built-in Function
> Return a unique temporary file name as a string.
>
> If *prefix* is omitted, a value of "oct-" is used. If *dir* is also omitted, the default directory for temporary files is used. If *dir* is provided, it must exist, otherwise the default directory for temporary files is used. Since the named file is not opened, by tmpnam, it is possible (though relatively unlikely) that it will not be available by the time your program attempts to open it.
>
> **See also:** tmpfile, mkstemp, P_tmpdir.

## 14.2.18 End of File and Errors

Once a file has been opened its status can be acquired. As an example the feof functions determines if the end of the file has been reached. This can be very useful when reading small parts of a file at a time. The following example shows how to read one line at a time from a file until the end has been reached.

```
filename = "myfile.txt";
fid = fopen (filename, "r");
while (! feof (fid) )
  text_line = fgetl (fid);
endwhile
fclose (fid);
```

Note that in some situations it is more efficient to read the entire contents of a file and then process it, than it is to read it line by line. This has the potential advantage of removing the loop in the above code.

feof (*fid*)  Built-in Function
> Return 1 if an end-of-file condition has been encountered for a given file and 0 otherwise. Note that it will only return 1 if the end of the file has already been encountered, not if the next read operation will result in an end-of-file condition.
>
> **See also:** fread, fopen, fclose.

ferror (*fid*)  Built-in Function
> Return 1 if an error condition has been encountered for a given file and 0 otherwise. Note that it will only return 1 if an error has already been encountered, not if the next operation will result in an error condition.

freport ()  Built-in Function
> Print a list of which files have been opened, and whether they are open for reading, writing, or both. For example,
>
> ```
> freport ()
> ```
>
> | | number | mode | name |
> |---|---|---|---|
> | ⊣ | | | |
> | ⊣ | | | |
> | ⊣ | 0 | r | stdin |
> | ⊣ | 1 | w | stdout |
> | ⊣ | 2 | w | stderr |
> | ⊣ | 3 | r | myfile |

## 14.2.19 File Positioning

Three functions are available for setting and determining the position of the file pointer for a given file.

`ftell` (*fid*)       Built-in Function

Return the position of the file pointer as the number of characters from the beginning of the file *fid*.

**See also:** fseek, fopen, fclose.

`fseek` (*fid*, *offset*, *origin*)       Built-in Function

Set the file pointer to any location within the file *fid*.

The pointer is positioned *offset* characters from the *origin*, which may be one of the predefined variables SEEK_CUR (current position), SEEK_SET (beginning), or SEEK_END (end of file) or strings "cof", "bof" or "eof". If *origin* is omitted, SEEK_SET is assumed. The offset must be zero, or a value returned by ftell (in which case *origin* must be SEEK_SET).

Return 0 on success and -1 on error.

**See also:** ftell, fopen, fclose.

`SEEK_SET ()`       Built-in Function
`SEEK_CUR ()`       Built-in Function
`SEEK_END ()`       Built-in Function

Return the value required to request that `fseek` perform one of the following actions:

SEEK_SET
    Position file relative to the beginning.

SEEK_CUR
    Position file relative to the current position.

SEEK_END
    Position file relative to the end.

`frewind` (*fid*)       Built-in Function

Move the file pointer to the beginning of the file *fid*, returning 0 for success, and -1 if an error was encountered. It is equivalent to `fseek (fid, 0, SEEK_SET)`.

The following example stores the current file position in the variable marker, moves the pointer to the beginning of the file, reads four characters, and then returns to the original position.

```
marker = ftell (myfile);
frewind (myfile);
fourch = fgets (myfile, 4);
fseek (myfile, marker, SEEK_SET);
```

# 15 Plotting

## 15.1 Plotting Basics

Octave makes it easy to create many different types of two- and three-dimensional plots using a few high-level functions.

If you need finer control over graphics, see Section 15.2 [Advanced Plotting], page 217.

### 15.1.1 Two-Dimensional Plots

The plot function allows you to create simple x-y plots with linear axes. For example,

```
x = -10:0.1:10;
plot (x, sin (x));
```

displays a sine wave shown in Figure 15.1. On most systems, this command will open a separate plot window to display the graph.

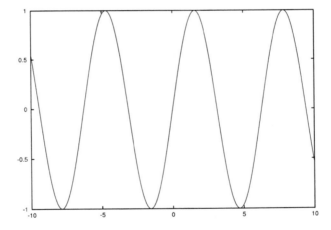

Figure 15.1: Simple Two-Dimensional Plot.

The function fplot also generates two-dimensional plots with linear axes using a function name and limits for the range of the x-coordinate instead of the x and y data. For example,

```
fplot (@sin, [-10, 10], 201);
```

produces a plot that is equivalent to the one above, but also includes a legend displaying the name of the plotted function.

| | |
|---|---|
| plot (y) | Function File |
| plot (x, y) | Function File |
| plot (x, y, property, value, ...) | Function File |
| plot (x, y, fmt) | Function File |
| plot (h, ...) | Function File |

Produces two-dimensional plots. Many different combinations of arguments are possible. The simplest form is

  plot (y)

where the argument is taken as the set of y coordinates and the x coordinates are taken to be the indices of the elements, starting with 1.

To save a plot, in one of several image formats such as PostScript or PNG, use the print command.

If more than one argument is given, they are interpreted as

  plot (y, property, value, ...)

or

  plot (x, y, property, value, ...)

or

  plot (x, y, fmt, ...)

and so on. Any number of argument sets may appear. The x and y values are interpreted as follows:

- If a single data argument is supplied, it is taken as the set of y coordinates and the x coordinates are taken to be the indices of the elements, starting with 1.
- If the x is a vector and y is a matrix, then the columns (or rows) of y are plotted versus x. (using whichever combination matches, with columns tried first.)
- If the x is a matrix and y is a vector, y is plotted versus the columns (or rows) of x. (using whichever combination matches, with columns tried first.)
- If both arguments are vectors, the elements of y are plotted versus the elements of x.
- If both arguments are matrices, the columns of y are plotted versus the columns of x. In this case, both matrices must have the same number of rows and columns and no attempt is made to transpose the arguments to make the number of rows match.

If both arguments are scalars, a single point is plotted.

Multiple property-value pairs may be specified, but they must appear in pairs. These arguments are applied to the lines drawn by plot.

If the fmt argument is supplied, it is interpreted as follows. If fmt is missing, the default gnuplot line style is assumed.

-    Set lines plot style (default).
- .   Set dots plot style.
- ˜   Set impulses plot style.

Chapter 15: Plotting                                                                  195

L   Set steps plot style.

n   Interpreted as the plot color if *n* is an integer in the range 1 to 6.

nm  If *nm* is a two digit integer and *m* is an integer in the range 1 to 6, *m* is interpreted as the point style. This is only valid in combination with the @ or -@ specifiers.

c   If *c* is one of "k" (black), "r" (red), "g" (green), "b" (blue), "m" (magenta), "c" (cyan), or "w" (white), it is interpreted as the line plot color.

";title;"
Here "title" is the label for the key.

+
*
o
x   Used in combination with the points or linespoints styles, set the point style.

The *fmt* argument may also be used to assign key titles. To do so, include the desired title between semi-colons after the formatting sequence described above, e.g. "+3;Key Title;" Note that the last semi-colon is required and will generate an error if it is left out.

Here are some plot examples:

    plot (x, y, "@12", x, y2, x, y3, "4", x, y4, "+")

This command will plot y with points of type 2 (displayed as +) and color 1 (red), y2 with lines, y3 with lines of color 4 (magenta) and y4 with points displayed as +.

    plot (b, "*", "markersize", 3)

This command will plot the data in the variable b, with points displayed as * with a marker size of 3.

    t = 0:0.1:6.3;
    plot (t, cos(t), "-;cos(t);", t, sin(t), "+3;sin(t);");

This will plot the cosine and sine functions and label them accordingly in the key.

If the first argument is an axis handle, then plot into these axes, rather than the current axis handle returned by gca.

**See also:** semilogx, semilogy, loglog, polar, mesh, contour, bar, stairs, errorbar, xlabel, ylabel, title, print.

fplot (*fn, limits*)                                          Function File
fplot (*fn, limits, tol*)                                     Function File
fplot (*fn, limits, n*)                                       Function File
fplot (..., *fmt*)                                            Function File

Plot a function *fn*, within the defined limits. *fn* an be either a string, a function handle or an inline function. The limits of the plot are given by *limits* of the form [*xlo, xhi*] or [*xlo, xhi, ylo, yhi*]. *tol* is the default tolerance to use for the plot, and if *tol* is an integer it is assumed that it defines the number points to use in the plot. The *fmt* argument is passed to the plot command.

```
fplot ("cos", [0, 2*pi])
fplot ("[cos(x), sin(x)]", [0, 2*pi])
```

**See also:** plot.

The functions `semilogx`, `semilogy`, and `loglog` are similar to the `plot` function, but produce plots in which one or both of the axes use log scales.

semilogx (*args*)                                                              Function File
    Produce a two-dimensional plot using a log scale for the x axis. See the description of `plot` for a description of the arguments that `semilogx` will accept.

    **See also:** plot, semilogy, loglog.

semilogy (*args*)                                                              Function File
    Produce a two-dimensional plot using a log scale for the y axis. See the description of `plot` for a description of the arguments that `semilogy` will accept.

    **See also:** plot, semilogx, loglog.

loglog (*args*)                                                                Function File
    Produce a two-dimensional plot using log scales for both axes. See the description of `plot` for a description of the arguments that `loglog` will accept.

    **See also:** plot, semilogx, semilogy.

The functions `bar`, `barh`, `stairs`, and `stem` are useful for displaying discrete data. For example,

```
hist (randn (10000, 1), 30);
```

produces the histogram of 10,000 normally distributed random numbers shown in Figure 15.2.

# Chapter 15: Plotting

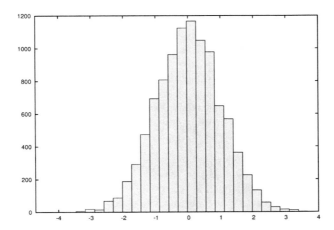

Figure 15.2: Histogram.

| | |
|---|---:|
| bar $(x, y)$ | Function File |
| bar $(y)$ | Function File |
| bar $(x, y, w)$ | Function File |
| bar $(x, y, w, style)$ | Function File |
| $h$ = bar $(\ldots, prop, val)$ | Function File |
| bar $(h, \ldots)$ | Function File |

Produce a bar graph from two vectors of x-y data.

If only one argument is given, it is taken as a vector of y-values and the x coordinates are taken to be the indices of the elements.

The default width of 0.8 for the bars can be changed using $w$.

If $y$ is a matrix, then each column of $y$ is taken to be a separate bar graph plotted on the same graph. By default the columns are plotted side-by-side. This behavior can be changed by the *style* argument, which can take the values "grouped" (the default), or "stacked".

The optional return value $h$ provides a handle to the patch object. Whereas the option input handle $h$ allows an axis handle to be passed. Properties of the patch graphics object can be changed using *prop*, *val* pairs.

See also: barh, plot.

| | |
|---|---:|
| barh $(x, y)$ | Function File |
| barh $(y)$ | Function File |
| barh $(x, y, w)$ | Function File |
| barh $(x, y, w, style)$ | Function File |
| $h$ = barh $(\ldots, prop, val)$ | Function File |
| barh $(h, \ldots)$ | Function File |

Produce a horizontal bar graph from two vectors of x-y data.

If only one argument is given, it is taken as a vector of y-values and the x coordinates are taken to be the indices of the elements.

The default width of 0.8 for the bars can be changed using w.

If y is a matrix, then each column of y is taken to be a separate bar graph plotted on the same graph. By default the columns are plotted side-by-side. This behavior can be changed by the *style* argument, which can take the values "grouped" (the default), or "stacked".

The optional return value h provides a handle to the patch object. Whereas the option input handle h allows an axis handle to be passed. Properties of the patch graphics object can be changed using *prop*, *val* pairs.

**See also:** bar, plot.

hist (*y*, *x*, *norm*)                                                            Function File
Produce histogram counts or plots.

With one vector input argument, plot a histogram of the values with 10 bins. The range of the histogram bins is determined by the range of the data.

Given a second scalar argument, use that as the number of bins.

Given a second vector argument, use that as the centers of the bins, with the width of the bins determined from the adjacent values in the vector.

If third argument is provided, the histogram is normalised such that the sum of the bars is equal to *norm*.

Extreme values are lumped in the first and last bins.

With two output arguments, produce the values *nn* and *xx* such that bar (*xx*, *nn*) will plot the histogram.

**See also:** bar.

stairs (*x*, *y*)                                                                Function File
Produce a stairstep plot. The arguments may be vectors or matrices.

If only one argument is given, it is taken as a vector of y-values and the x coordinates are taken to be the indices of the elements.

If two output arguments are specified, the data are generated but not plotted. For example,

```
    stairs (x, y);
```
and
```
    [xs, ys] = stairs (x, y);
    plot (xs, ys);
```
are equivalent.

**See also:** plot, semilogx, semilogy, loglog, polar, mesh, contour, bar, xlabel, ylabel, title.

Chapter 15: Plotting                                                            199

h = stem (x, y, linespec)                                          Function File
Plot a stem graph and return the handles of the line and marker objects used to draw the stems. The default color is "r" (red). The default line style is "-" and the default marker is "o".

For example,
```
x = 1:10;
stem (x);
```
plots 10 stems with heights from 1 to 10;
```
x = 1:10;
y = ones (1, length (x))*2.*x;
stem (x, y);
```
plots 10 stems with heights from 2 to 20;
```
x = 1:10;
y = ones (size (x))*2.*x;
h = stem (x, y, "b");
```
plots 10 bars with heights from 2 to 20 (the color is blue, and h is a 2-by-10 array of handles in which the first row holds the line handles and the second row holds the marker handles);
```
x = 1:10;
y = ones (size (x))*2.*x;
h = stem (x, y, "-.k");
```
plots 10 stems with heights from 2 to 20 (the color is black, line style is "-.", and h is a 2-by-10 array of handles in which the first row holds the line handles and the second row holds the marker handles);
```
x = 1:10;
y = ones (size (x))*2.*x;
h = stem (x, y, "-.k.");
```
plots 10 stems with heights from 2 to 20 (the color is black, line style is "-." and the marker style is ".", and h is a 2-by-10 array of handles in which the first row holds the line handles and the second row holds the marker handles);
```
x = 1:10;
y = ones (size (x))*2.*x;
h = stem (x, y, "fill");
```
plots 10 stems with heights from 2 to 20 (the color is rgb-triple defined, the line style is "-", the marker style is "o", and h is a 2-by-10 array of handles in which the first row holds the line handles and the second row holds the marker handles).

Color definitions with rgb-triples are not valid!

**See also:** bar, barh, plot.

The contour and contourc functions produce two-dimensional contour plots from three dimensional data.

contour (z)                                                                                                                Function File
contour (z, vn)                                                Function File
contour (x, y, z)                                            Function File
contour (x, y, z, vn)                                Function File
contour (..., style)                              Function File
contour (h, ...)                                      Function File
[c, h] = contour (...)                         Function File

Plot level curves (contour lines) of the matrix $z$, using the contour matrix $c$ computed by contourc from the same arguments; see the latter for their interpretation. The set of contour levels, $c$, is only returned if requested. For example:

```
x = 0:2;
y = x;
z = x' * y;
contour (x, y, z, 2:3)
```

The style to use for the plot can be defined with a line style *style* in a similar manner to the line styles used with the plot command. Any markers defined by *style* are ignored.

The optional input and output argument $h$ allows an axis handle to be passed to contour and the handles to the contour objects to be returned.

**See also:** contourc, patch, plot.

[c, lev] = contourc (x, y, z, vn)                      Function File
Compute isolines (contour lines) of the matrix $z$. Parameters $x$, $y$ and $vn$ are optional.

The return value *lev* is a vector of the contour levels. The return value $c$ is a 2 by $n$ matrix containing the contour lines in the following format

$$c = [lev1, x1, x2, \ldots, levn, x1, x2, \ldots$$
$$\phantom{c = [}len1, y1, y2, \ldots, lenn, y1, y2, \ldots]$$

in which contour line $n$ has a level (height) of *levn* and length of *lenn*.

If $x$ and $y$ are omitted they are taken as the row/column index of $z$. $vn$ is either a scalar denoting the number of lines to compute or a vector containing the values of the lines. If only one value is wanted, set $vn$ = [val, val]; If $vn$ is omitted it defaults to 10.

For example,

```
x = 0:2;
y = x;
z = x' * y;
contourc (x, y, z, 2:3)
⇒  2.0000   2.0000   1.0000   3.0000   1.5000   2.0000
   2.0000   1.0000   2.0000   2.0000   2.0000   1.5000
```

**See also:** contour.

Chapter 15: Plotting                                                                  201

The errorbar, semilogxerr, semilogyerr, and loglogerr functions produce plots with error bar markers. For example,

```
x = 0:0.1:10;
y = sin (x);
yp =  0.1 .* randn (size (x));
ym = -0.1 .* randn (size (x));
errorbar (x, sin (x), ym, yp);
```
produces the figure shown in Figure 15.3.

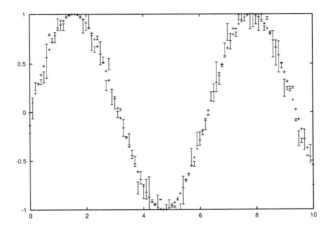

Figure 15.3: Errorbar plot.

errorbar (args)                                                              Function File
This function produces two-dimensional plots with errorbars. Many different combinations of arguments are possible. The simplest form is

   errorbar (y, ey)

where the first argument is taken as the set of y coordinates and the second argument ey is taken as the errors of the y values. x coordinates are taken to be the indices of the elements, starting with 1.

If more than two arguments are given, they are interpreted as

   errorbar (x, y, ..., fmt, ...)

where after x and y there can be up to four error parameters such as ey, ex, ly, uy etc., depending on the plot type. Any number of argument sets may appear, as long as they are separated with a format string fmt.

If y is a matrix, x and error parameters must also be matrices having same dimensions. The columns of y are plotted versus the corresponding columns of x and errorbars are drawn from the corresponding columns of error parameters.

If fmt is missing, yerrorbars ("~") plot style is assumed.

If the *fmt* argument is supplied, it is interpreted as in normal plots. In addition the following plot styles are supported by errorbar:

~     Set yerrorbars plot style (default).

&gt;     Set xerrorbars plot style.

~&gt;    Set xyerrorbars plot style.

\#     Set boxes plot style.

\#~    Set boxerrorbars plot style.

\#~&gt;

       Set boxxyerrorbars plot style.

Examples:

     errorbar (x, y, ex, "&gt;")

produces an xerrorbar plot of *y* versus *x* with *x* errorbars drawn from *x-ex* to *x+ex*.

     errorbar (x, y1, ey, "~",
     x, y2, ly, uy)

produces yerrorbar plots with *y1* and *y2* versus *x*. Errorbars for *y1* are drawn from *y1-ey* to *y1+ey*, errorbars for *y2* from *y2-ly* to *y2+uy*.

     errorbar (x, y, lx, ux,
     ly, uy, "~&gt;")

produces an xyerrorbar plot of *y* versus *x* in which *x* errorbars are drawn from *x-lx* to *x+ux* and *y* errorbars from *y-ly* to *y+uy*.

**See also:** semilogxerr, semilogyerr, loglogerr.

**semilogxerr** (*args*)                                                               *Function File*
Produce two-dimensional plots on a semilogarithm axis with errorbars. Many different combinations of arguments are possible. The most used form is

     semilogxerr (x, y, ey, fmt)

which produces a semi-logarithm plot of *y* versus *x* with errors in the *y*-scale defined by *ey* and the plot format defined by *fmt*. See errorbar for available formats and additional information.

**See also:** errorbar, loglogerr semilogyerr.

**semilogyerr** (*args*)                                                               *Function File*
Produce two-dimensional plots on a semilogarithm axis with errorbars. Many different combinations of arguments are possible. The most used form is

     semilogyerr (x, y, ey, fmt)

which produces a semi-logarithm plot of *y* versus *x* with errors in the *y*-scale defined by *ey* and the plot format defined by *fmt*. See errorbar for available formats and additional information.

**See also:** errorbar, loglogerr semilogxerr.

**loglogerr** (*args*)                                           Function File
Produce two-dimensional plots on double logarithm axis with errorbars. Many different combinations of arguments are possible. The most used form is

    loglogerr (x, y, ey, fmt)

which produces a double logarithm plot of *y* versus *x* with errors in the *y*-scale defined by *ey* and the plot format defined by *fmt*. See errorbar for available formats and additional information.

**See also:** errorbar, semilogxerr, semilogyerr.

Finally, the `polar` function allows you to easily plot data in polar coordinates. However, the display coordinates remain rectangular and linear. For example,

    polar (0:0.1:10*pi, 0:0.1:10*pi);

produces the spiral plot shown in Figure 15.4.

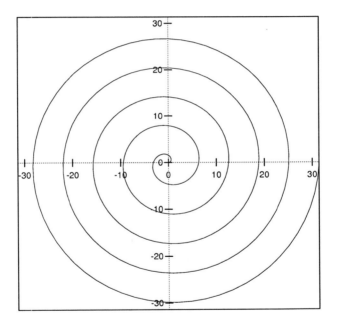

Figure 15.4: Polar plot.

**polar** (*theta, rho, fmt*)                                    Function File
Make a two-dimensional plot given the polar coordinates *theta* and *rho*. The optional third argument specifies the line type.

**See also:** plot.

| | |
|---|---|
| pie (y) | Function File |
| pie (y, explode) | Function File |
| pie (..., labels) | Function File |
| pie (h, ...); | Function File |
| h = pie (...); | Function File |

Produce a pie chart.

Called with a single vector argument, produces a pie chart of the elements in x, with the size of the slice determined by percentage size of the values of x.

The variable *explode* is a vector of the same length as x that if non zero 'explodes' the slice from the pie chart.

If given *labels* is a cell array of strings of the same length as x, giving the labels of each of the slices of the pie chart.

The optional return value h provides a handle to the patch object.

**See also:** bar, stem.

| | |
|---|---|
| quiver (u, v) | Function File |
| quiver (x, y, u, v) | Function File |
| quiver (..., s) | Function File |
| quiver (..., style) | Function File |
| quiver (..., 'filled') | Function File |
| quiver (h, ...) | Function File |
| h = quiver (...) | Function File |

Plot the (u, v) components of a vector field in an (x, y) meshgrid. If the grid is uniform, you can specify x and y as vectors.

If x and y are undefined they are assumed to be (1:m, 1:n) where [m, n] = size(u).

The variable s is a scalar defining a scaling factor to use for the arrows of the field relative to the mesh spacing. A value of 0 disables all scaling. The default value is 1.

The style to use for the plot can be defined with a line style *style* in a similar manner to the line styles used with the plot command. If a marker is specified then markers at the grid points of the vectors are printed rather than arrows. If the argument 'filled' is given then the markers as filled.

The optional return value h provides a list of handles to the parts of the vector field (body, arrow and marker).

```
[x, y] = meshgrid (1:2:20);
quiver (x, y, sin (2*pi*x/10), sin (2*pi*y/10));
```

**See also:** plot.

# Chapter 15: Plotting

pcolor (*x*, *y*, *c*)     Function File
pcolor (*c*)     Function File
: Density plot for given matrices *x*, and *y* from meshgrid and a matrix *c* corresponding to the *x* and *y* coordinates of the mesh. If *x* and *y* are vectors, then a typical vertex is (x(j), y(i), c(i,j)). Thus, columns of *c* correspond to different *x* values and rows of *c* correspond to different *y* values.

**See also:** meshgrid, contour.

area (*x*, *y*)     Function File
area (*x*, *y*, *lvl*)     Function File
area (..., *prop*, *val*, ...)     Function File
area (*y*, ...)     Function File
area (*h*, ...)     Function File
h = area (...)     Function File
: Area plot of cumulative sum of the columns of *y*. This shows the contributions of a value to a sum, and is functionally similar to plot (x, cumsum (y, 2)), except that the area under the curve is shaded.

If the *x* argument is omitted it is assumed to be given by 1 : rows (y). A value *lvl* can be defined that determines where the base level of the shading under the curve should be defined.

Additional arguments to the area function are passed to the patch. The optional return value *h* provides a handle to the list of patch objects.

**See also:** plot, patch.

The axis function may be used to change the axis limits of an existing plot.

axis (*limits*)     Function File
: Set axis limits for plots.

The argument *limits* should be a 2, 4, or 6 element vector. The first and second elements specify the lower and upper limits for the x axis. The third and fourth specify the limits for the y axis, and the fifth and sixth specify the limits for the z axis.

Without any arguments, axis turns autoscaling on.

With one output argument, x=axis returns the current axes

The vector argument specifying limits is optional, and additional string arguments may be used to specify various axis properties. For example,

    axis ([1, 2, 3, 4], "square");

forces a square aspect ratio, and

    axis ("labely", "tic");

turns tic marks on for all axes and tic mark labels on for the y-axis only.

The following options control the aspect ratio of the axes.

"square"
: Force a square aspect ratio.

"equal"
: Force x distance to equal y-distance.

"normal"
: Restore the balance.

The following options control the way axis limits are interpreted.

"auto"
: Set the specified axes to have nice limits around the data or all if no axes are specified.

"manual"
: Fix the current axes limits.

"tight"
: Fix axes to the limits of the data (not implemented).

The option "image" is equivalent to "tight" and "equal".
The following options affect the appearance of tic marks.

"on"
: Turn tic marks and labels on for all axes.

"off"
: Turn tic marks off for all axes.

"tic[xyz]"
: Turn tic marks on for all axes, or turn them on for the specified axes and off for the remainder.

"label[xyz]"
: Turn tic labels on for all axes, or turn them on for the specified axes and off for the remainder.

"nolabel"
: Turn tic labels off for all axes.

Note, if there are no tic marks for an axis, there can be no labels.
The following options affect the direction of increasing values on the axes.

"ij"
: Reverse y-axis, so lower values are nearer the top.

"xy"
: Restore y-axis, so higher values are nearer the top.

If an axes handle is passed as the first argument, then operate on this axes rather than the current axes.

Similarly the axis limits of the colormap can be changed with the caxis function.

caxis (*limits*)                                         Function File
caxis (*h*, ...)                                         Function File
   Set color axis limits for plots.

   The argument *limits* should be a 2 element vector specifying the lower and upper limits to assign to the first and last value in the colormap. Values outside this range are clamped to the first and last colormap entries.

   If *limits* is 'auto', then automatic colormap scaling is applied, whereas if *limits* is 'manual' the colormap scaling is set to manual.

   Called without any arguments to current color axis limits are returned.

   If an axes handle is passed as the first argument, then operate on this axes rather than the current axes.

### 15.1.2 Three-Dimensional Plotting

The function mesh produces mesh surface plots. For example,

```
tx = ty = linspace (-8, 8, 41)';
[xx, yy] = meshgrid (tx, ty);
r = sqrt (xx .^ 2 + yy .^ 2) + eps;
tz = sin (r) ./ r;
mesh (tx, ty, tz);
```

produces the familiar "sombrero" plot shown in Figure 15.5. Note the use of the function meshgrid to create matrices of X and Y coordinates to use for plotting the Z data. The ndgrid function is similar to meshgrid, but works for N-dimensional matrices.

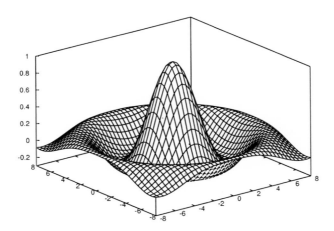

Figure 15.5: Mesh plot.

The meshc function is similar to mesh, but also produces a plot of contours for the surface.

The plot3 function displays arbitrary three-dimensional data, without requiring it to form a surface. For example

```
t = 0:0.1:10*pi;
r = linspace (0, 1, numel (t));
z = linspace (0, 1, numel (t));
plot3 (r.*sin(t), r.*cos(t), z);
```

displays the spiral in three dimensions shown in Figure 15.6.

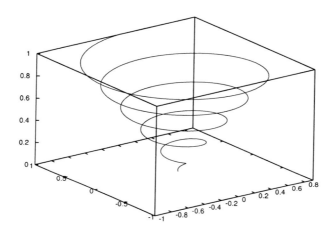

Figure 15.6: Three dimensional spiral.

Finally, the view function changes the viewpoint for three-dimensional plots.

mesh (x, y, z)                                                    Function File
    Plot a mesh given matrices x, and y from meshgrid and a matrix z corresponding to the x and y coordinates of the mesh. If x and y are vectors, then a typical vertex is $(x(j), y(i), z(i,j))$. Thus, columns of z correspond to different x values and rows of z correspond to different y values.

    **See also:** meshgrid, contour.

meshc (x, y, z)                                                   Function File
    Plot a mesh and contour given matrices x, and y from meshgrid and a matrix z corresponding to the x and y coordinates of the mesh. If x and y are vectors, then a typical vertex is $(x(j), y(i), z(i,j))$. Thus, columns of z correspond to different x values and rows of z correspond to different y values.

    **See also:** meshgrid, mesh, contour.

## Chapter 15: Plotting

hidden (*mode*)     Function File
hidden ()     Function File
    Manipulation the mesh hidden line removal. Called with no argument the hidden line removal is toggled. The argument *mode* can be either 'on' or 'off' and the set of the hidden line removal is set accordingly.

    **See also:** mesh, meshc, surf.

surf (*x, y, z*)     Function File
    Plot a surface given matrices *x*, and *y* from meshgrid and a matrix *z* corresponding to the *x* and *y* coordinates of the mesh. If *x* and *y* are vectors, then a typical vertex is ($x$(j), $y$(i), $z$(i,j)). Thus, columns of *z* correspond to different *x* values and rows of *z* correspond to different *y* values.

    **See also:** mesh, surface.

surfc (*x, y, z*)     Function File
    Plot a surface and contour given matrices *x*, and *y* from meshgrid and a matrix *z* corresponding to the *x* and *y* coordinates of the mesh. If *x* and *y* are vectors, then a typical vertex is ($x$(j), $y$(i), $z$(i,j)). Thus, columns of *z* correspond to different *x* values and rows of *z* correspond to different *y* values.

    **See also:** meshgrid, surf, contour.

[*xx, yy, zz*] = meshgrid (*x, y, z*)     Function File
[*xx, yy*] = meshgrid (*x, y*)     Function File
[*xx, yy*] = meshgrid (*x*)     Function File
    Given vectors of *x* and *y* and *z* coordinates, and returning 3 arguments, return three dimensional arrays corresponding to the *x*, *y*, and *z* coordinates of a mesh. When returning only 2 arguments, return matrices corresponding to the *x* and *y* coordinates of a mesh. The rows of *xx* are copies of *x*, and the columns of *yy* are copies of *y*. If *y* is omitted, then it is assumed to be the same as *x*, and *z* is assumed the same as *y*.

    **See also:** mesh, contour.

[*y1, y2, ..., yn*] = ndgrid (*x1, x2, ..., xn*)     Function File
[*y1, y2, ..., yn*] = ndgrid (*x*)     Function File
    Given n vectors *x1*, ... *xn*, ndgrid returns n arrays of dimension n. The elements of the i-th output argument contains the elements of the vector *xi* repeated over all dimensions different from the i-th dimension. Calling ndgrid with only one input argument *x* is equivalent of calling ndgrid with all n input arguments equal to *x*:

[*y1, y2, ..., yn*] = ndgrid (*x, ..., x*)

    **See also:** meshgrid.

plot3 (*args*)                                                                  Function File
    Produce three-dimensional plots. Many different combinations of arguments are possible. The simplest form is
        plot3 (*x, y, z*)
    in which the arguments are taken to be the vertices of the points to be plotted in three dimensions. If all arguments are vectors of the same length, then a single continuous line is drawn. If all arguments are matrices, then each column of the matrices is treated as a separate line. No attempt is made to transpose the arguments to make the number of rows match.
    If only two arguments are given, as
        plot3 (*x, c*)
    the real and imaginary parts of the second argument are used as the *y* and *z* coordinates, respectively.
    If only one argument is given, as
        plot3 (*c*)
    the real and imaginary parts of the argument are used as the *y* and *z* values, and they are plotted versus their index.
    Arguments may also be given in groups of three as
        plot3 (*x1, y1, z1, x2, y2, z2,* ...)
    in which each set of three arguments is treated as a separate line or set of lines in three dimensions.
    To plot multiple one- or two-argument groups, separate each group with an empty format string, as
        plot3 (*x1, c1,* "", *c2,* "", ...)
    An example of the use of plot3 is
        z = [0:0.05:5];
        plot3 (cos(2*pi*z), sin(2*pi*z), z, ";helix;");
        plot3 (z, exp(2i*pi*z), ";complex sinusoid;");
    **See also:** plot.

view (*azimuth, elevation*)                                                      Function File
view (*dims*)                                                                    Function File
[*azimuth, elevation*] = view ()                                                 Function File
    Set or get the viewpoint for the current axes.

shading (*type*)                                                                 Function File
shading (*ax,* ...)                                                              Function File
    Set the shading of surface or patch graphic objects. Valid arguments for *type* are "flat", "interp", or "faceted". If *ax* is given the shading is applied to axis *ax* instead of the current axis.

### 15.1.3 Plot Annotations

You can add titles, axis labels, legends, and arbitrary text to an existing plot. For example,

```
x = -10:0.1:10;
plot (x, sin (x));
title ("sin(x) for x = -10:0.1:10");
xlabel ("x");
ylabel ("sin (x)");
text (pi, 0.7, "arbitrary text");
legend ("sin (x)");
```

The functions grid and box may also be used to add grid and border lines to the plot. By default, the grid is off and the border lines are on.

title (*title*)                  Function File
    Create a title object and return a handle to it.

legend (*st1, st2, ...*)                  Function File
legend (*st1, st2, ...*, "location", *pos*)          Function File
legend (*matstr*)                  Function File
legend (*matstr*, "location", *pos*)          Function File
legend (*cell*)                  Function File
legend (*cell*, "location", *pos*)          Function File
legend ('*func*')                  Function File

Display a legend for the current axes using the specified strings as labels. Legend entries may be specified as individual character string arguments, a character array, or a cell array of character strings. Legend works on line graphs, bar graphs, etc. A plot must exist before legend is called.

The optional parameter *pos* specifies the location of the legend as follows:

| | |
|---|---|
| north | center top |
| south | center bottom |
| east | right center |
| west | left center |
| northeast | right top (default) |
| northwest | left top |
| southeast | right bottom |
| southwest | left bottom |
| outside | can be appended to any location string |

Some specific functions are directly available using *func*:

"show"
    Show legends from the plot

"hide"
"off"
    Hide legends from the plot

"boxon"
    Draw a box around legends

"boxoff"
: Withdraw the box around legends

"left"
: Text is to the left of the keys

"right"
: Text is to the right of the keys

h = text (x, y, label)                        *Function File*
h = text (x, y, z, label)                     *Function File*
h = text (x, y, label, p1, v1, ...)           *Function File*
h = text (x, y, z, label, p1, v1, ...)        *Function File*
: Create a text object with text *label* at position x, y, z on the current axes. Property-value pairs following *label* may be used to specify the appearance of the text.

xlabel (*string*)                             *Function File*
ylabel (*string*)                             *Function File*
zlabel (*string*)                             *Function File*
xlabel (h, *string*)                          *Function File*
: Specify x, y, and z axis labels for the current figure. If h is specified then label the axis defined by h.

**See also:** plot, semilogx, semilogy, loglog, polar, mesh, contour, bar, stairs, ylabel, title.

box (*arg*)                                   *Function File*
box (h, ...)                                  *Function File*
: Control the display of a border around the plot. The argument may be either "on" or "off". If it is omitted, the current box state is toggled.

**See also:** grid.

grid (*arg*)                                  *Function File*
grid ("minor", *arg2*)                        *Function File*
: Force the display of a grid on the plot. The argument may be either "on" or "off". If it is omitted, the current grid state is toggled.

If *arg* is "minor" then the minor grid is toggled. When using a minor grid a second argument *arg2* is allowed, which can be either "on" or "off" to explicitly set the state of the minor grid.

**See also:** plot.

## 15.1.4 Multiple Plots on One Page

Octave can display more than one plot in a single figure. The simplest way to do this is to use the subplot function to divide the plot area into a series of subplot windows that are indexed by an integer. For example,

```
subplot (2, 1, 1)
fplot (@sin, [-10, 10]);
subplot (2, 1, 2)
fplot (@cos, [-10, 10]);
```

creates a figure with two separate axes, one displaying a sine wave and the other a cosine wave. The first call to subplot divides the figure into two plotting areas (two rows and one column) and makes the first plot area active. The grid of plot areas created by subplot is numbered in column-major order (top to bottom, left to right).

subplot (*rows, cols, index*)            Function File
subplot (*rcn*)            Function File

Set up a plot grid with *cols* by *rows* subwindows and plot in location given by *index*.

If only one argument is supplied, then it must be a three digit value specifying the location in digits 1 (rows) and 2 (columns) and the plot index in digit 3.

The plot index runs row-wise. First all the columns in a row are filled and then the next row is filled.

For example, a plot with 2 by 3 grid will have plot indices running as follows:

| 1 | 2 | 3 |
|---|---|---|
| 4 | 5 | 6 |

See also: plot.

## 15.1.5 Multiple Plot Windows

You can open multiple plot windows using the figure function. For example

```
figure (1);
fplot (@sin, [-10, 10]);
figure (2);
fplot (@cos, [-10, 10]);
```

creates two figures, with the first displaying a sine wave and the second a cosine wave. Figure numbers must be positive integers.

figure (*n*)                                                                           Function File
figure (*n, property, value,* ...)                                                     Function File
    Set the current plot window to plot window *n*. If no arguments are specified, the next available window number is chosen.

    Multiple property-value pairs may be specified for the figure, but they must appear in pairs.

### 15.1.6 Printing Plots

The print command allows you to save plots in a variety of formats. For example,

    print -deps foo.eps

writes the current figure to an encapsulated PostScript file called 'foo.eps'.

print (*filename, options*)                                                            Function File
    Print a graph, or save it to a file

    *filename* defines the file name of the output file. If no filename is specified, output is sent to the printer.

    *options*:

-P*printer*
    Set the *printer* name to which the graph is sent if no *filename* is specified.

-color
-mono
    Monochrome or colour lines.

-solid
-dashed
    Solid or dashed lines.

-portrait
-landscape
    Plot orientation, as returned by "orient".

-d*device*
    Output device, where *device* is one of:

    ps
    ps2
    psc
    psc2
        Postscript (level 1 and 2, mono and color)

    eps
    eps2
    epsc
    epsc2
        Encapsulated postscript (level 1 and 2, mono and color)

## Chapter 15: Plotting

**tex**
**epslatex**
**epslatexstandalone**
**pstex**
**pslatex**
: Generate a LaTeX (or TeX) file for labels, and eps/ps for graphics. The file produced by `epslatexstandalone` can be processed directly by LaTeX. The other formats are intended to be included in a LaTeX (or TeX) document. The `tex` device is the same as the `epslatex` device.

**ill**
**aifm**
: Adobe Illustrator

**cdr**
**corel**
: CorelDraw

**dxf**
: AutoCAD

**emf**
: Microsoft Enhanced Metafile

**fig**
: XFig. If this format is selected the additional options -textspecial or -textnormal can be used to control whether the special flag should be set for the text in the figure (default is -textnormal).

**hpgl**
: HP plotter language

**mf** Metafont

**png**
: Portable network graphics

**jpg**
**jpeg**
: JPEG image

**gif**
: GIF image

**pbm**
: PBMplus

**svg**
: Scalable vector graphics

**pdf**
: Portable document format

Other devices are supported by "convert" from ImageMagick. Type system("convert") to see what formats are available.

If the device is omitted, it is inferred from the file extension, or if there is no filename it is sent to the printer as postscript.

-S*xsize,ysize*
Plot size in pixels for PNG and SVG. If using the command form of the print function, you must quote the *xsize,ysize* option. For example, by writing "-S640,480".

-F*fontname*
-F*fontname*:*size*
-F:*size*
*fontname* set the postscript font (for use with postscript, aifm, corel and fig). By default, 'Helvetica' is set for PS/Aifm, and 'Switzerland-Light' for Corel. It can also be 'Times-Roman'. *size* is given in points. *fontname* is ignored for the fig device.

The filename and options can be given in any order.

orient (*orientation*)                                                     Function File
Set the default print orientation. Valid values for *orientation* include "landscape" and "portrait". If called with no arguments, return the default print orientation.

### 15.1.7 Test Plotting Functions

The functions sombrero and peaks provide a way to check that plotting is working. Typing either sombrero or peaks at the Octave prompt should display a three dimensional plot.

sombrero (*n*)                                                                   Function File
Produce the familiar three-dimensional sombrero plot using *n* grid lines. If *n* is omitted, a value of 41 is assumed.
The function plotted is

      z = sin (sqrt (x^2 + y^2)) / (sqrt (x^2 + y^2))

See also: surf, meshgrid, mesh.

peaks ()                                                                                           Function File
peaks (*n*)                                                                             Function File
peaks (*x*, *y*)                                                                   Function File
z = peaks (...)                                                           Function File
[*x*, *y*, *z*] = peaks (...)                                         Function File
Generate a function with lots of local maxima and minima. The function has the form

$$f(x,y) = 3(1-x)^2 e^{\left(-x^2-(y+1)^2\right)} - 10\left(\frac{x}{5} - x^3 - y^5\right)) - \frac{1}{3}e^{\left(-(x+1)^2-y^2\right)}$$

Called without a return argument, peaks plots the surface of the above function using mesh. If *n* is a scalar, the peaks returns the values of the above function on a *n*-by-*n* mesh over the range [-3,3]. The default value for *n* is 49.

If *n* is a vector, then it represents the *x* and *y* values of the grid on which to calculate the above function. The *x* and *y* values can be specified separately.

See also: surf, mesh, meshgrid.

## 15.2 Advanced Plotting

### 15.2.1 Graphics Objects

Plots in Octave are constructed from the following *graphics objects*. Each graphics object has a set of properties that define its appearance and may also contain links to other graphics objects. Graphics objects are only referenced by a numeric index, or *handle*.

root figure
: The parent of all figure objects. The index for the root figure is defined to be 0.

figure
: A figure window.

axes
: An set of axes. This object is a child of a figure object and may be a parent of line, text, image, patch, or surface objects.

line
: A line in two or three dimensions.

text
: Text annotations.

image
: A bitmap image.

patch
: A filled polygon, currently limited to two dimensions.

surface
: A three-dimensional surface.

To determine whether an object is a graphics object index or a figure index, use the functions ishandle and isfigure.

ishandle (*h*)       Built-in Function
: Return true if *h* is a graphics handle and false otherwise.

isfigure (*h*)      Function File
: Return true if *h* is a graphics handle that contains a figure object and false otherwise.

The function gcf returns an index to the current figure object, or creates one if none exists. Similarly, gca returns the current axes object, or creates one (and its parent figure object) if none exists.

gcf ()                                                                                          Function File

Return the current figure handle. If a figure does not exist, create one and return its handle. The handle may then be used to examine or set properties of the figure. For example,

```
fplot (@sin, [-10, 10]);
fig = gcf ();
set (fig, "visible", "off");
```

plots a sine wave, finds the handle of the current figure, and then makes that figure invisible. Setting the visible property of the figure to "on" will cause it to be displayed again.

**See also:** get, set.

gca ()                                                                                          Function File

Return a handle to the current axis object. If no axis object exists, create one and return its handle. The handle may then be used to examine or set properties of the axes. For example,

```
ax = gca ();
set (ax, "position", [0.5, 0.5, 0.5, 0.5]);
```

creates an empty axes object, then changes its location and size in the figure window.

**See also:** get, set.

The get and set functions may be used to examine and set properties for graphics objects. For example,

```
get (0)
⇒ ans =
  {
    type = root figure
    currentfigure = [](0x0)
    children = [](0x0)
    visible = on
  }
```

returns a structure containing all the properties of the root figure. As with all functions in Octave, the structure is returned by value, so modifying it will not modify the internal root figure plot object. To do that, you must use the set function. Also, note that in this case, the currentfigure property is empty, which indicates that there is no current figure window.

The get function may also be used to find the value of a single property. For example,

```
get (gca (), "xlim")
⇒ [ 0 1 ]
```
returns the range of the x-axis for the current axes object in the current figure.

To set graphics object properties, use the set function. For example,
```
set (gca (), "xlim", [-10, 10]);
```
sets the range of the x-axis for the current axes object in the current figure to [-10, 10]. Additionally, calling set with a graphics object index as the only argument returns a structure containing the default values for all the properties for the given object type. For example,
```
set (gca ())
```
returns a structure containing the default property values for axes objects.

*get (h, p)*                                                             Built-in Function
    Return the named property $p$ from the graphics handle $h$. If $p$ is omitted, return the complete property list for $h$. If $h$ is a vector, return a cell array including the property values or lists respectively.

*set (h, p, v, ...)*                                         Built-in Function
    Set the named property value or vector $p$ to the value $v$ for the graphics handle $h$.

*parent = ancestor (h, type)*                           Function File
*parent = ancestor (h, type, 'toplevel')*                Function File
    Return the first ancestor of handle object $h$ whose type matches *type*, where *type* is a character string. If *type* is a cell array of strings, return the first parent whose type matches any of the given type strings.

    If the handle object $h$ is of type *type*, return $h$.

    If "toplevel" is given as a 3rd argument, return the highest parent in the object hierarchy that matches the condition, instead of the first (nearest) one.

    **See also:** get, set.

You can create axes, line, and patch objects directly using the `axes`, `line`, and `patch` functions. These objects become children of the current axes object.

*axes ()*                                                                 Function File
*axes (property, value, ...)*                                      Function File
*axes (h)*                                                            Function File
    Create an axes object and return a handle to it.

line ()                                              Function File
line (*x*, *y*)                                      Function File
line (*x*, *y*, *z*)                                 Function File
line (*x*, *y*, *z*, *property*, *value*, ...)       Function File
  Create line object from *x* and *y* and insert in current axes object. Return a handle (or vector of handles) to the line objects created.

  Multiple property-value pairs may be specified for the line, but they must appear in pairs.

patch ()                                             Function File
patch (*x*, *y*, *c*)                                Function File
patch (*x*, *y*, *c*, *opts*)                        Function File
patch ('Faces', *f*, 'Vertices', *v*, ...)           Function File
patch (..., *prop*, *val*)                           Function File
patch (*h*, ...)                                     Function File
*h* = patch (...)                                    Function File
  Create patch object from *x* and *y* with color *c* and insert in the current axes object. Return handle to patch object.

  For a uniform colored patch, *c* can be given as an RGB vector, scalar value referring to the current colormap, or string value (for example, "r" or "red").

surface (*x*, *y*, *z*, *c*)                         Function File
surface (*x*, *y*, *z*)                              Function File
surface (*z*, *c*)                                   Function File
surface (*z*)                                        Function File
surface (..., *prop*, *val*)                         Function File
surface (*h*, ...)                                   Function File
*h* = surface (...)                                  Function File
  Plot a surface graphic object given matrices *x*, and *y* from meshgrid and a matrix *z* corresponding to the *x* and *y* coordinates of the surface. If *x* and *y* are vectors, then a typical vertex is $(x(j), y(i), z(i,j))$. Thus, columns of *z* correspond to different *x* values and rows of *z* correspond to different *y* values. If *x* and *y* are missing, they are constructed from size of the matrix *z*.

  Any additional properties passed are assigned to the surface.

  **See also:** surf, mesh, patch, line.

By default, Octave refreshes the plot window when a prompt is printed, or when waiting for input. To force an update at other times, call the drawnow function.

drawnow ()                                           Function File
  Update and display the current graphics.

  Octave automatically calls drawnow just before printing a prompt, when sleep or pause is called, or while waiting for command-line input.

# Chapter 15: Plotting

Normally, high-level plot functions like `plot` or `mesh` call `newplot` to initialize the state of the current axes so that the next plot is drawn in a blank window with default property settings. To have two plots superimposed over one another, call the `hold` function. For example,

```
hold ("on");
x = -10:0.1:10;
plot (x, sin (x));
plot (x, cos (x));
hold ("off");
```

displays sine and cosine waves on the same axes. If the hold state is off, consecutive plotting commands like this will only display the last plot.

**newplot ()**  *Function File*
Prepare graphics engine to produce a new plot. This function should be called at the beginning of all high-level plotting functions.

**hold** *args*  *Function File*
Tell Octave to 'hold' the current data on the plot when executing subsequent plotting commands. This allows you to execute a series of plot commands and have all the lines end up on the same figure. The default is for each new plot command to clear the plot device first. For example, the command

    hold on

turns the hold state on. An argument of `"off"` turns the hold state off, and hold with no arguments toggles the current hold state.

**ishold**  *Function File*
Return true if the next line will be added to the current plot, or false if the plot device will be cleared before drawing the next line.

To clear the current figure, call the `clf` function. To bring it to the top of the window stack, call the `shg` function. To delete a graphics object, call `delete` on its index. To close the figure window, call the `close` function.

**clf ()**  *Function File*
Clear the current figure.

**See also:** close, delete.

**shg**  *Function File*
Show the graph window. Currently, this is the same as executing `drawnow`.

**See also:** drawnow, figure.

**delete** (*file*)  *Function File*
**delete** (*h*)  *Function File*
Delete the named file or figure handle.

close                                                              Command
close (n)                                                          Command
close all                                                          Command
close all hidden                                                   Command
   Close figure window(s) by calling the function specified by the "closerequestfcn" property for each figure. By default, the function closereq is used.

   See also: closereq.

closereq ()                                                   Function File
   Close the current figure and delete all graphics objects associated with it.

   See also: close, delete.

### 15.2.2 Graphics Object Properties

#### 15.2.2.1 Root Figure Properties

currentfigure
   Index to graphics object for the current figure.

#### 15.2.2.2 Figure Properties

nextplot
   May be one of

   "new"
   "add"
   "replace"
   "replacechildren"

closerequestfcn
   Handle of function to call when a figure is closed.

currentaxes
   Index to graphics object of current axes.

colormap
   An N-by-3 matrix containing the color map for the current axes.

visible
   Either "on" or "off" to toggle display of the figure.

paperorientation
   Indicates the orientation for printing. Either "landscape" or "portrait".

### 15.2.2.3 Axes Properties

position
: A four-element vector specifying the coordinates of the lower left corner and width and height of the plot, in normalized units. For example, [0.2, 0.3, 0.4, 0.5] sets the lower left corner of the axes at (0.2, 0.3) and the width and height to be 0.4 and 0.5 respectively.

title
: Index of text object for the axes title.

box
: Either "on" or "off" to toggle display of the box around the axes.

key
: Either "on" or "off" to toggle display of the legend. Note that this property is not compatible with MATLAB and may be removed in a future version of Octave.

keybox
: Either "on" or "off" to toggle display of a box around the legend. Note that this property is not compatible with MATLAB and may be removed in a future version of Octave.

keypos
: An integer from 1 to 4 specifying the position of the legend. 1 indicates upper right corner, 2 indicates upper left, 3 indicates lower left, and 4 indicates lower right. Note that this property is not compatible with MATLAB and may be removed in a future version of Octave.

dataaspectratio
: A two-element vector specifying the relative height and width of the data displayed in the axes. Setting dataaspectratio to 1, 2] causes the length of one unit as displayed on the y axis to be the same as the length of 2 units on the x axis. Setting dataaspectratio also forces the dataaspectratiomode property to be set to "manual".

dataaspectratiomode
: Either "manual" or "auto".

xlim
ylim
zlim
clim
: Two-element vectors defining the limits for the x, y, and z axes and the Setting one of these properties also forces the corresponding mode property to be set to "manual".

xlimmode
ylimmode
zlimmode
climmode
: Either "manual" or "auto".

xlabel
ylabel
zlabel
> Indices to text objects for the axes labels.

xgrid
ygrid
zgrid
> Either "on" or "off" to toggle display of grid lines.

xminorgrid
yminorgrid
zminorgrid
> Either "on" or "off" to toggle display of minor grid lines.

xtick
ytick
ztick
> Setting one of these properties also forces the corresponding mode property to be set to "manual".

xtickmode
ytickmode
ztickmode
> Either "manual" or "auto".

xticklabel
yticklabel
zticklabel
> Setting one of these properties also forces the corresponding mode property to be set to "manual".

xticklabelmode
yticklabelmode
zticklabelmode
> Either "manual" or "auto".

xscale
yscale
zscale
> Either "linear" or "log".

xdir
ydir
zdir
> Either "forward" or "reverse".

xaxislocation
yaxislocation
> Either "top" or "bottom" for the x axis and "left" or "right" for the y axis.

Chapter 15: Plotting                                                                 225

`view`
: A three element vector specifying the view point for three-dimensional plots.

`visible`
: Either "on" or "off" to toggle display of the axes.

`nextplot`
: May be one of

    `"new"`
    `"add"`
    `"replace"`
    `"replacechildren"`

`outerposition`
: A four-element vector specifying the coordinates of the lower left corner and width and height of the plot, in normalized units. For example, [0.2, 0.3, 0.4, 0.5] sets the lower left corner of the axes at (0.2, 0.3) and the width and height to be 0.4 and 0.5 respectively.

### 15.2.2.4 Line Properties

`xdata`
`ydata`
`zdata`
`ldata`
`udata`
`xldata`
`xudata`
: The data to be plotted. The `ldata` and `udata` elements are for errorbars in the y direction, and the `xldata` and `xudata` elements are for errorbars in the x direction.

`color`
: The RGB color of the line, or a color name. See Section 15.2.4 [Colors], page 231.

`linestyle`
`linewidth`
: See Section 15.2.5 [Line Styles], page 232.

`marker`
`markeredgecolor`
`markerfacecolor`
`markersize`
: See Section 15.2.6 [Marker Styles], page 232.

`keylabel`
: The text of the legend entry corresponding to this line. Note that this property is not compatible with MATLAB and may be removed in a future version of Octave.

## 15.2.2.5 Text Properties

string
: The character string contained by the text object.

units
: May be "normalized" or "graph".

position
: The coordinates of the text object.

rotation
: The angle of rotation for the displayed text, measured in degrees.

horizontalalignment
: May be "left", "center", or "right".

color
: The color of the text. See Section 15.2.4 [Colors], page 231.

fontname
: The font used for the text.

fontsize
: The size of the font, in points to use.

fontangle
: Flag whether the font is italic or normal. Valid values are 'normal', 'italic' and 'oblique'.

fontweight
: Flag whether the font is bold, etc. Valid values are 'normal', 'bold', 'demi' or 'light'.

interpreter
: Determines how the text is rendered. Valid values are 'none', 'tex' or 'latex'.

All text objects, including titles, labels, legends, and text, include the property 'interpreter', this property determines the manner in which special control sequences in the text are rendered. If the interpreter is set to 'none', then no rendering occurs. At this point the 'latex' option is not implemented and so the 'latex' interpreter also does not interpret the text.

The 'tex' option implements a subset of TEX functionality in the rendering of the text. This allows the insertion of special characters such as Greek or mathematical symbols within the text. The special characters are also inserted with a code starting with the back-slash (\) character, as in the table Table 15.1.

In addition, the formatting of the text can be changed within the string with the codes

| | |
|---|---|
| \bf | Bold font |
| \it | Italic font |
| \sl | Oblique Font |
| \rm | Normal font |

## Chapter 15: Plotting

These are be used in conjunction with the { and } characters to limit the change in the font to part of the string. For example

```
xlabel ('{\bf H} = a {\bf V}')
```

where the character 'a' will not appear in a bold font. Note that to avoid having Octave interpret the backslash characters in the strings, the strings should be in single quotes.

It is also possible to change the fontname and size within the text

|  |  |
|---|---|
| \fontname{*fontname*} | Specify the font to use |
| \fontsize{*size*} | Specify the size of the font to use |

Finally, the superscript and subscripting can be controlled with the '^' and '_' characters. If the '^' or '_' is followed by a { character, then all of the block surrounded by the { } pair is super- or sub-scripted. Without the { } pair, only the character immediately following the '^' or '_' is super- or sub-scripted.

| Code | Sym | Code | Sym | Code | Sym |
|---|---|---|---|---|---|
| \forall | ∀ | \exists | ∃ | \ni | ∋ |
| \cong | ≅ | \Delta | Δ | \Phi | Φ |
| \Gamma | Γ | \vartheta | ϑ | \Lambda | Λ |
| \Pi | Π | \Theta | Θ | \Sigma | Σ |
| \varsigma | ς | \Omega | Ω | \Xi | Ξ |
| \Psi | Ψ | \perp | ⊥ | \alpha | α |
| \beta | β | \chi | χ | \delta | δ |
| \epsilon | ε | \phi | φ | \gamma | γ |
| \eta | η | \iota | ι | \varphi | φ |
| \kappa | κ | \lambda | λ | \mu | μ |
| \nu | ν | \o | ≤ | \pi | π |
| \theta | θ | \rho | ρ | \sigma | σ |
| \tau | τ | \upsilon | υ | \varpi | ϖ |
| \omega | ω | \xi | ξ | \psi | ψ |
| \zeta | ζ | \sim | ~ | \Upsilon | Υ |
| \prime | ′ | \leq | ≤ | \infty | ∞ |
| \clubsuit | ♣ | \diamondsuit | ♦ | \heartsuit | ♡ |
| \spadesuit | ♠ | \leftrightarrow | ↔ | \leftarrow | ← |
| \uparrow | ↑ | \rightarrow | → | \downarrow | ↓ |
| \circ | ∘ | \pm | ± | \geq | ≥ |
| \times | × | \propto | ∝ | \partial | ∂ |
| \bullet | • | \div | ÷ | \neq | ≠ |
| \equiv | ≡ | \approx | ≈ | \ldots | ... |
| \mid | \| | \aleph | ℵ | \Im | ℑ |
| \Re | ℜ | \wp | ℘ | \otimes | ⊗ |
| \oplus | ⊕ | \oslash | ⊘ | \cap | ∩ |
| \cup | ∪ | \supset | ⊃ | \supseteq | ⊇ |
| \subset | ⊂ | \subseteq | ⊆ | \in | ∈ |
| \notin | ∉ | \angle | ∠ | \bigtriangledown | ▽ |
| \langle | ⟨ | \rangle | ⟩ | \nabla | ∇ |
| \prod | ∏ | \surd | √ | \cdot | · |
| \neg | ¬ | \wedge | ∧ | \vee | ∨ |
| \Leftrightarrow | ⇔ | \Leftarrow | ⇐ | \Uparrow | ⇑ |
| \Rightarrow | ⇒ | \Downarrow | ⇓ | \diamond | ◇ |
| \copyright | © | \rfloor | ⌋ | \lceil | ⌈ |
| \lfloor | ⌊ | \rceil | ⌉ | \int | ∫ |

Table 15.1: Available special characters in TEX mode

A complete example showing the capabilities of the extended text is

## Chapter 15: Plotting

```
x = 0:0.01:3;
plot(x,erf(x));
hold on;
plot(x,x,"r");
axis([0, 3, 0, 1]);
text(0.65, 0.6175, strcat('\leftarrow x = {2/\surd\pi}',
  ' {\fontsize{16}\int_{\fontsize{8}0}^{\fontsize{8}x}}',
  ' e^{-t^2} dt} = 0.6175'))
```
The result of which can be seen in Figure 15.7

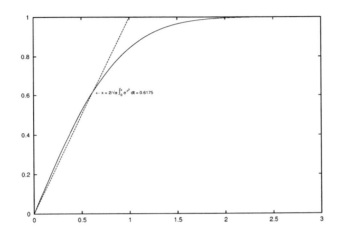

Figure 15.7: Example of inclusion of text with the TEX interpreter

### 15.2.2.6 Image Properties

cdata
> The data for the image. Each pixel of the image corresponds to an element of cdata. The value of an element of cdata specifies the row-index into the colormap of the axes object containing the image. The color value found in the color map for the given index determines the color of the pixel.

xdata
ydata
> Two-element vectors specifying the range of the x- and y- coordinates for the image.

### 15.2.2.7 Patch Properties

cdata  
xdata  
ydata  
zdata  
    Data defining the patch object.

facecolor  
    The fill color of the patch. See Section 15.2.4 [Colors], page 231.

facealpha  
    A number in the range [0, 1] indicating the transparency of the patch.

edgecolor  
    The color of the line defining the patch. See Section 15.2.4 [Colors], page 231.

linestyle  
linewidth  
    See Section 15.2.5 [Line Styles], page 232.

marker  
markeredgecolor  
markerfacecolor  
markersize  
    See Section 15.2.6 [Marker Styles], page 232.

### 15.2.2.8 Surface Properties

xdata  
ydata  
zdata  
    The data determining the surface. The xdata and ydata elements are vectors and zdata must be a matrix.

keylabel  
    The text of the legend entry corresponding to this surface. Note that this property is not compatible with MATLAB and may be removed in a future version of Octave.

## 15.2.3 Managing Default Properties

Object properties have two classes of default values, *factory defaults* (the initial values) and *user-defined defaults*, which may override the factory defaults.

Although default values may be set for any object, they are set in parent objects and apply to child objects. For example,

    set (0, "defaultlinecolor", "green");

sets the default line color for all objects. The rule for constructing the property name to set a default value is default + *object-type* + *property-name*

This rule can lead to some strange looking names, for example defaultlinelinewidth" specifies the default linewidth property for line objects.

The example above used the root figure object, 0, so the default property value will apply to all line objects. However, default values are hierarchical, so defaults set in a figure objects override those set in the root figure object. Likewise, defaults set in axes objects override those set in figure or root figure objects. For example,

```
subplot (2, 1, 1);
set (0, "defaultlinecolor", "red");
set (1, "defaultlinecolor", "green");
set (gca (), "defaultlinecolor", "blue");
line (1:10, rand (1, 10));
subplot (2, 1, 2);
line (1:10, rand (1, 10));
figure (2)
line (1:10, rand (1, 10));
```

produces two figures. The line in first subplot window of the first figure is blue because it inherits its color from its parent axes object. The line in the second subplot window of the first figure is green because it inherits its color from its parent figure object. The line in the second figure window is red because it inherits its color from the global root figure parent object.

To remove a user-defined default setting, set the default property to the value "remove". For example,

```
set (gca (), "defaultlinecolor", "remove");
```

removes the user-defined default line color setting from the current axes object.

Getting the "default" property of an object returns a list of user-defined defaults set for the object. For example,

```
get (gca (), "default");
```

returns a list of user-defined default values for the current axes object.

Factory default values are stored in the root figure object. The command

```
get (0, "factory");
```

returns a list of factory defaults.

## 15.2.4 Colors

Colors may be specified as RGB triplets with values ranging from zero to one, or by name. Recognized color names include "blue", "black", "cyan", "green", "magenta", "red", "white", and "yellow".

### 15.2.5 Line Styles

Line styles are specified by the following properties:

`linestyle`
: May be one of

    `"-"`
    : Solid lines.

    `"--"`
    : Dashed lines.

    `":"`
    : Points.

    `"-."`
    : A dash-dot line.

`linewidth`
: A number specifying the width of the line. The default is 1. A value of 2 is twice as wide as the default, etc.

### 15.2.6 Marker Styles

Marker styles are specified by the following properties:

`marker`
: A character indicating a plot marker to be place at each data point, or "none", meaning no markers should be displayed.

`markeredgecolor`
: The color of the edge around the marker, or "auto", meaning that the edge color is the same as the face color. See Section 15.2.4 [Colors], page 231.

`markerfacecolor`
: The color of the marker, or "none" to indicate that the marker should not be filled. See Section 15.2.4 [Colors], page 231.

`markersize`
: A number specifying the size of the marker. The default is 1. A value of 2 is twice as large as the default, etc.

### 15.2.7 Interaction with gnuplot

*val* = gnuplot_binary ()  *Loadable Function*
*old_val* = gnuplot_binary (*new_val*)  *Loadable Function*

Query or set the name of the program invoked by the plot command. The default value "gnuplot". See Appendix E [Installation], page 511.

*val* = gnuplot_use_title_option ()                             Loadable Function
*old_val* = gnuplot_use_title_option (*new_val*)            Loadable Function

If enabled, append -title "Figure NN" to the gnuplot command. By default, this feature is enabled if the DISPLAY environment variable is set when Octave starts.

**This function is obsolete and will be removed from a future version of Octave.**

# 16 Matrix Manipulation

There are a number of functions available for checking to see if the elements of a matrix meet some condition, and for rearranging the elements of a matrix. For example, Octave can easily tell you if all the elements of a matrix are finite, or are less than some specified value. Octave can also rotate the elements, extract the upper- or lower-triangular parts, or sort the columns of a matrix.

## 16.1 Finding Elements and Checking Conditions

The functions any and all are useful for determining whether any or all of the elements of a matrix satisfy some condition. The find function is also useful in determining which elements of a matrix meet a specified condition.

any (*x*, *dim*)  Built-in Function

For a vector argument, return 1 if any element of the vector is nonzero.

For a matrix argument, return a row vector of ones and zeros with each element indicating whether any of the elements of the corresponding column of the matrix are nonzero. For example,

> any (eye (2, 4))
> ⇒ [ 1, 1, 0, 0 ]

If the optional argument *dim* is supplied, work along dimension *dim*. For example,

> any (eye (2, 4), 2)
> ⇒ [ 1; 1 ]

all (*x*, *dim*)  Built-in Function

The function all behaves like the function any, except that it returns true only if all the elements of a vector, or all the elements along dimension *dim* of a matrix, are nonzero.

Since the comparison operators (see Section 8.4 [Comparison Ops], page 100) return matrices of ones and zeros, it is easy to test a matrix for many things, not just whether the elements are nonzero. For example,

> all (all (rand (5) < 0.9))
> ⇒ 0

tests a random 5 by 5 matrix to see if all of its elements are less than 0.9.

Note that in conditional contexts (like the test clause of if and while statements) Octave treats the test as if you had typed all (all (condition)).

xor (*x*, *y*)  Mapping Function

Return the 'exclusive or' of the entries of *x* and *y*. For boolean expressions *x* and *y*, xor (*x*, *y*) is true if and only if *x* or *y* is true, but not if both *x* and *y* are true.

is_duplicate_entry (x)                                            Function File
    Return non-zero if any entries in x are duplicates of one another.

diff (x, k, dim)                                                 Function File
    If x is a vector of length n, diff (x) is the vector of first differences $x_2 - x_1, \ldots, x_n - x_{n-1}$.

    If x is a matrix, diff (x) is the matrix of column differences along the first non-singleton dimension.

    The second argument is optional. If supplied, diff (x, k), where k is a non-negative integer, returns the k-th differences. It is possible that k is larger than then first non-singleton dimension of the matrix. In this case, diff continues to take the differences along the next non-singleton dimension.

    The dimension along which to take the difference can be explicitly stated with the optional variable dim. In this case the k-th order differences are calculated along this dimension. In the case where k exceeds size (x, dim) then an empty matrix is returned.

isinf (x)                                                              Mapping Function
    Return 1 for elements of x that are infinite and zero otherwise. For example,
```
        isinf ([13, Inf, NA, NaN])
             ⇒ [ 0, 1, 0, 0 ]
```

isnan (x)                                                             Mapping Function
    Return 1 for elements of x that are NaN values and zero otherwise. NA values are also considered NaN values. For example,
```
        isnan ([13, Inf, NA, NaN])
             ⇒ [ 0, 0, 1, 1 ]
```

finite (x)                                                         Mapping Function
    Return 1 for elements of x that are finite values and zero otherwise. For example,
```
        finite ([13, Inf, NA, NaN])
             ⇒ [ 1, 0, 0, 0 ]
```

find (x)                                                                 Loadable Function
find (x, n)                                                          Loadable Function
find (x, n, direction)                                          Loadable Function
    Return a vector of indices of nonzero elements of a matrix, as a row if x is a row or as a column otherwise. To obtain a single index for each matrix element, Octave pretends that the columns of a matrix form one long vector (like Fortran arrays are stored). For example,
```
        find (eye (2))
             ⇒ [ 1; 4 ]
```
    If two outputs are requested, find returns the row and column indices of nonzero elements of a matrix. For example,

```
[i, j] = find (2 * eye (2))
    ⇒ i = [ 1; 2 ]
    ⇒ j = [ 1; 2 ]
```
If three outputs are requested, find also returns a vector containing the nonzero values. For example,
```
[i, j, v] = find (3 * eye (2))
    ⇒ i = [ 1; 2 ]
    ⇒ j = [ 1; 2 ]
    ⇒ v = [ 3; 3 ]
```
If two inputs are given, $n$ indicates the number of elements to find from the beginning of the matrix or vector.

If three inputs are given, *direction* should be one of "first" or "last" indicating that it should start counting found elements from the first or last element.

[*err*, *y1*, ...] = common_size (*x1*, ...)                                     Function File
Determine if all input arguments are either scalar or of common size. If so, *err* is zero, and *yi* is a matrix of the common size with all entries equal to *xi* if this is a scalar or *xi* otherwise. If the inputs cannot be brought to a common size, errorcode is 1, and *yi* is *xi*. For example,
```
[errorcode, a, b] = common_size ([1 2; 3 4], 5)
    ⇒ errorcode = 0
    ⇒ a = [ 1, 2; 3, 4 ]
    ⇒ b = [ 5, 5; 5, 5 ]
```
This is useful for implementing functions where arguments can either be scalars or of common size.

## 16.2 Rearranging Matrices

fliplr (*x*)                                                                     Function File
Return a copy of *x* with the order of the columns reversed. For example,
```
fliplr ([1, 2; 3, 4])
    ⇒   2   1
        4   3
```
Note that fliplr only work with 2-D arrays. To flip N-d arrays use flipdim instead.

**See also:** flipud, flipdim, rot90, rotdim.

flipud (*x*)                                                                     Function File
Return a copy of *x* with the order of the rows reversed. For example,

```
flipud ([1, 2; 3, 4])
⇒  3  4
   1  2
```

Due to the difficulty of defining which axis about which to flip the matrix flipud only work with 2-d arrays. To flip N-d arrays use flipdim instead.

See also: fliplr, flipdim, rot90, rotdim.

flipdim (*x*, *dim*)     Function File
Return a copy of *x* flipped about the dimension *dim*. For example

```
flipdim ([1, 2; 3, 4], 2)
⇒  2  1
   4  3
```

See also: fliplr, flipud, rot90, rotdim.

rot90 (*x*, *n*)     Function File
Return a copy of *x* with the elements rotated counterclockwise in 90-degree increments. The second argument is optional, and specifies how many 90-degree rotations are to be applied (the default value is 1). Negative values of *n* rotate the matrix in a clockwise direction. For example,

```
rot90 ([1, 2; 3, 4], -1)
⇒  3  1
   4  2
```

rotates the given matrix clockwise by 90 degrees. The following are all equivalent statements:

```
rot90 ([1, 2; 3, 4], -1)
rot90 ([1, 2; 3, 4], 3)
rot90 ([1, 2; 3, 4], 7)
```

Due to the difficulty of defining an axis about which to rotate the matrix rot90 only work with 2-D arrays. To rotate N-d arrays use rotdim instead.

See also: rotdim, flipud, fliplr, flipdim.

rotdim (*x*, *n*, *plane*)     Function File
Return a copy of *x* with the elements rotated counterclockwise in 90-degree increments. The second argument is optional, and specifies how many 90-degree rotations are to be applied (the default value is 1). The third argument is also optional and defines the plane of the rotation. As such *plane* is a two element vector containing two different valid dimensions of the matrix. If *plane* is not given Then the first two non-singleton dimensions are used.

Negative values of *n* rotate the matrix in a clockwise direction. For example,

```
rotdim ([1, 2; 3, 4], -1, [1, 2])
⇒  3  1
   4  2
```

rotates the given matrix clockwise by 90 degrees. The following are all equivalent statements:

Chapter 16: Matrix Manipulation                                         239

         rotdim ([1, 2; 3, 4], -1, [1, 2])
         rotdim ([1, 2; 3, 4], 3, [1, 2])
         rotdim ([1, 2; 3, 4], 7, [1, 2])
   See also: rot90, flipud, fliplr, flipdim.

cat (*dim*, *array1*, *array2*, ..., *arrayN*)                      Built-in Function
   Return the concatenation of N-d array objects, *array1*, *array2*, ..., *arrayN*
   along dimension *dim*.
         A = ones (2, 2);
         B = zeros (2, 2);
         cat (2, A, B)
      ⇒ ans =

            1 1 0 0
            1 1 0 0
   Alternatively, we can concatenate *A* and *B* along the second dimension the
   following way:
         [A, B].
   *dim* can be larger than the dimensions of the N-d array objects and the
   result will thus have *dim* dimensions as the following example shows:
         cat (4, ones(2, 2), zeros (2, 2))
      ⇒ ans =

            ans(:,:,1,1) =

            1 1
            1 1

            ans(:,:,1,2) =
            0 0
            0 0
   See also: horzcat, vertcat.

horzcat (*array1*, *array2*, ..., *arrayN*)                         Built-in Function
   Return the horizontal concatenation of N-d array objects, *array1*, *array2*,
   ..., *arrayN* along dimension 2.
   See also: cat, vertcat.

vertcat (*array1*, *array2*, ..., *arrayN*)                         Built-in Function
   Return the vertical concatenation of N-d array objects, *array1*, *array2*, ...,
   *arrayN* along dimension 1.
   See also: cat, horzcat.

**permute (a, *perm*)**  Built-in Function
Return the generalized transpose for an N-d array object a. The permutation vector *perm* must contain the elements 1:ndims(a) (in any order, but each element must appear just once).

See also: ipermute.

**ipermute (a, *iperm*)**  Built-in Function
The inverse of the permute function. The expression

    ipermute (permute (a, perm), perm)

returns the original array a.

See also: permute.

**reshape (a, *m*, *n*, ...)**  Built-in Function
**reshape (a, *size*)**  Built-in Function
Return a matrix with the given dimensions whose elements are taken from the matrix a. The elements of the matrix are accessed in column-major order (like Fortran arrays are stored).

For example,

    reshape ([1, 2, 3, 4], 2, 2)
    ⇒  1  3
       2  4

Note that the total number of elements in the original matrix must match the total number of elements in the new matrix.

A single dimension of the return matrix can be unknown and is flagged by an empty argument.

**y = circshift (*x*, *n*)**  Function File
Circularly shifts the values of the array x. n must be a vector of integers no longer than the number of dimensions in x. The values of n can be either positive or negative, which determines the direction in which the values or x are shifted. If an element of n is zero, then the corresponding dimension of x will not be shifted. For example

    x = [1, 2, 3; 4, 5, 6; 7, 8, 9];
    circshift (x, 1)
    ⇒  7, 8, 9
       1, 2, 3
       4, 5, 6
    circshift (x, -2)
    ⇒  7, 8, 9
       1, 2, 3
       4, 5, 6
    circshift (x, [0,1])
    ⇒  3, 1, 2
       6, 4, 5
       9, 7, 8

See also: permute, ipermute, shiftdim.

## Chapter 16: Matrix Manipulation

y = shiftdim (x, n)      Function File
[y, ns] = shiftdim (x)      Function File

Shifts the dimension of x by n, where n must be an integer scalar. When n is positive, the dimensions of x are shifted to the left, with the leading dimensions circulated to the end. If n is negative, then the dimensions of x are shifted to the right, with n leading singleton dimensions added.

Called with a single argument, shiftdim, removes the leading singleton dimensions, returning the number of dimensions removed in the second output argument ns.

For example

```
x = ones (1, 2, 3);
size (shiftdim (x, -1))
⇒ [1, 1, 2, 3]
size (shiftdim (x, 1))
⇒ [2, 3]
[b, ns] = shiftdim (x);
⇒ b = [1, 1, 1; 1, 1, 1]
⇒ ns = 1
```

See also: reshape, permute, ipermute, circshift, squeeze.

shift (x, b)      Function File
shift (x, b, dim)      Function File

If x is a vector, perform a circular shift of length b of the elements of x.

If x is a matrix, do the same for each column of x. If the optional *dim* argument is given, operate along this dimension

[s, i] = sort (x)      Loadable Function
[s, i] = sort (x, dim)      Loadable Function
[s, i] = sort (x, mode)      Loadable Function
[s, i] = sort (x, dim, mode)      Loadable Function

Return a copy of x with the elements arranged in increasing order. For matrices, sort orders the elements in each column.

For example,

```
sort ([1, 2; 2, 3; 3, 1])
⇒  1  1
   2  2
   3  3
```

The sort function may also be used to produce a matrix containing the original row indices of the elements in the sorted matrix. For example,

```
[s, i] = sort ([1, 2; 2, 3; 3, 1])
⇒ s = 1  1
      2  2
      3  3
⇒ i = 1  3
      2  1
      3  2
```

If the optional argument *dim* is given, then the matrix is sorted along the dimension defined by *dim*. The optional argument mode defines the order in which the values will be sorted. Valid values of mode are 'ascend' or 'descend'.

For equal elements, the indices are such that the equal elements are listed in the order that appeared in the original list.

The sort function may also be used to sort strings and cell arrays of strings, in which case the dictionary order of the strings is used.

The algorithm used in sort is optimized for the sorting of partially ordered lists.

sortrows (a, c)                                                              Function File

Sort the rows of the matrix a according to the order of the columns specified in c. If c is omitted, a lexicographical sort is used. By default ascending order is used however if elements of c are negative then the corresponding column is sorted in descending order.

Since the sort function does not allow sort keys to be specified, it can't be used to order the rows of a matrix according to the values of the elements in various columns[1] in a single call. Using the second output, however, it is possible to sort all rows based on the values in a given column. Here's an example that sorts the rows of a matrix based on the values in the second column.

```
a = [1, 2; 2, 3; 3, 1];
[s, i] = sort (a (:, 2));
a (i, :)
    ⇒  3  1
       1  2
       2  3
```

swap (*inputs*)                                                                           Function File
   [a1,b1] = swap(a,b)
   interchange a and b

swapcols (*inputs*)                                                              Function File
   function B = swapcols(A)
   permute columns of A into reverse order

swaprows (*inputs*)                                                             Function File
   function B = swaprows(A)
   permute rows of A into reverse order

---

[1] For example, to first sort based on the values in column 1, and then, for any values that are repeated in column 1, sort based on the values found in column 2, etc.

## Chapter 16: Matrix Manipulation

**tril (a, k)**           Function File
**triu (a, k)**           Function File

Return a new matrix formed by extracting the lower (tril) or upper (triu) triangular part of the matrix a, and setting all other elements to zero. The second argument is optional, and specifies how many diagonals above or below the main diagonal should also be set to zero.

The default value of $k$ is zero, so that triu and tril normally include the main diagonal as part of the result matrix.

If the value of $k$ is negative, additional elements above (for tril) or below (for triu) the main diagonal are also selected.

The absolute value of $k$ must not be greater than the number of sub- or super-diagonals.

For example,

```
tril (ones (3), -1)
⇒  0  0  0
   1  0  0
   1  1  0
```

and

```
tril (ones (3), 1)
⇒  1  1  0
   1  1  1
   1  1  1
```

**See also:** triu, diag.

**vec (x)**           Function File

Return the vector obtained by stacking the columns of the matrix x one above the other.

**vech (x)**           Function File

Return the vector obtained by eliminating all supradiagonal elements of the square matrix x and stacking the result one column above the other.

**prepad (x, l, c)**           Function File
**postpad (x, l, c)**           Function File
**postpad (x, l, c, dim)**           Function File

Prepends (appends) the scalar value c to the vector x until it is of length l. If the third argument is not supplied, a value of 0 is used.

If length (x) > l, elements from the beginning (end) of x are removed until a vector of length l is obtained.

If x is a matrix, elements are prepended or removed from each row.

If the optional dim argument is given, then operate along this dimension.

**blkdiag (a, b, c, ...)**           Function File

Build a block diagonal matrix from a, b, c, .... All the arguments must be numeric and are two-dimensional matrices or scalars.

**See also:** diag, horzcat, vertcat.

## 16.3 Applying a Function to an Array

a = arrayfun (*name, c*)                                       Function File
a = arrayfun (*func, c*)                                       Function File
a = arrayfun (*func, c, d*)                                  Function File
a = arrayfun (*func, c, options*)                          Function File
[a, b, ...] = arrayfun (*func, c, ...*)                  Function File

    Execute a function on each element of an array. This is useful for functions that do not accept array arguments. If the function does accept array arguments it is better to call the function directly.

    See cellfun for complete usage instructions.

    **See also:** cellfun.

bsxfun (*f, a, b*)                                                 Loadable Function

    Applies a binary function f element-wise to two matrix arguments a and b. The function f must be capable of accepting two column vector arguments of equal length, or one column vector argument and a scalar.

    The dimensions of a and b must be equal or singleton. The singleton dimensions of the matrices will be expanded to the same dimensionality as the other matrix.

    **See also:** arrayfun, cellfun.

## 16.4 Special Utility Matrices

eye (*x*)                                                         Built-in Function
eye (*n, m*)                                                 Built-in Function
eye (..., *class*)                                           Built-in Function

    Return an identity matrix. If invoked with a single scalar argument, eye returns a square matrix with the dimension specified. If you supply two scalar arguments, eye takes them to be the number of rows and columns. If given a vector with two elements, eye uses the values of the elements as the number of rows and columns, respectively. For example,

        eye (3)
            ⇒   1  0  0
                 0  1  0
                 0  0  1

The following expressions all produce the same result:
        eye (2)
        ≡
        eye (2, 2)
        ≡
        eye (size ([1, 2; 3, 4]))

The optional argument *class*, allows eye to return an array of the specified type, like

Chapter 16: Matrix Manipulation                                                245

>     val = zeros (n,m, "uint8")

Calling eye with no arguments is equivalent to calling it with an argument of 1. This odd definition is for compatibility with MATLAB.

ones (x)                                                      Built-in Function
ones (n, m)                                                   Built-in Function
ones (n, m, k, ...)                                           Built-in Function
ones (..., class)                                             Built-in Function

Return a matrix or N-dimensional array whose elements are all 1. The arguments are handled the same as the arguments for eye.

If you need to create a matrix whose values are all the same, you should use an expression like

>     val_matrix = val * ones (n, m)

The optional argument *class*, allows ones to return an array of the specified type, for example

>     val = ones (n,m, "uint8")

zeros (x)                                                     Built-in Function
zeros (n, m)                                                  Built-in Function
zeros (n, m, k, ...)                                          Built-in Function
zeros (..., class)                                            Built-in Function

Return a matrix or N-dimensional array whose elements are all 0. The arguments are handled the same as the arguments for eye.

The optional argument *class*, allows zeros to return an array of the specified type, for example

>     val = zeros (n,m, "uint8")

repmat (A, m, n)                                              Function File
repmat (A, [m n])                                             Function File
repmat (A, [m n p ...])                                       Function File

Form a block matrix of size m by n, with a copy of matrix A as each element. If n is not specified, form an m by m block matrix.

diag (v, k)                                                   Built-in Function

Return a diagonal matrix with vector v on diagonal k. The second argument is optional. If it is positive, the vector is placed on the k-th super-diagonal. If it is negative, it is placed on the -k-th sub-diagonal. The default value of k is 0, and the vector is placed on the main diagonal. For example,

>     diag ([1, 2, 3], 1)
    ⇒  0  1  0  0
       0  0  2  0
       0  0  0  3
       0  0  0  0

Given a matrix argument, instead of a vector, diag extracts the k-th diagonal of the matrix.

The functions linspace and logspace make it very easy to create vectors with evenly or logarithmically spaced elements. See Section 4.2 [Ranges], page 38.

linspace (*base*, *limit*, *n*)                                                Built-in Function
Return a row vector with *n* linearly spaced elements between *base* and *limit*. If the number of elements is greater than one, then the *base* and *limit* are always included in the range. If *base* is greater than *limit*, the elements are stored in decreasing order. If the number of points is not specified, a value of 100 is used.

The linspace function always returns a row vector.

For compatibility with MATLAB, return the second argument if fewer than two values are requested.

logspace (*base*, *limit*, *n*)                                                  Function File
Similar to linspace except that the values are logarithmically spaced from $10^{base}$ to $10^{limit}$.

If *limit* is equal to $\pi$, the points are between $10^{base}$ and $\pi$, *not* $10^{base}$ and $10^{\pi}$, in order to be compatible with the corresponding MATLAB function.

Also for compatibility, return the second argument if fewer than two values are requested.

See also: linspace.

## 16.5 Random Matrices

This section describes the basic generators for random matrices in Octave. Additional random distributions can be found in Section 24.7 [Random Number Generation], page 376.

rand (*x*)                                                                                   Loadable Function
rand (*n*, *m*)                                                                     Loadable Function
rand ("state", *x*)                                                        Loadable Function
rand ("seed", *x*)                                                          Loadable Function
Return a matrix with random elements uniformly distributed on the interval (0, 1). The arguments are handled the same as the arguments for eye.

You can query the state of the random number generator using the form

    v = rand ("state")

This returns a column vector *v* of length 625. Later, you can restore the random number generator to the state *v* using the form

    rand ("state", v)

You may also initialize the state vector from an arbitrary vector *v* of length 625 or less. This new state will be a hash based on the value of *v*, not *v* itself.

By default, the generator is initialized from /dev/urandom if it is available, otherwise from cpu time, wall clock time and the current fraction of a second.

## Chapter 16: Matrix Manipulation

To compute the pseudo-random sequence, **rand** uses the Mersenne Twister with a period of $2^{19937} - 1$.[2] Do **not** use for cryptography without securely hashing several returned values together, otherwise the generator state can be learned after reading 624 consecutive values.

Older versions of Octave used a different random number generator. The new generator is now used by default as it is faster and produces random numbers with a significantly longer cycle time. However, in some circumstances it might be desirable to obtain the random sequences produced by the old generator. The keyword "seed" specifies that the old generators should be used, as in

    rand ("seed", val)

which sets the seed of the generator to *val*. The seed of the generator can be queried with

    s = rand ("seed")

However, it should be noted that querying the seed will not cause **rand** to use the old generators, only setting the seed will. To cause **rand** to once again use the new generators, the keyword "state" should be used to reset the state of the rand.

See also: randn, rande, randg, randp.

| | |
|---|---|
| randn (*x*) | Loadable Function |
| randn (*n*, *m*) | Loadable Function |
| randn ("state", *x*) | Loadable Function |
| randn ("seed", *x*) | Loadable Function |

Return a matrix with normally distributed random elements. The arguments are handled the same as the arguments for rand.

By default, randn uses the Marsaglia and Tsang "Ziggurat technique" to transform from a uniform to a normal distribution.[3]

See also: rand, rande, randg, randp.

| | |
|---|---|
| rande (*x*) | Loadable Function |
| rande (*n*, *m*) | Loadable Function |
| rande ("state", *x*) | Loadable Function |
| rande ("seed", *x*) | Loadable Function |

Return a matrix with exponentially distributed random elements. The arguments are handled the same as the arguments for rand.

By default, rande uses the Marsaglia and Tsang Ziggurat technique to transform a uniform distribution to an exponential distribution.

See also: rand, randn, randg, randp.

---

[2] See M. Matsumoto and T. Nishimura, *Mersenne Twister: A 623-dimensionally equidistributed uniform pseudorandom number generator*, ACM Trans. on Modeling and Computer Simulation Vol. 8, No. 1, January pp.3-30 1998, http://www.math.sci.hiroshima-u.ac.jp/~m-mat/MT/emt.html

[3] G. Marsaglia and W.W. Tsang, *Ziggurat method for generating random variables*, J. Statistical Software, vol 5, 2000, http://www.jstatsoft.org/v05/i08/

randp (*l*, *x*)     Loadable Function
randp (*l*, *n*, *m*)     Loadable Function
randp ("state", *x*)     Loadable Function
randp ("seed", *x*)     Loadable Function

    Return a matrix with Poisson distributed random elements with mean value parameter given by the first argument, *l*. The arguments are handled the same as the arguments for rand, except for the argument *l*.

    **See also:** rand, randn, rande, randg.

randg (*a*, *x*)     Loadable Function
randg (*a*, *n*, *m*)     Loadable Function
randg ("state", *x*)     Loadable Function
randg ("seed", *x*)     Loadable Function

    Return a matrix with gamma(a,1) distributed random elements. The arguments are handled the same as the arguments for rand, except for the argument a.

    **See also:** rand, randn, rande, randp.

    The generators operate in the new or old style together, it is not possible to mix the two. Initializing any generator with "state" or "seed" causes the others to switch to the same style for future calls.

    The state of each generator is independent and calls to different generators can be interleaved without affecting the final result. For example,

```
rand ("state", [11, 22, 33]);
randn ("state", [44, 55, 66]);
u = rand (100, 1);
n = randn (100, 1);
```

and

```
rand ("state", [11, 22, 33]);
randn ("state", [44, 55, 66]);
u = zeros (100, 1);
n = zeros (100, 1);
for i = 1:100
  u(i) = rand ();
  n(i) = randn ();
end
```

produce equivalent results. When the generators are initialized in the old style with "seed" only rand and randn are independent, because the old rande, randg and randp generators make calls to rand and randn.

    The generators are initialized with random states at start-up, so that the sequences of random numbers are not the same each time you run Octave.[4] If you really do need to reproduce a sequence of numbers exactly, you can set the state or seed to a specific value.

---

[4] The old versions of rand and randn obtain their initial seeds from the system clock.

Chapter 16: Matrix Manipulation                                                        249

If invoked without arguments, rand and randn return a single element of a random sequence.

The original rand and randn functions use Fortran code from RANLIB, a library of fortran routines for random number generation, compiled by Barry W. Brown and James Lovato of the Department of Biomathematics at The University of Texas, M.D. Anderson Cancer Center, Houston, TX 77030.

randperm (n)                                                              Function File
    Return a row vector containing a random permutation of the integers from 1 to n.

## 16.6 Famous Matrices

The following functions return famous matrix forms.

hadamard (n)                                                              Function File
    Construct a Hadamard matrix Hn of size n-by-n. The size n must be of the form 2 ^ k * p in which p is one of 1, 12, 20 or 28. The returned matrix is normalized, meaning Hn(:,1) == 1 and H(1,:) == 1.

Some of the properties of Hadamard matrices are:

- kron (Hm, Hn) is a Hadamard matrix of size m-by-n.
- Hn * Hn' == n * eye (n).
- The rows of Hn are orthogonal.
- det (A) <= det (Hn) for all A with abs (A (i, j)) <= 1.
- Multiply any row or column by -1 and still have a Hadamard matrix.

hankel (c, r)                                                             Function File
    Return the Hankel matrix constructed given the first column c, and (optionally) the last row r. If the last element of c is not the same as the first element of r, the last element of c is used. If the second argument is omitted, it is assumed to be a vector of zeros with the same size as c.

A Hankel matrix formed from an m-vector c, and an n-vector r, has the elements

$$H(i,j) = \begin{cases} c_{i+j-1}, & i+j-1 \leq m; \\ r_{i+j-m}, & \text{otherwise.} \end{cases}$$

See also: vander, sylvester_matrix, hilb, invhilb, toeplitz.

hilb (n)                                                                  Function File
    Return the Hilbert matrix of order n. The i, j element of a Hilbert matrix is defined as

$$H(i,j) = \frac{1}{(i+j-1)}$$

See also: hankel, vander, sylvester_matrix, invhilb, toeplitz.

**invhilb (n)**  Function File
Return the inverse of a Hilbert matrix of order n. This can be computed exactly using

$$A_{ij} = -1^{i+j}(i+j-1)\binom{n+i-1}{n-j}\binom{n+j-1}{n-i}\binom{i+j-2}{i-2}^2$$
$$= \frac{p(i)p(j)}{(i+j-1)}$$

where

$$p(k) = -1^k \binom{k+n-1}{k-1}\binom{n}{k}$$

The validity of this formula can easily be checked by expanding the binomial coefficients in both formulas as factorials. It can be derived more directly via the theory of Cauchy matrices: see J. W. Demmel, Applied Numerical Linear Algebra, page 92.

Compare this with the numerical calculation of inverse (hilb (n)), which suffers from the ill-conditioning of the Hilbert matrix, and the finite precision of your computer's floating point arithmetic.

See also: hankel, vander, sylvester_matrix, hilb, toeplitz.

**magic (n)**  Function File
Create an n-by-n magic square. Note that magic (2) is undefined since there is no 2-by-2 magic square.

**pascal (n, t)**  Function File
Return the Pascal matrix of order n if t = 0. t defaults to 0. Return lower triangular Cholesky factor of the Pascal matrix if t = 1. This matrix is its own inverse, that is pascal (n, 1) ^ 2 == eye (n). If t = 2, return a transposed and permuted version of pascal (n, 1), which is the cube-root of the identity matrix. That is pascal (n, 2) ^ 3 == eye (n).

See also: hankel, vander, sylvester_matrix, hilb, invhilb, toeplitz hadamard, wilkinson, compan, rosser.

**rosser ()**  Function File
Returns the Rosser matrix. This is a difficult test case used to test eigenvalue algorithms.

See also: hankel, vander, sylvester_matrix, hilb, invhilb, toeplitz hadamard, wilkinson, compan, pascal.

**sylvester_matrix (k)**  Function File
Return the Sylvester matrix of order $n = 2^k$.

See also: hankel, vander, hilb, invhilb, toeplitz.

# Chapter 16: Matrix Manipulation

**toeplitz ($c$, $r$)** — Function File
Return the Toeplitz matrix constructed given the first column $c$, and (optionally) the first row $r$. If the first element of $c$ is not the same as the first element of $r$, the first element of $c$ is used. If the second argument is omitted, the first row is taken to be the same as the first column.

A square Toeplitz matrix has the form:

$$\begin{bmatrix} c_0 & r_1 & r_2 & \cdots & r_n \\ c_1 & c_0 & r_1 & \cdots & r_{n-1} \\ c_2 & c_1 & c_0 & \cdots & r_{n-2} \\ \vdots & \vdots & \vdots & \ddots & \vdots \\ c_n & c_{n-1} & c_{n-2} & \cdots & c_0 \end{bmatrix}$$

**See also:** hankel, vander, sylvester_matrix, hilb, invhilb.

**vander ($c$)** — Function File
Return the Vandermonde matrix whose next to last column is $c$.

A Vandermonde matrix has the form:

$$\begin{bmatrix} c_1^{n-1} & \cdots & c_1^2 & c_1 & 1 \\ c_2^{n-1} & \cdots & c_2^2 & c_2 & 1 \\ \vdots & \ddots & \vdots & \vdots & \vdots \\ c_n^{n-1} & \cdots & c_n^2 & c_n & 1 \end{bmatrix}$$

**See also:** hankel, sylvester_matrix, hilb, invhilb, toeplitz.

**wilkinson ($n$)** — Function File
Return the Wilkinson matrix of order $n$.

**See also:** hankel, vander, sylvester_matrix, hilb, invhilb, toeplitz, hadamard, rosser, compan, pascal.

# 17 Arithmetic

Unless otherwise noted, all of the functions described in this chapter will work for real and complex scalar or matrix arguments. Functions described as *mapping functions* apply the given operation to each element when given a matrix argument.

## 17.1 Utility Functions

ceil (*x*)                                                                  Mapping Function
    Return the smallest integer not less than *x*. If *x* is complex, return ceil (real (*x*)) + ceil (imag (*x*)) * I.

cplxpair (*z, tol, dim*)                                                        Function File
    Sort the numbers *z* into complex conjugate pairs ordered by increasing real part. With identical real parts, order by increasing imaginary magnitude. Place the negative imaginary complex number first within each pair. Place all the real numbers after all the complex pairs (those with abs (imag (*z*)) / *z*) < *tol*), where the default value of *tol* is 100 * *eps*.

    By default the complex pairs are sorted along the first non-singleton dimension of *z*. If *dim* is specified, then the complex pairs are sorted along this dimension.

    Signal an error if some complex numbers could not be paired. Requires all complex numbers to be exact conjugates within tol, or signals an error. Note that there are no guarantees on the order of the returned pairs with identical real parts but differing imaginary parts.

        cplxpair (exp(2i*pi*[0:4]'/5)) == exp(2i*pi*[3; 2; 4; 1; 0]/5)

*d* = del2 (*m*)                                                                Function File
*d* = del2 (*m, h*)                                                             Function File
*d* = del2 (*m, dx, dy, ...*)                                                   Function File
    Calculates the discrete Laplace operator. If *m* is a matrix this is defined as

$$d = \frac{1}{4}\left(\frac{d^2}{dx^2}M(x,y) + \frac{d^2}{dy^2}M(x,y)\right)$$

where the derivatives are calculated as finite difference approximations. The formula above is extended to N-dimensional arrays by calculating the second derivative over the higher dimensions.

The spacing between evaluation points may be defined by *h*, which is a scalar defining the spacing in all dimensions. Or alternatively, the spacing in each dimension may be defined separately by *dx*, *dy*, etc. Scalar spacing values give equidistant spacing, whereas vector spacing values can be used to specify variable spacing. The length of the vectors must match the respective dimension of *m*. The default spacing value is 1.

You need at least 3 data points for each dimension. Boundary points are calculated as the linear extrapolation of the interior points.

**See also:** gradient, diff.

exp (*x*)          Mapping Function
Compute the exponential of *x*. To compute the matrix exponential, see Chapter 18 [Linear Algebra], page 271.

*p* = factor (*q*)          Function File
[*p*, *n*] = factor (*q*)          Function File
Return prime factorization of *q*. That is prod (*p*) == *q*. If *q* == 1, returns 1.

With two output arguments, returns the unique primes *p* and their multiplicities. That is prod (*p* .^ *n*) == *q*.

factorial (*n*)          Function File
Return the factorial of *n*. If *n* is scalar, this is equivalent to prod (1:*n*). If *n* is an array, the factorial of the elements of the array are returned.

fix (*x*)          Mapping Function
Truncate *x* toward zero. If *x* is complex, return fix (real (*x*)) + fix (imag (*x*)) * I.

floor (*x*)          Mapping Function
Return the largest integer not greater than *x*. If *x* is complex, return floor (real (*x*)) + floor (imag (*x*)) * I.

fmod (*x*, *y*)          Mapping Function
Compute the floating point remainder of dividing *x* by *y* using the C library function fmod. The result has the same sign as *x*. If *y* is zero, the result is implementation-defined.

*g* = gcd (*a1*, ...)          Loadable Function
[*g*, *v1*, ...] = gcd (*a1*, ...)          Loadable Function
If a single argument is given then compute the greatest common divisor of the elements of this argument. Otherwise if more than one argument is given all arguments must be the same size or scalar. In this case the greatest common divisor is calculated for element individually. All elements must be integers. For example,

    gcd ([15, 20])
       ⇒ 5

and

## Chapter 17: Arithmetic

        gcd ([15, 9], [20 18])
        ⇒  5  9

Optional return arguments *v1*, etc, contain integer vectors such that,

$g = v_1 a_1 + v_2 a_2 + \cdots$

For backward compatibility with previous versions of this function, when all arguments are scalar, a single return argument *v1* containing all of the values of *v1*, ... is acceptable.

**See also:** lcm, min, max, ceil, floor.

| | |
|---|---:|
| *x* = gradient (*M*) | Function File |
| [*x*, *y*, ...] = gradient (*M*) | Function File |
| [...] = gradient (*M*, *s*) | Function File |
| [...] = gradient (*M*, *dx*, *dy*, ...) | Function File |

Calculates the gradient. *x* = gradient (*M*) calculates the one dimensional gradient if *M* is a vector. If *M* is a matrix the gradient is calculated for each row.

[*x*, *y*] = gradient (*M*) calculates the one dimensional gradient for each direction if *M* if *M* is a matrix. Additional return arguments can be use for multi-dimensional matrices.

Spacing values between two points can be provided by the *dx*, *dy* or *h* parameters. If *h* is supplied it is assumed to be the spacing in all directions. Otherwise, separate values of the spacing can be supplied by the *dx*, etc variables. A scalar value specifies an equidistant spacing, while a vector value can be used to specify a variable spacing. The length must match their respective dimension of *M*.

At boundary points a linear extrapolation is applied. Interior points are calculated with the first approximation of the numerical gradient

        y'(i) = 1/(x(i+1)-x(i-1)) *(y(i-1)-y(i+1)).

| | |
|---|---:|
| lcm (*x*, ...) | Mapping Function |

Compute the least common multiple of the elements of *x*, or the list of all the arguments. For example,

        lcm (a1, ..., ak)

is the same as

        lcm ([a1, ..., ak]).

All elements must be the same size or scalar.

**See also:** gcd, min, max, ceil, floor.

| | |
|---|---:|
| log (*x*) | Mapping Function |

Compute the natural logarithm for each element of *x*. To compute the matrix logarithm, see Chapter 18 [Linear Algebra], page 271.

**See also:** log2, log10, logspace, exp.

log10 (x)                                                                              Mapping Function

Compute the base-10 logarithm for each element of x.

See also: log, log2, logspace, exp.

log2 (x)                                                                                Mapping Function

[f, e] = log2 (x)                                                Mapping Function

Compute the base-2 logarithm of x. With two outputs, returns $f$ and $e$ such that $1/2 <= |f| < 1$ and $x = f \cdot 2^e$.

See also: log, log10, logspace, exp.

max (x, y, dim)                                              Mapping Function

[w, iw] = max (x)                                     Mapping Function

For a vector argument, return the maximum value. For a matrix argument, return the maximum value from each column, as a row vector, or over the dimension *dim* if defined. For two matrices (or a matrix and scalar), return the pair-wise maximum. Thus,

    max (max (x))

returns the largest element of x, and

    max (2:5, pi)
       ⇒   3.1416   3.1416   4.0000   5.0000

compares each element of the range 2:5 with pi, and returns a row vector of the maximum values.

For complex arguments, the magnitude of the elements are used for comparison.

If called with one input and two output arguments, max also returns the first index of the maximum value(s). Thus,

    [x, ix] = max ([1, 3, 5, 2, 5])
       ⇒   x = 5
           ix = 3

min (x, y, dim)                                              Mapping Function

[w, iw] = min (x)                                     Mapping Function

For a vector argument, return the minimum value. For a matrix argument, return the minimum value from each column, as a row vector, or over the dimension *dim* if defined. For two matrices (or a matrix and scalar), return the pair-wise minimum. Thus,

    min (min (x))

returns the smallest element of x, and

    min (2:5, pi)
       ⇒   2.0000   3.0000   3.1416   3.1416

compares each element of the range 2:5 with pi, and returns a row vector of the minimum values.

For complex arguments, the magnitude of the elements are used for comparison.

## Chapter 17: Arithmetic

If called with one input and two output arguments, min also returns the first index of the minimum value(s). Thus,

[x, ix] = min ([1, 3, 0, 2, 5])
⇒  x = 0
    ix = 3

mod (*x, y*)  *Mapping Function*
Compute modulo function. Conceptually this is given by

x - y .* floor (x ./ y)

and is written in a manner that the correct modulus is returned for integer types. This function handles negative values correctly. That is mod (-1, 3) is 2, not -1 as rem (-1, 3) returns. Also, mod (x, 0) returns x.

An error message is printed if the dimensions of the arguments do not agree, or if either of the arguments is complex.

See also: rem, round.

nextpow2 (*x*)  *Function File*
If x is a scalar, returns the first integer n such that $2^n \geq |x|$.
If x is a vector, return nextpow2 (length (x)).

See also: pow2.

nthroot (*x, n*)  *Function File*
Compute the n-th root of x, returning real results for real components of x. For example

nthroot (-1, 3)
⇒ -1
(-1) ^ (1 / 3)
⇒ 0.50000 - 0.86603i

pow2 (*x*)  *Mapping Function*
pow2 (*f, e*)  *Mapping Function*
With one argument, computes $2^x$ for each element of x. With two arguments, returns $f \cdot 2^e$.

See also: nextpow2.

primes (*n*)  *Function File*
Return all primes up to n.

Note that if you need a specific number of primes, you can use the fact the distance from one prime to the next is on average proportional to the logarithm of the prime. Integrating, you find that there are about $k$ primes less than $k \log(5k)$.

The algorithm used is called the Sieve of Erastothenes.

**rem** (*x, y*)     *Mapping Function*
Return the remainder of *x* / *y*, computed using the expression

    `x - y .* fix (x ./ y)`

An error message is printed if the dimensions of the arguments do not agree, or if either of the arguments is complex.

See also: mod, round.

**round** (*x*)     *Mapping Function*
Return the integer nearest to *x*. If *x* is complex, return round (real (*x*)) + round (imag (*x*)) * I.

See also: rem.

**sign** (*x*)     *Mapping Function*
Compute the *signum* function, which is defined as

$$\text{sign}(x) = \begin{cases} 1, & x > 0; \\ 0, & x = 0; \\ -1, & x < 0. \end{cases}$$

For complex arguments, sign returns x ./ abs (x).

**sqrt** (*x*)     *Mapping Function*
Compute the square root of *x*. If *x* is negative, a complex result is returned. To compute the matrix square root, see Chapter 18 [Linear Algebra], page 271.

## 17.2 Complex Arithmetic

The following functions are available for working with complex numbers. Each expects a single argument. Given a matrix they work on an element by element basis. In the descriptions of the following functions, $z$ is the complex number $x + iy$, where $i$ is defined as $\sqrt{-1}$.

**abs** (*z*)     *Mapping Function*
Compute the magnitude of *z*, defined as $|z| = \sqrt{x^2 + y^2}$.
For example,

    `abs (3 + 4i)`
    ⇒ 5

**arg** (*z*)     *Mapping Function*
**angle** (*z*)     *Mapping Function*
Compute the argument of *z*, defined as $\theta = \tan^{-1}(y/x)$. in radians.
For example,

    `arg (3 + 4i)`
    ⇒ 0.92730

conj (z)                                                    Mapping Function
    Return the complex conjugate of z, defined as $\bar{z} = x - iy$.

    See also: real, imag.

imag (z)                                                    Mapping Function
    Return the imaginary part of z as a real number.

    See also: real, conj.

real (z)                                                    Mapping Function
    Return the real part of z.

    See also: imag, conj.

## 17.3 Trigonometry

Octave provides the following trigonometric functions. Angles are specified in radians. To convert from degrees to radians multiply by $\pi/180$ (e.g. sin (30 * pi/180) returns the sine of 30 degrees).

sin (x)                                                     Mapping Function
    Compute the sine of each element of x.

cos (x)                                                     Mapping Function
    Compute the cosine of each element of x.

tan (z)                                                     Mapping Function
    Compute tangent of each element of x.

sec (x)                                                     Mapping Function
    Compute the secant of each element of x.

csc (x)                                                     Mapping Function
    Compute the cosecant of each element of x.

cot (x)                                                     Mapping Function
    Compute the cotangent of each element of x.

asin (x)                                                    Mapping Function
    Compute the inverse sine of each element of x.

acos (x)                                                    Mapping Function
    Compute the inverse cosine of each element of x.

atan (x)                                                    Mapping Function
    Compute the inverse tangent of each element of x.

`asec (x)` — Mapping Function
  Compute the inverse secant of each element of x.

`acsc (x)` — Mapping Function
  Compute the inverse cosecant of each element of x.

`acot (x)` — Mapping Function
  Compute the inverse cotangent of each element of x.

`sinh (x)` — Mapping Function
  Compute the hyperbolic sine of each element of x.

`cosh (x)` — Mapping Function
  Compute the hyperbolic cosine of each element of x.

`tanh (x)` — Mapping Function
  Compute hyperbolic tangent of each element of x.

`sech (x)` — Mapping Function
  Compute the hyperbolic secant of each element of x.

`csch (x)` — Mapping Function
  Compute the hyperbolic cosecant of each element of x.

`coth (x)` — Mapping Function
  Compute the hyperbolic cotangent of each element of x.

`asinh (x)` — Mapping Function
  Compute the inverse hyperbolic sine of each element of x.

`acosh (x)` — Mapping Function
  Compute the inverse hyperbolic cosine of each element of x.

`atanh (x)` — Mapping Function
  Compute the inverse hyperbolic tangent of each element of x.

`asech (x)` — Mapping Function
  Compute the inverse hyperbolic secant of each element of x.

`acsch (x)` — Mapping Function
  Compute the inverse hyperbolic cosecant of each element of x.

`acoth (x)` — Mapping Function
  Compute the inverse hyperbolic cotangent of each element of x.

Chapter 17: Arithmetic       261

Each of these functions expects a single argument. For matrix arguments, they work on an element by element basis. For example,

```
sin ([1, 2; 3, 4])
⇒  0.84147   0.90930
   0.14112  -0.75680
```

atan2 (*y*, *x*)  *Mapping Function*
Compute atan (*y* / *x*) for corresponding elements of *y* and *x*. The result is in the range -pi to pi.

In addition to the trigonometric functions that work with radians, Octave also provides the following functions which work with degrees.

sind (*x*)  *Function File*
Compute the sine of each element of *x*. Returns zero in elements for which *x*/180 is an integer.

**See also:** sin, cosd, tand, acosd, asind, atand.

cosd (*x*)  *Function File*
Compute the cosine of an angle in degrees. Returns zero in elements for which (*x*-90)/180 is an integer.

**See also:** cos, sind, tand, acosd, asind, atand.

tand (*x*)  *Function File*
Compute the tangent of an angle in degrees. Returns zero for elements of for which *x*/180 is an integer and Inf for elements where (*x*-90)/180 is an integer.

**See also:** tan, cosd, sind, acosd, asind, atand.

secd (*x*)  *Function File*
Compute the secant of an angle in degrees.

**See also:** sec, cscd, sind, cosd.

cscd (*x*)  *Function File*
Compute the cosecant of an angle in degrees.

**See also:** csc, secd, sind, cosd.

cotd (*x*)  *Function File*
Compute the cotangent of an angle in degrees.

**See also:** cot, tand.

asind (*x*)  *Function File*
Compute the inverse sine of an angle in degrees.

**See also:** asin, sind, acosd.

acosd (x)　　　　　　　　　　　　　　　　　　　　　　　　Function File
　　Compute the inverse cosine of an angle in degrees.
　　See also: acos, cosd, asecd.

atand (x)　　　　　　　　　　　　　　　　　　　　　　　　Function File
　　Compute the inverse tangent of an angle in degrees.
　　See also: acot, tand.

asecd (x)　　　　　　　　　　　　　　　　　　　　　　　　Function File
　　Compute inverse secant in degrees.
　　See also: asec, secd, acscd.

acscd (x)　　　　　　　　　　　　　　　　　　　　　　　　Function File
　　Compute the inverse cosecant of an angle in degrees.
　　See also: acsc, cscd, asecd.

acotd (x)　　　　　　　　　　　　　　　　　　　　　　　　Function File
　　Compute the inverse cotangent of an angle in degrees.
　　See also: atan, tand.

## 17.4 Sums and Products

sum (x, dim)　　　　　　　　　　　　　　　　　　　　　Built-in Function
sum (..., 'native')　　　　　　　　　　　　　　　　　Built-in Function
　　Sum of elements along dimension dim. If dim is omitted, it defaults to 1 (column-wise sum).

　　As a special case, if x is a vector and dim is omitted, return the sum of the elements.

　　If the optional argument 'native' is given, then the sum is performed in the same type as the original argument, rather than in the default double type. For example
```
         sum ([true, true])
            ⇒ 2
         sum ([true, true], 'native')
            ⇒ true
```

prod (x, dim)　　　　　　　　　　　　　　　　　　　　　Built-in Function
　　Product of elements along dimension dim. If dim is omitted, it defaults to 1 (column-wise products).

　　As a special case, if x is a vector and dim is omitted, return the product of the elements.

Chapter 17: Arithmetic                                                                 263

cumsum (x, dim)                                                    Built-in Function
Cumulative sum of elements along dimension *dim*. If *dim* is omitted, it defaults to 1 (column-wise cumulative sums).

As a special case, if *x* is a vector and *dim* is omitted, return the cumulative sum of the elements as a vector with the same orientation as *x*.

cumprod (x, dim)                                                   Built-in Function
Cumulative product of elements along dimension *dim*. If *dim* is omitted, it defaults to 1 (column-wise cumulative products).

As a special case, if *x* is a vector and *dim* is omitted, return the cumulative product of the elements as a vector with the same orientation as *x*.

sumsq (x, dim)                                                     Built-in Function
Sum of squares of elements along dimension *dim*. If *dim* is omitted, it defaults to 1 (column-wise sum of squares).

As a special case, if *x* is a vector and *dim* is omitted, return the sum of squares of the elements.

This function is conceptually equivalent to computing
        sum (x .* conj (x), dim)
but it uses less memory and avoids calling conj if *x* is real.

accumarray (*subs, vals, sz, fun, fillval, issparse*)                  Function File
accumarray (*csubs, vals, . . .* )                                     Function File
Create an array by accumulating the elements of a vector into the positions defined by their subscripts. The subscripts are defined by the rows of the matrix *subs* and the values by *vals*. Each row of *subs* corresponds to one of the values in *vals*.

The size of the matrix will be determined by the subscripts themselves. However, if *sz* is defined it determines the matrix size. The length of *sz* must correspond to the number of columns in *subs*.

The default action of accumarray is to sum the elements with the same subscripts. This behavior can be modified by defining the *fun* function. This should be a function or function handle that accepts a column vector and returns a scalar. The result of the function should not depend on the order of the subscripts.

The elements of the returned array that have no subscripts associated with them are set to zero. Defining *fillval* to some other value allows these values to be defined.

By default accumarray returns a full matrix. If *issparse* is logically true, then a sparse matrix is returned instead.

An example of the use of accumarray is:
        accumarray ([1,1,1;2,1,2;2,3,2;2,1,2;2,3,2], 101:105)
        ⇒ ans(:,:,1) = [101, 0, 0; 0, 0, 0]
           ans(:,:,2) = [0, 0, 0; 206, 0, 208]

## 17.5 Special Functions

| | |
|---|---|
| [j, ierr] = besselj (alpha, x, opt) | Loadable Function |
| [y, ierr] = bessely (alpha, x, opt) | Loadable Function |
| [i, ierr] = besseli (alpha, x, opt) | Loadable Function |
| [k, ierr] = besselk (alpha, x, opt) | Loadable Function |
| [h, ierr] = besselh (alpha, k, x, opt) | Loadable Function |

Compute Bessel or Hankel functions of various kinds:

besselj
> Bessel functions of the first kind. If the argument *opt* is supplied, the result is multiplied by exp(-abs(imag(x))).

bessely
> Bessel functions of the second kind. If the argument *opt* is supplied, the result is multiplied by exp(-abs(imag(x))).

besseli
> Modified Bessel functions of the first kind. If the argument *opt* is supplied, the result is multiplied by exp(-abs(real(x))).

besselk
> Modified Bessel functions of the second kind. If the argument *opt* is supplied, the result is multiplied by exp(x).

besselh
> Compute Hankel functions of the first ($k = 1$) or second ($k = 2$) kind. If the argument *opt* is supplied, the result is multiplied by exp (-I*x) for $k = 1$ or exp (I*x) for $k = 2$.

If *alpha* is a scalar, the result is the same size as *x*. If *x* is a scalar, the result is the same size as *alpha*. If *alpha* is a row vector and *x* is a column vector, the result is a matrix with length (x) rows and length (alpha) columns. Otherwise, *alpha* and *x* must conform and the result will be the same size.

The value of *alpha* must be real. The value of *x* may be complex.

If requested, *ierr* contains the following status information and is the same size as the result.

0. Normal return.
1. Input error, return NaN.
2. Overflow, return Inf.
3. Loss of significance by argument reduction results in less than half of machine accuracy.
4. Complete loss of significance by argument reduction, return NaN.
5. Error—no computation, algorithm termination condition not met, return NaN.

[a, ierr] = airy (k, z, opt)     Loadable Function
Compute Airy functions of the first and second kind, and their derivatives. If the optional argument *opt* is non-zero the value is multiplied by the scale factor given in the table below.

| k | Function | Scale factor (if 'opt' is supplied) |
|---|---|---|
| 0 | $A_i(Z)$ | $\exp(\frac{2}{3}Z\sqrt{Z})$ |
| 1 | $\frac{dA_i(Z)}{dZ}$ | $\exp(\frac{2}{3}Z\sqrt{Z})$ |
| 2 | $B_i(Z)$ | $\exp(-|\text{Re}(\frac{2}{3}Z\sqrt{Z})|)$ |
| 3 | $\frac{dB_i(Z)}{dZ}$ | $\exp(-|\text{Re}(\frac{2}{3}Z\sqrt{Z})|)$ |

The function call airy (z) is equivalent to airy (0, z).

The result is the same size as z.

If requested, *ierr* contains the following status information and is the same size as the result.

0. Normal return.
1. Input error, return NaN.
2. Overflow, return Inf.
3. Loss of significance by argument reduction results in less than half of machine accuracy.
4. Complete loss of significance by argument reduction, return NaN.
5. Error—no computation, algorithm termination condition not met, return NaN.

beta (a, b)     Mapping Function
Return the Beta function,

$$B(a,b) = \frac{\Gamma(a)\Gamma(b)}{\Gamma(a+b)}.$$

betainc (x, a, b)     Mapping Function
Return the incomplete Beta function,

$$\beta(x,a,b) = B(a,b)^{-1} \int_0^x t^{(a-z)}(1-t)^{(b-1)} dt.$$

If x has more than one component, both a and b must be scalars. If x is a scalar, a and b must be of compatible dimensions.

**betaln (a, b)**     *Mapping Function*
    Return the log of the Beta function,

$$B(a,b) = \log \frac{\Gamma(a)\Gamma(b)}{\Gamma(a+b)}.$$

    **See also:** beta, betai, gammaln.

**bincoeff (n, k)**     *Mapping Function*
    Return the binomial coefficient of $n$ and $k$, defined as

$$\binom{n}{k} = \frac{n(n-1)(n-2)\cdots(n-k+1)}{k!}$$

    For example,
```
     bincoeff (5, 2)
     ⇒ 10
```

**erf (z)**     *Mapping Function*
    Computes the error function,

$$\operatorname{erf}(z) = \frac{2}{\sqrt{\pi}} \int_0^z e^{-t^2} dt$$

    **See also:** erfc, erfinv.

**erfc (z)**     *Mapping Function*
    Computes the complementary error function, $1 - \operatorname{erf}(z)$.
    **See also:** erf, erfinv.

**erfinv (z)**     *Mapping Function*
    Computes the inverse of the error function.
    **See also:** erf, erfc.

**gamma (z)**     *Mapping Function*
    Computes the Gamma function,

$$\Gamma(z) = \int_0^\infty t^{z-1} e^{-t} dt.$$

    **See also:** gammai, lgamma.

Chapter 17: Arithmetic                                                     267

**gammainc (x, a)**                                            Mapping Function
Compute the normalized incomplete gamma function,

$$\gamma(x, a) = \frac{\int_0^x e^{-t} t^{a-1} dt}{\Gamma(a)}$$

with the limiting value of 1 as x approaches infinity. The standard notation is $P(a, x)$, e.g. Abramowitz and Stegun (6.5.1).

If a is scalar, then gammainc (x, a) is returned for each element of x and vice versa.

If neither x nor a is scalar, the sizes of x and a must agree, and *gammainc* is applied element-by-element.

See also: gamma, lgamma.

**L = legendre (n, X)**                                          Function File
Computes the Legendre Function of degree n and order m where all values for $m = 0 \ldots n$ are returned. n must be a scalar in the range $[0 \ldots 255]$. The return value has one dimension more than x.

The Legendre Function of degree $n$ and order $m$ is defined as

$$P_n^m(x) = (-1)^m (1-x^2)^{m/2} \frac{d^m}{dx^m} P_n(x)$$

where $P_n(x)$ is the Legendre polynomial of degree $n$

$$P_n(x) = \frac{1}{2^n n!} \left[ \frac{d^n}{dx^n} (x^2 - 1)^n \right]$$

For example, legendre(3,[-1.0 -0.9 -0.8]) returns a matrix of results corresponding to the following values,

```
    x  |    -1.0   |    -0.9   |    -0.8
-----------------------------------------
   m=0 |  -1.00000 |  -0.47250 |  -0.08000
   m=1 |   0.00000 |  -1.99420 |  -1.98000
   m=2 |   0.00000 |  -2.56500 |  -4.32000
   m=3 |   0.00000 |  -1.24229 |  -3.24000
```

**lgamma (x)**                                                Mapping Function
**gammaln (x)**                                               Mapping Function
Return the natural logarithm of the absolute value of the gamma function of x.

See also: gamma, gammai.

cross (x, y, dim)                                            Function File
Computes the vector cross product of the two 3-dimensional vectors x and
y.

>     cross ([1,1,0], [0,1,1])
>     ⇒ [ 1; -1; 1 ]

If x and y are matrices, the cross product is applied along the first dimension
with 3 elements. The optional argument dim is used to force the cross
product to be calculated along the dimension defined by dim.

commutation_matrix (m, n)                                    Function File
Return the commutation matrix $K_{m,n}$ which is the unique $mn \times mn$ matrix
such that $K_{m,n} \cdot \text{vec}(A) = \text{vec}(A^T)$ for all $m \times n$ matrices $A$.

If only one argument m is given, $K_{m,m}$ is returned.

See Magnus and Neudecker (1988), Matrix differential calculus with applications in statistics and econometrics.

duplication_matrix (n)                                       Function File
Return the duplication matrix $D_n$ which is the unique $n^2 \times n(n+1)/2$ matrix
such that $D_n * \text{vech}(A) = \text{vec}(A)$ for all symmetric $n \times n$ matrices $A$.

See Magnus and Neudecker (1988), Matrix differential calculus with applications in statistics and econometrics.

## 17.6 Coordinate Transformations

[theta, r] = cart2pol (x, y)                                 Function File
[theta, r, z] = cart2pol (x, y, z)                           Function File
Transform cartesian to polar or cylindrical coordinates. x, y (and z) must
be of same shape. theta describes the angle relative to the x-axis. r is the
distance to the z-axis (0, 0, z).

See also: pol2cart, cart2sph, sph2cart.

[x, y] = pol2cart (theta, r)                                 Function File
[x, y, z] = pol2cart (theta, r, z)                           Function File
Transform polar or cylindrical to cartesian coordinates. theta, r (and z)
must be of same shape. theta describes the angle relative to the x-axis. r is
the distance to the z-axis (0, 0, z).

See also: cart2pol, cart2sph, sph2cart.

[theta, phi, r] = cart2sph (x, y, z)                         Function File
Transform cartesian to spherical coordinates. x, y and z must be of same
shape. theta describes the angle relative to the x-axis. phi is the angle
relative to the xy-plane. r is the distance to the origin (0, 0, 0).

See also: pol2cart, cart2pol, sph2cart.

Chapter 17: Arithmetic            269

[x, y, z] = sph2cart (*theta*, *phi*, *r*)            Function File
  Transform spherical to cartesian coordinates. x, y and z must be of same shape. *theta* describes the angle relative to the x-axis. *phi* is the angle relative to the xy-plane. *r* is the distance to the origin (0, 0, 0).

  **See also:** pol2cart, cart2pol, cart2sph.

## 17.7 Mathematical Constants

I (*x*)            Built-in Function
I (*n*, *m*)            Built-in Function
I (*n*, *m*, *k*, ...)            Built-in Function
I (..., *class*)            Built-in Function
  Return a matrix or N-dimensional array whose elements are all equal to the pure imaginary unit, defined as $\sqrt{-1}$. Since I (also i, J, and j) is a function, you can use the name(s) for other purposes.

Inf (*x*)            Built-in Function
Inf (*n*, *m*)            Built-in Function
Inf (*n*, *m*, *k*, ...)            Built-in Function
Inf (..., *class*)            Built-in Function
  Return a matrix or N-dimensional array whose elements are all Infinity. The arguments are handled the same as the arguments for eye. The optional argument *class* may be either "single" or "double". The default is "double".

NaN (*x*)            Built-in Function
NaN (*n*, *m*)            Built-in Function
NaN (*n*, *m*, *k*, ...)            Built-in Function
NaN (..., *class*)            Built-in Function
  Return a matrix or N-dimensional array whose elements are all NaN (Not a Number). The value NaN is the result of an operation like 0/0, or $\infty - \infty$, or any operation with a NaN.

  Note that NaN always compares not equal to NaN. This behavior is specified by the IEEE standard for floating point arithmetic. To find NaN values, you must use the isnan function.

  The arguments are handled the same as the arguments for eye. The optional argument *class* may be either "single" or "double". The default is "double".

pi (*x*)            Built-in Function
pi (*n*, *m*)            Built-in Function
pi (*n*, *m*, *k*, ...)            Built-in Function
pi (..., *class*)            Built-in Function
  Return a matrix or N-dimensional array whose elements are all equal to the ratio of the circumference of a circle to its diameter. Internally, pi is computed as 4.0 * atan (1.0).

e (x)  Built-in Function
e (n, m)  Built-in Function
e (n, m, k, ...)  Built-in Function
e (..., class)  Built-in Function
    Return a matrix or N-dimensional array whose elements are all equal to the base of natural logarithms. The constant e satisfies the equation $\log(e) = 1$.

eps (x)  Built-in Function
eps (n, m)  Built-in Function
eps (n, m, k, ...)  Built-in Function
eps (..., class)  Built-in Function
    Return a matrix or N-dimensional array whose elements are all eps, the machine precision. More precisely, eps is the largest relative spacing between any two adjacent numbers in the machine's floating point system. This number is obviously system-dependent. On machines that support 64-bit IEEE floating point arithmetic, eps is approximately $2.2204 \times 10^{-16}$.

realmax (x)  Built-in Function
realmax (n, m)  Built-in Function
realmax (n, m, k, ...)  Built-in Function
realmax (..., class)  Built-in Function
    Return a matrix or N-dimensional array whose elements are all equal to the largest floating point number that is representable. The actual value is system-dependent. On machines that support 64-bit IEEE floating point arithmetic, realmax is approximately $1.7977 \times 10^{308}$.

    **See also:** realmin.

realmin (x)  Built-in Function
realmin (n, m)  Built-in Function
realmin (n, m, k, ...)  Built-in Function
realmin (..., class)  Built-in Function
    Return a matrix or N-dimensional array whose elements are all equal to the smallest normalized floating point number that is representable. The actual value is system-dependent. On machines that support 64-bit IEEE floating point arithmetic, realmin is approximately $2.2251 \times 10^{-308}$.

    **See also:** realmax.

# 18 Linear Algebra

This chapter documents the linear algebra functions of Octave. Reference material for many of these functions may be found in Golub and Van Loan, *Matrix Computations, 2nd Ed.*, Johns Hopkins, 1989, and in the LAPACK *Users' Guide*, SIAM, 1992.

## 18.1 Techniques used for Linear Algebra

Octave uses a polymorphic solver which selects an appropriate matrix factorization depending on the properties of the matrix itself. Generally, the cost of determining the matrix type is small relative to the cost of factorizing the matrix, and the type is cached once it is calculated, so that it is not re-determined each time the matrix is used.

The selection tree for solving a linear system or computing a matrix inverse is as follows:

1. If the matrix is upper or lower triangular sparse, try a forward or backward substitution using the LAPACK xTRTRS function, and goto 4.
2. If the matrix is square, hermitian with a real positive diagonal, attempt Cholesky factorization using the LAPACK xPOTRF function.
3. If the Cholesky factorization failed or the matrix is not hermitian with a real positive diagonal, and the matrix is square, factorize using the LAPACK xGETRF function.
4. If the matrix is not square, or any of the previous solvers flags a singular or near singular matrix, find a least squares solution using the LAPACK xGELSD function.

The user can force the type of the matrix with the `matrix_type` function. This overcomes the cost of discovering the type of the matrix. However, it should be noted that identifying the type of the matrix incorrectly will lead to unpredictable results, and so `matrix_type` should be used with care.

## 18.2 Basic Matrix Functions

aa = balance (a, *opt*)  Loadable Function
[dd, aa] = balance (a, *opt*)  Loadable Function
[cc, dd, aa, bb] = balance (a, b, *opt*)  Loadable Function

  Compute aa = dd \ a * dd in which aa is a matrix whose row and column norms are roughly equal in magnitude, and dd = p * d, in which p is a permutation matrix and d is a diagonal matrix of powers of two. This allows the equilibration to be computed without roundoff. Results of eigenvalue calculation are typically improved by balancing first.

  If four output values are requested, compute aa = cc*a*dd and bb = cc*b*dd), in which aa and bb have non-zero elements of approximately the same magnitude and cc and dd are permuted diagonal matrices as in dd for the algebraic eigenvalue problem.

The eigenvalue balancing option opt may be one of:

"N", "n"
: No balancing; arguments copied, transformation(s) set to identity.

"P", "p"
: Permute argument(s) to isolate eigenvalues where possible.

"S", "s"
: Scale to improve accuracy of computed eigenvalues.

"B", "b"
: Permute and scale, in that order. Rows/columns of a (and b) that are isolated by permutation are not scaled. This is the default behavior.

Algebraic eigenvalue balancing uses standard LAPACK routines.

Generalized eigenvalue problem balancing uses Ward's algorithm (SIAM Journal on Scientific and Statistical Computing, 1981).

**cond (a,p)**                                                        *Function File*

Compute the p-norm condition number of a matrix. cond (a) is defined as norm (a, p) * norm (inv (a), p). By default p=2 is used which implies a (relatively slow) singular value decomposition. Other possible selections are p= 1, Inf, inf, 'Inf', 'fro' which are generally faster.

See also: norm, inv, det, svd, rank.

**[d, rcond] = det (a)**                                           *Loadable Function*

Compute the determinant of a using LAPACK. Return an estimate of the reciprocal condition number if requested.

**dmult (a, b)**                                                       *Function File*

If a is a vector of length rows (b), return diag (a) * b (but computed much more efficiently).

**dot (x, y, dim)**                                                    *Function File*

Computes the dot product of two vectors. If x and y are matrices, calculate the dot-product along the first non-singleton dimension. If the optional argument dim is given, calculate the dot-product along this dimension.

**lambda = eig (a)**                                              *Loadable Function*
**[v, lambda] = eig (a)**                                         *Loadable Function*

The eigenvalues (and eigenvectors) of a matrix are computed in a several step process which begins with a Hessenberg decomposition, followed by a Schur decomposition, from which the eigenvalues are apparent. The eigenvectors, when desired, are computed by further manipulations of the Schur decomposition.

The eigenvalues returned by eig are not ordered.

g = givens (x, y)                           Loadable Function
[c, s] = givens (x, y)                      Loadable Function
   Return a 2 × 2 orthogonal matrix

$$G = \begin{bmatrix} c & s \\ -s' & c \end{bmatrix}$$

such that

$$G \begin{bmatrix} x \\ y \end{bmatrix} = \begin{bmatrix} * \\ 0 \end{bmatrix}$$

with $x$ and $y$ scalars.
For example,
   givens (1, 1)
      ⇒    0.70711    0.70711
          -0.70711    0.70711

[x, rcond] = inv (a)                        Loadable Function
[x, rcond] = inverse (a)                    Loadable Function
   Compute the inverse of the square matrix a. Return an estimate of the reciprocal condition number if requested, otherwise warn of an ill-conditioned matrix if the reciprocal condition number is small.

type = matrix_type (a)                      Loadable Function
a = matrix_type (a, type)                   Loadable Function
a = matrix_type (a, 'upper', perm)          Loadable Function
a = matrix_type (a, 'lower', perm)          Loadable Function
a = matrix_type (a, 'banded', nl, nu)       Loadable Function
   Identify the matrix type or mark a matrix as a particular type. This allows rapid for solutions of linear equations involving a to be performed. Called with a single argument, matrix_type returns the type of the matrix and caches it for future use. Called with more than one argument, matrix_type allows the type of the matrix to be defined.

   The possible matrix types depend on whether the matrix is full or sparse, and can be one of the following

   'unknown'
      Remove any previously cached matrix type, and mark type as unknown

   'full'
      Mark the matrix as full.

   'positive definite'
      Probable full positive definite matrix.

   'diagonal'
      Diagonal Matrix. (Sparse matrices only)

'permuted diagonal'
> Permuted Diagonal matrix. The permutation does not need to be specifically indicated, as the structure of the matrix explicitly gives this. (Sparse matrices only)

'upper'
> Upper triangular. If the optional third argument *perm* is given, the matrix is assumed to be a permuted upper triangular with the permutations defined by the vector *perm*.

'lower'
> Lower triangular. If the optional third argument *perm* is given, the matrix is assumed to be a permuted lower triangular with the permutations defined by the vector *perm*.

'banded'
'banded positive definite'
> Banded matrix with the band size of *nl* below the diagonal and *nu* above it. If *nl* and *nu* are 1, then the matrix is tridiagonal and treated with specialized code. In addition the matrix can be marked as probably a positive definite (Sparse matrices only)

'singular'
> The matrix is assumed to be singular and will be treated with a minimum norm solution

Note that the matrix type will be discovered automatically on the first attempt to solve a linear equation involving a. Therefore matrix_type is only useful to give Octave hints of the matrix type. Incorrectly defining the matrix type will result in incorrect results from solutions of linear equations, and so it is entirely the responsibility of the user to correctly identify the matrix type.

Also the test for positive definiteness is a low-cost test for a hermitian matrix with a real positive diagonal. This does not guarantee that the matrix is positive definite, but only that it is a probable candidate. When such a matrix is factorized, a Cholesky factorization is first attempted, and if that fails the matrix is then treated with an LU factorization. Once the matrix has been factorized, matrix_type will return the correct classification of the matrix.

norm (a, p)                                                          Function File
Compute the p-norm of the matrix a. If the second argument is missing, p = 2 is assumed.

If a is a matrix:

$p = 1$
> 1-norm, the largest column sum of the absolute values of a.

$p = 2$
> Largest singular value of a.

## Chapter 18: Linear Algebra

$p = \text{Inf}$ or `"inf"`
Infinity norm, the largest row sum of the absolute values of a.

$p = $ `"fro"`
Frobenius norm of a, sqrt (sum (diag (a' * a))).

If a is a vector or a scalar:

$p = \text{Inf}$ or `"inf"`
max (abs (a)).

$p = -\text{Inf}$
min (abs (a)).

$p = $ `"fro"`
Frobenius norm of a, sqrt (sumsq (abs (a))).

other
p-norm of a, (sum (abs (a) .^ p)) ^ (1/p).

See also: cond, svd.

**null (a, *tol*)**                                     Function File
Return an orthonormal basis of the null space of a.
The dimension of the null space is taken as the number of singular values of a not greater than *tol*. If the argument *tol* is missing, it is computed as

max (size (a)) * max (svd (a)) * eps

**orth (a, *tol*)**                                     Function File
Return an orthonormal basis of the range space of a.
The dimension of the range space is taken as the number of singular values of a greater than *tol*. If the argument *tol* is missing, it is computed as

max (size (a)) * max (svd (a)) * eps

**pinv (x, *tol*)**                                     Loadable Function
Return the pseudoinverse of x. Singular values less than *tol* are ignored.
If the second argument is omitted, it is assumed that

tol = max (size (x)) * sigma_max (x) * eps,

where sigma_max (x) is the maximal singular value of x.

**rank (a, *tol*)**                                     Function File
Compute the rank of a, using the singular value decomposition. The rank is taken to be the number of singular values of a that are greater than the specified tolerance *tol*. If the second argument is omitted, it is taken to be

tol = max (size (a)) * sigma(1) * eps;

where eps is machine precision and sigma(1) is the largest singular value of a.

trace (a)  Function File
    Compute the trace of a, sum (diag (a)).

[r, k] = rref (a, *tol*)  Function File
    Returns the reduced row echelon form of a. *tol* defaults to eps * max (size (a)) * norm (a, inf).

    Called with two return arguments, k returns the vector of "bound variables", which are those columns on which elimination has been performed.

## 18.3 Matrix Factorizations

chol (a)  Loadable Function
    Compute the Cholesky factor, r, of the symmetric positive definite matrix a, where $R^T R = A$.

    See also: cholinv, chol2inv.

cholinv (a)  Loadable Function
    Use the Cholesky factorization to compute the inverse of the symmetric positive definite matrix a.

    See also: chol, chol2inv.

chol2inv (u)  Loadable Function
    Invert a symmetric, positive definite square matrix from its Cholesky decomposition, u. Note that u should be an upper-triangular matrix with positive diagonal elements. chol2inv (u) provides inv (u'*u) but it is much faster than using inv.

    See also: chol, cholinv.

h = hess (a)  Loadable Function
[p, h] = hess (a)  Loadable Function
    Compute the Hessenberg decomposition of the matrix a.

    The Hessenberg decomposition is usually used as the first step in an eigenvalue computation, but has other applications as well (see Golub, Nash, and Van Loan, IEEE Transactions on Automatic Control, 1979). The Hessenberg decomposition is

$$A = PHP^T$$

    where $P$ is a square unitary matrix $(P^H P = I)$, and $H$ is upper Hessenberg $(H_{i,j} = 0, \forall i \geq j+1)$.

# Chapter 18: Linear Algebra

[l, u, p] = lu (a)                                                                                           Loadable Function

Compute the LU decomposition of a, using subroutines from LAPACK. The result is returned in a permuted form, according to the optional return value p. For example, given the matrix a = [1, 2; 3, 4],

    [l, u, p] = lu (a)

returns

    l =

        1.00000   0.00000
        0.33333   1.00000

    u =

        3.00000   4.00000
        0.00000   0.66667

    p =

        0  1
        1  0

The matrix is not required to be square.

[q, r, p] = qr (a)                                                                           Loadable Function

Compute the QR factorization of a, using standard LAPACK subroutines. For example, given the matrix a = [1, 2; 3, 4],

    [q, r] = qr (a)

returns

    q =

      -0.31623   -0.94868
      -0.94868    0.31623

    r =

      -3.16228   -4.42719
       0.00000   -0.63246

The qr factorization has applications in the solution of least squares problems

$$\min_x \|Ax - b\|_2$$

for overdetermined systems of equations (i.e., $A$ is a tall, thin matrix). The QR factorization is $QR = A$ where $Q$ is an orthogonal matrix and $R$ is upper triangular.

The permuted QR factorization [q, r, p] = qr (a) forms the QR factorization such that the diagonal entries of r are decreasing in magnitude order. For example, given the matrix a = [1, 2; 3, 4],

```
    [q, r, p] = qr(a)
returns
    q =

        -0.44721  -0.89443
        -0.89443   0.44721

    r =

        -4.47214  -3.13050
         0.00000   0.44721

    p =

        0  1
        1  0
```

The permuted qr factorization [q, r, p] = qr (a) factorization allows the construction of an orthogonal basis of span (a).

lambda = qz (a, b)   Loadable Function
Generalized eigenvalue problem $Ax = sBx$, $QZ$ decomposition. There are three ways to call this function:

1. lambda = qz(A,B)

   Computes the generalized eigenvalues $\lambda$ of $(A - sB)$.

2. [AA, BB, Q, Z, V, W, lambda] = qz (A, B)

   Computes qz decomposition, generalized eigenvectors, and generalized eigenvalues of $(A - sB)$

   $$AV = BV \text{diag}(\lambda)$$

   $$W^T A = \text{diag}(\lambda) W^T B$$

   $$AA = Q^T AZ, BB = Q^T BZ$$

   with $Q$ and $Z$ orthogonal (unitary)$= I$

3. [AA,BB,Z{, lambda}] = qz(A,B,opt)

   As in form [2], but allows ordering of generalized eigenpairs for (e.g.) solution of discrete time algebraic Riccati equations. Form 3 is not available for complex matrices, and does not compute the generalized eigenvectors $V$, $W$, nor the orthogonal matrix $Q$.

   opt
   > for ordering eigenvalues of the GEP pencil. The leading block of the revised pencil contains all eigenvalues that satisfy:
   >
   > "N"
   > > = unordered (default)

# Chapter 18: Linear Algebra

"S"
: = small: leading block has all |lambda| <=1

"B"
: = big: leading block has all |lambda| >= 1

"-"
: = negative real part: leading block has all eigenvalues in the open left half-plane

"+"
: = nonnegative real part: leading block has all eigenvalues in the closed right half-plane

Note: qz performs permutation balancing, but not scaling (see balance). Order of output arguments was selected for compatibility with MATLAB

See also: balance, dare, eig, schur.

[aa, bb, q, z] = qzhess (a, b)  *Function File*
Compute the Hessenberg-triangular decomposition of the matrix pencil (a, b), returning aa = q * a * z, bb = q * b * z, with q and z orthogonal. For example,

```
[aa, bb, q, z] = qzhess ([1, 2; 3, 4], [5, 6; 7, 8])
⇒ aa = [ -3.02244, -4.41741;  0.92998,  0.69749 ]
⇒ bb = [ -8.60233, -9.99730;  0.00000, -0.23250 ]
⇒ q  = [ -0.58124, -0.81373; -0.81373,  0.58124 ]
⇒ z  = [ 1, 0; 0, 1 ]
```

The Hessenberg-triangular decomposition is the first step in Moler and Stewart's QZ decomposition algorithm.

Algorithm taken from Golub and Van Loan, *Matrix Computations*, 2nd edition.

s = schur (a)  *Loadable Function*
[u, s] = schur (a, opt)  *Loadable Function*

The Schur decomposition is used to compute eigenvalues of a square matrix, and has applications in the solution of algebraic Riccati equations in control. schur always returns $S = U^T A U$ where $U$ is a unitary matrix ($U^T U$ is identity) and $S$ is upper triangular. The eigenvalues of $A$ (and $S$) are the diagonal elements of $S$. If the matrix $A$ is real, then the real Schur decomposition is computed, in which the matrix $U$ is orthogonal and $S$ is block upper triangular with blocks of size at most 2 × 2 along the diagonal. The diagonal elements of $S$ (or the eigenvalues of the 2 × 2 blocks, when appropriate) are the eigenvalues of $A$ and $S$.

The eigenvalues are optionally ordered along the diagonal according to the value of opt. opt = "a" indicates that all eigenvalues with negative real parts should be moved to the leading block of $S$ (used in are), opt = "d" indicates that all eigenvalues with magnitude less than one should be moved to the leading block of $S$ (used in dare), and opt = "u", the default, indicates that no ordering of eigenvalues should occur. The leading $k$ columns of $U$ always span the $A$-invariant subspace corresponding to the $k$ leading eigenvalues of $S$.

s = svd (a)     Loadable Function
[u, s, v] = svd (a)     Loadable Function
    Compute the singular value decomposition of a

$$A = USV^H$$

The function svd normally returns the vector of singular values. If asked for three return values, it computes $U$, $S$, and $V$. For example,

    svd (hilb (3))

returns

```
ans =

  1.4083189
  0.1223271
  0.0026873
```

and

    [u, s, v] = svd (hilb (3))

returns

```
u =

  -0.82704   0.54745   0.12766
  -0.45986  -0.52829  -0.71375
  -0.32330  -0.64901   0.68867

s =

  1.40832  0.00000  0.00000
  0.00000  0.12233  0.00000
  0.00000  0.00000  0.00269

v =

  -0.82704   0.54745   0.12766
  -0.45986  -0.52829  -0.71375
  -0.32330  -0.64901   0.68867
```

If given a second argument, svd returns an economy-sized decomposition, eliminating the unnecessary rows or columns of u or v.

## Chapter 18: Linear Algebra

[housv, beta, zer] = housh (x, j, z)      Function File

Compute Householder reflection vector *housv* to reflect *x* to be the j-th column of identity, i.e.,

(I - beta*housv*housv')x = norm(x)*e(j) if x(1) < 0,
(I - beta*housv*housv')x = -norm(x)*e(j) if x(1) >= 0

Inputs

- *x*      vector
- *j*      index into vector
- *z*      threshold for zero (usually should be the number 0)

Outputs (see Golub and Van Loan):

*beta*
     If beta = 0, then no reflection need be applied (zer set to 0)

*housv*
     householder vector

[u, h, nu] = krylov (a, v, k, eps1, pflg)      Function File

Construct an orthogonal basis *u* of block Krylov subspace

[v a*v a^2*v ... a^(k+1)*v]

Using Householder reflections to guard against loss of orthogonality.

If *v* is a vector, then *h* contains the Hessenberg matrix such that a*u == u*h+rk*ek', in which rk = a*u(:,k)-u*h(:,k), and ek' is the vector [0, 0, ..., 1] of length k. Otherwise, *h* is meaningless.

If *v* is a vector and *k* is greater than length(A)-1, then *h* contains the Hessenberg matrix such that a*u == u*h.

The value of *nu* is the dimension of the span of the krylov subspace (based on *eps1*).

If *b* is a vector and *k* is greater than *m-1*, then *h* contains the Hessenberg decomposition of *a*.

The optional parameter *eps1* is the threshold for zero. The default value is 1e-12.

If the optional parameter *pflg* is nonzero, row pivoting is used to improve numerical behavior. The default value is 0.

Reference: Hodel and Misra, "Partial Pivoting in the Computation of Krylov Subspaces", to be submitted to Linear Algebra and its Applications

## 18.4 Functions of a Matrix

expm (a)                                                                             Loadable Function

Return the exponential of a matrix, defined as the infinite Taylor series

$$\exp(A) = I + A + \frac{A^2}{2!} + \frac{A^3}{3!} + \cdots$$

The Taylor series is *not* the way to compute the matrix exponential; see Moler and Van Loan, *Nineteen Dubious Ways to Compute the Exponential of a Matrix*, SIAM Review, 1978. This routine uses Ward's diagonal Padé approximation method with three step preconditioning (SIAM Journal on Numerical Analysis, 1977). Diagonal Padé approximations are rational polynomials of matrices $D_q(a)^{-1} N_q(a)$ whose Taylor series matches the first $2q+1$ terms of the Taylor series above; direct evaluation of the Taylor series (with the same preconditioning steps) may be desirable in lieu of the Padé approximation when $D_q(a)$ is ill-conditioned.

logm (a)                                                                                   Function File

Compute the matrix logarithm of the square matrix a. Note that this is currently implemented in terms of an eigenvalue expansion and needs to be improved to be more robust.

[result, error_estimate] = sqrtm (a)                    Loadable Function

Compute the matrix square root of the square matrix a.

Ref: Nicholas J. Higham. A new sqrtm for MATLAB. Numerical Analysis Report No. 336, Manchester Centre for Computational Mathematics, Manchester, England, January 1999.

See also: expm, logm, funm.

kron (a, b)                                                                Loadable Function

Form the kronecker product of two matrices, defined block by block as

    x = [a(i, j) b]

For example,

```
    kron (1:4, ones (3, 1))
       ⇒  1  2  3  4
          1  2  3  4
          1  2  3  4
```

x = syl (a, b, c)                                                 Loadable Function

Solve the Sylvester equation

$$AX + XB + C = 0$$

using standard LAPACK subroutines. For example,

```
    syl ([1, 2; 3, 4], [5, 6; 7, 8], [9, 10; 11, 12])
       ⇒ [ -0.50000, -0.66667; -0.66667, -0.50000 ]
```

# 19 Nonlinear Equations

Octave can solve sets of nonlinear equations of the form

$$f(x) = 0$$

using the function `fsolve`, which is based on the MINPACK subroutine hybrd. This is an iterative technique so a starting point must be provided, and convergence is not guaranteed even if a solution exists.

[x, fval, info] = fsolve (fcn, x0)  Loadable Function
Given *fcn*, the name of a function of the form f (x) and an initial starting point *x0*, fsolve solves the set of equations such that f(x) == 0.

On return, *fval* contains the value of the function *fcn* evaluated at *x*, and *info* may be one of the following values:

-2   Invalid input parameters.

-1   Error in user-supplied function.

1    Relative error between two consecutive iterates is at most the specified tolerance (see fsolve_options).

3    Algorithm failed to converge.

4    Limit on number of function calls reached.

If *fcn* is a two-element string array, or a two element cell array containing either the function name or inline or function handle. The first element names the function *f* described above, and the second element names a function of the form j (x) to compute the Jacobian matrix with elements

$$J = \frac{\partial f_i}{\partial x_j}$$

You can use the function fsolve_options to set optional parameters for fsolve.

fsolve_options (opt, val)  Loadable Function
When called with two arguments, this function allows you to set options parameters for the function fsolve. Given one argument, fsolve_options returns the value of the corresponding option. If no arguments are supplied, the names of all the available options and their current values are displayed.

Options include

"tolerance"
    Nonnegative relative tolerance.

Here is a complete example. To solve the set of equations

$$-2x^2 + 3xy + 4\sin(y) - 6 = 0$$
$$3x^2 - 2xy^2 + 3\cos(x) + 4 = 0$$

you first need to write a function to compute the value of the given function. For example:

```
function y = f (x)
   y(1) = -2*x(1)^2 + 3*x(1)*x(2)   + 4*sin(x(2)) - 6;
   y(2) =  3*x(1)^2 - 2*x(1)*x(2)^2 + 3*cos(x(1)) + 4;
endfunction
```

Then, call fsolve with a specified initial condition to find the roots of the system of equations. For example, given the function f defined above,

```
[x, fval, info] = fsolve (@f, [1; 2])
```

results in the solution

```
x =

  0.57983
  2.54621

fval =

  -5.7184e-10
   5.5460e-10

info = 1
```

A value of info = 1 indicates that the solution has converged.

The function perror may be used to print English messages corresponding to the numeric error codes. For example,

```
perror ("fsolve", 1)
    -| solution converged to requested tolerance
```

When no Jacobian is supplied (as in the example above) it is approximated numerically. This requires more function evaluations, and hence is less efficient. In the example above we could derive the Jacobian analytically as

$$\begin{bmatrix} \frac{\partial f_1}{\partial x_1} & \frac{\partial f_1}{\partial x_2} \\ \frac{\partial f_2}{\partial x_1} & \frac{\partial f_2}{\partial x_2} \end{bmatrix} = \begin{bmatrix} 3x_2 - 4x_1 & 4\cos(x_2) + 3x_1 \\ -2x_2^2 - 3\sin(x_1) + 6x_1 & -4x_1 x_2 \end{bmatrix}$$

and compute it with the following Octave function

```
function J = jacobian(x)
   J(1,1) =  3*x(2) - 4*x(1);
   J(1,2) =  4*cos(x(2)) + 3*x(1);
   J(2,1) = -2*x(2)^2 - 3*sin(x(1)) + 6*x(1);
   J(2,2) = -4*x(1)*x(2);
endfunction
```

The Jacobian can then be used with the following call to fsolve:

```
[x, fval, info] = fsolve ({@f, @jacobian}, [1; 2]);
```
which gives the same solution as before.

# 20 Sparse Matrices

This chapter describes the use of *sparse matrices* in Octave. For matrices where a large number of the elements are zero, Octave provides a sparse matrix type which stores only the non-zero elements. This reduces the amount of memory required and allows unnecessary operations involving empty elements to be avoided.

## 20.1 Basics

### 20.1.1 Storage of Sparse Matrices

It is not necessary to know how sparse matrices are stored by Octave in order to use them. However, knowledge of the storage technique is helpful in estimating the memory requirements (and costs) of operations involving sparse matrices.[1]

In general, there are many techniques for storing sparse matrix data, each having different tradeoffs for different classes of operations. A good summary of the available methods is given by Saad[2]. One obvious way to store the elements of a sparse matrix $M$ is as a set of triplets containing the row and column of each non-zero element and its value, $(i, j, M_{ij})$. This is conceptually simple but typically requires more memory than is actually needed.

The storage format used within Octave is *compressed column format*. In this format, the non-zero values from each column and their row positions are stored sequentially in memory. A separate array records the locations which correspond to the start of each column.

As an example, consider the matrix

```
1  2  0  0
0  0  0  3
0  0  0  4
```

The non-zero elements of this matrix are

(1, 1)  ⇒ 1
(1, 2)  ⇒ 2
(2, 4)  ⇒ 3
(3, 4)  ⇒ 4

This will be stored as three vectors *cidx*, *ridx* and *data*, representing the column indexing, row indexing and data respectively. The contents of these three vectors for the above matrix will be

---

[1] Knowledge of the storage technique is also necessary for those wishing to create their own opt-files.

[2] Youcef Saad "SPARSKIT: A basic toolkit for sparse matrix computation", 1994. http://www-users.cs.umn.edu/~saad/software/SPARSKIT/paper.ps

```
cidx = [0, 1, 2, 2, 4] # column start index in ridx & data
ridx = [0, 0, 1, 2]
data = [1, 2, 3, 4]
```

Note that this is the C representation, with the first row and column assumed to start at zero (in Octave the row and column indexing starts at one). Thus the number of elements in the $i$-th column is given by `cidx (i + 1) - cidx (i)`.

For simplicity, the column index contains one more entry than the number of columns, to record the position of the final element. The first element is always zero. This avoids any need to the handle the first and last columns as a special case.

The code to access the elements of a sparse matrix in C looks like this:

```
for (j = 0; j < nc; j++)
    for (i = cidx [j]; i < cidx[j+1]; i++)
        printf ("non-zero element (%i,%i) is %d\n",
            ridx[i], j, data[i]);
```

Note that Octave uses compressed column format, rather than compressed row format, for consistency with the column-major ordering of dense matrices. This makes operations that involve both sparse and dense matrices more efficient.

In the storage format used by Octave, sparse matrix elements are stored in increasing order of their row index. This requires sorting them on creation. Allowing disordered elements would make some operations faster (such as concatenating two matrices together) but would add complexity and reduce speed elsewhere.

### 20.1.2 Creating Sparse Matrices

A sparse matrix can be created in the following ways:

Returned from a function
: Many functions return sparse matrices directly. These include speye, sprand, spdiag, etc.

Constructed from matrices or vectors
: The function sparse allows a sparse matrix to be constructed from three vectors representing the row, column and data. Alternatively, the function spconvert uses a three column matrix format to allow easy importation of data from elsewhere.

Created and then filled
: The function sparse or spalloc can be used to create an empty matrix that is then filled by the user.

From a user binary program
: The user can directly create the sparse matrix within an oct-file.

There are several basic functions to return specific sparse matrices. For example speye ($n$) or speye ($r$, $c$) returns an $n$-by-$n$ or $r$-by-$c$ sparse identity matrix. The function diag returns a sparse diagonal matrix when given a sparse input vector. For example,

```
s = diag (sparse(randn(1,n)), -1);
```
creates a sparse $(n+1)$-by-$(n+1)$ sparse matrix with a single diagonal defined. The function spdiags described below is a generalisation of diag.

y = speye ($m$)                                                                Function File
y = speye ($m$, $n$)                                                Function File
y = speye ($sz$)                                                  Function File

Returns a sparse identity matrix. This is significantly more efficient than sparse (eye ($m$)) as the full matrix is not constructed.

Called with a single argument a square matrix of size $m$ by $m$ is created. Otherwise a matrix of $m$ by $n$ is created. If called with a single vector argument, this argument is taken to be the size of the matrix to create.

y = spfun ($f$, $x$)                                                 Function File
Compute $f(x)$ for the non-zero values of $x$, This results in a sparse matrix with the same structure as $x$. The function $f$ can be passed as a string, a function handle or an inline function.

y = spones ($x$)                                                  Function File
Replace the non-zero entries of $x$ with ones. This creates a sparse matrix with the same structure as $x$.

spdiag ($v$, $k$)                                                Loadable Function
Return a diagonal matrix with the sparse vector $v$ on diagonal $k$. The second argument is optional. If it is positive, the vector is placed on the $k$-th super-diagonal. If it is negative, it is placed on the $-k$-th sub-diagonal. The default value of $k$ is 0, and the vector is placed on the main diagonal. For example,

```
spdiag ([1, 2, 3], 1)
ans =

Compressed Column Sparse (rows=4, cols=4, nnz=3)
  (1 , 2) -> 1
  (2 , 3) -> 2
  (3 , 4) -> 3
```

Given a matrix argument, instead of a vector, spdiag extracts the $k$-th diagonal of the sparse matrix.

See also: diag.

[b, c] = spdiags (a)                                  Function File
b = spdiags (a, c)                                    Function File
b = spdiags ($v$, c, a)                                  Function File
b = spdiags ($v$, c, $m$, $n$)                           Function File
A generalization of the function spdiag. Called with a single input argument, the non-zero diagonals $c$ of $A$ are extracted. With two arguments the diagonals to extract are given by the vector $c$.

The other two forms of spdiags modify the input matrix by replacing the diagonals. They use the columns of $v$ to replace the columns represented by the vector $c$. If the sparse matrix a is defined then the diagonals of this matrix are replaced. Otherwise a matrix of $m$ by $n$ is created with the diagonals given by $v$.

Negative values of $c$ representative diagonals below the main diagonal, and positive values of $c$ diagonals above the main diagonal.

For example

```
spdiags (reshape (1:12, 4, 3), [-1 0 1], 5, 4)
⇒      5 10  0  0
       1  6 11  0
       0  2  7 12
       0  0  3  8
       0  0  0  4
```

**s = sparse (a)**                                               Loadable Function
Create a sparse matrix from the full matrix a.

**s = sparse (i, j, sv, m, n, nzmax)**                           Loadable Function
Create a sparse matrix given integer index vectors $i$ and $j$, a 1-by-nnz vector of real of complex values sv, overall dimensions $m$ and $n$ of the sparse matrix. The argument nzmax is ignored but accepted for compatibility with MATLAB.

Note: if multiple values are specified with the same $i$, $j$ indices, the corresponding values in s will be added.

The following are all equivalent:

```
s = sparse (i, j, s, m, n)
s = sparse (i, j, s, m, n, "summation")
s = sparse (i, j, s, m, n, "sum")
```

**s = sparse (i, j, s, m, n, "unique")**                         Loadable Function
Same as above, except that if more than two values are specified for the same $i$, $j$ indices, the last specified value will be used.

**s = sparse (i, j, sv)**                                        Loadable Function
Uses $m$ = max $(i)$, $n$ = max $(j)$.

**s = sparse (m, n)**                                            Loadable Function
Equivalent to sparse ([], [], [], m, n, 0)

If any of sv, $i$ or $j$ are scalars, they are expanded to have a common size.

**See also:** full.

The recommended way to create a sparse matrix is to use the function sparse with three vectors containing the row index, column index and value to be stored. For example, the following commands create the 3-by-4 sparse matrix shown earlier:

```
r = 3; c = 4;
ri = [1, 1, 2, 3];
ci = [1, 2, 4, 4];
di = [1, 2, 3, 4];
s = sparse (ri, ci, di, r, c);
```
The indices do not need to be in any particular order as Octave will sort them prior to storing the data.[3]

x = spconvert (*m*)                        Function File

This function converts from a simple sparse matrix format easily produced by other programs into Octave's internal sparse format. The input *x* is either a three or four column real matrix, containing the row index, column index, real and imaginary parts of the elements of the sparse matrix. The matrix can contain zero elements and the elements can be sorted in any order. An element with a zero real and imaginary part can be used to force a particular matrix size.

For example,
```
s = spconvert ([1 2 3 4; 1 3 4 4; 1 2 3 0]')
⇒ Compressed Column Sparse (rows=4, cols=4, nnz=3)
    (1 , 1) -> 1
    (2 , 3) -> 2
    (3 , 4) -> 3
```

s = spalloc (*r, c, nz*)                      Function File

Returns an empty sparse matrix of size *r*-by-*c*. This function is provided only for compatibility reasons and the argument *nz* is ignored. Octave resizes the memory required for sparse matrices on demand whenever they are modified, so spalloc does not preassign any memory.

It should be noted that this means that code like
```
k = 5;
nz = r * k;
s = spalloc (r, c, nz)
for j = 1:c
    idx = randperm (r);
    s (:, j) = [zeros(r - k, 1); rand(k, 1)] (idx);
endfor
```
will reallocate memory at each step. It is therefore vitally important that such code should be vectorized as much as possible, to minimize the number of assignments and reduce the number of memory allocations.[4]

See also: sparse, nzmax.

---

[3] Pre-sorting the data will make the creation of the sparse matrix faster.

[4] The above problem of memory reallocation can be avoided in oct-files. However, the construction of a sparse matrix from an oct-file is more complex than can be discussed here.

*FM* = full (*SM*)      Loadable Function
returns a full storage matrix from a sparse one

See also: sparse.

sprand (*m*, *n*, *d*)      Function File
sprand (*s*)      Function File

Generate a random sparse matrix using uniform variates. The size of the matrix will be *m* by *n*, with a density of values given by *d*, which should be between 0 and 1. Values will be uniformly distributed between 0 and 1.

The actual density may be slightly lower than *d*, due to random collisions in the selection of indices. These are unlikely to occur for large, really sparse matrices.

If called with a single matrix argument, a sparse matrix is generated with random values wherever the matrix *S* is non-zero.

See also: sprandn.

sprandn (*m*, *n*, *d*)      Function File
sprandn (*s*)      Function File

Generate a random sparse matrix using normal variates. The size of the matrix will be *m* by *n*, with a density of values given by *d*, which should be between 0 and 1. Values will be normally distributed with mean of zero and variance 1.

The actual density may be slightly lower than *d*, due to random collisions in the selection of indices. These are unlikely to occur for large, really sparse matrices.

If called with a single matrix argument, a sparse matrix is generated with random values wherever the matrix *S* is non-zero.

See also: sprand.

sprandsym (*n*, *d*)      Function File
sprandsym (*s*)      Function File

Generate a symmetric random sparse matrix. The size of the matrix will be *n* by *n*, with a density of values given by *d*. *d* should be between 0 and 1. Values will be normally distributed with mean of zero and variance 1.

The actual density may be slightly lower than *d*, due to random collisions in the selection of indices. These are unlikely to occur for large, really sparse matrices.

If called with a single matrix argument, a sparse matrix is generated with random values wherever the matrix *S* is non-zero in its lower triangular part.

See also: sprand, sprandn.

### 20.1.3 Sparse Matrix Properties

There are a number of functions that return basic information concerning sparse matrix objects. The most basic of these is issparse which identifies whether a particular Octave object is a sparse matrix.

The function nnz returns the number of non-zero entries in a sparse matrix, and nzmax returns the amount of storage allocated to it.

The function spstats returns some basic statistics on the columns of a sparse matrix, including the number of elements, mean value and variance of each column.

issparse (*expr*)                                                         Loadable Function
    Return 1 if the value of the expression *expr* is a sparse matrix.

nonzeros (*s*)                                                         Function File
    Returns a vector of the non-zero values of the sparse matrix *s*.

*scalar* = nnz (*a*)                                             Built-in Function
    Returns the number of non zero elements in *a*.

    See also: sparse.

*scalar* = nzmax (*SM*)                                     Built-in Function
    Return the amount of storage allocated to the sparse matrix *SM*. Note that Octave tries to free unused memory at the first opportunity for sparse objects. The values returned by *nzmaz* and *nnz* will typically be the same, although they may differ for user-created sparse objects.

    See also: sparse, spalloc.

spmax (*x, y, dim*)                                        Mapping Function
[w, iw] = spmax (*x*)                                  Mapping Function
    For a vector argument, return the maximum value. For a matrix argument, return the maximum value from each column, as a row vector, or over the dimension *dim* if defined. For two matrices (or a matrix and scalar), return the pair-wise maximum. Thus,

        max (max (*x*))

returns the largest element of *x*, and

        max (2:5, pi)
            ⇒   3.1416   3.1416   4.0000   5.0000

compares each element of the range 2:5 with pi, and returns a row vector of the maximum values.

For complex arguments, the magnitude of the elements are used for comparison.

If called with one input and two output arguments, max also returns the first index of the maximum value(s). Thus,

        [x, ix] = max ([1, 3, 5, 2, 5])
            ⇒   x = 5
               ix = 3

spmin (*x, y, dim*)  Mapping Function
[*w, iw*] = spmin (*x*)  Mapping Function
  For a vector argument, return the minimum value. For a matrix argument, return the minimum value from each column, as a row vector, or over the dimension *dim* if defined. For two matrices (or a matrix and scalar), return the pair-wise minimum. Thus,

      min (min (x))

  returns the smallest element of *x*, and

      min (2:5, pi)
      ⇒  2.0000  3.0000  3.1416  3.1416

  compares each element of the range 2:5 with pi, and returns a row vector of the minimum values.

  For complex arguments, the magnitude of the elements are used for comparison.

  If called with one input and two output arguments, min also returns the first index of the minimum value(s). Thus,

      [x, ix] = min ([1, 3, 0, 2, 5])
      ⇒   x = 0
          ix = 3

[*count, mean, var*] = spstats (*s*)  Function File
[*count, mean, var*] = spstats (*s, j*)  Function File
  Return statistics for the non-zero elements of the sparse matrix *s*. *count* is the number of non-zeros in each column, *mean* is the mean of the non-zeros in each column, and *var* is the variance of the non-zeros in each column.

  Called with two input arguments, if *s* is the data and *j* is the bin number for the data, compute the statistics for each bin. In this case, bins can contain data values of zero, whereas with spstats (*s*) the zeros are not included.

### 20.1.4 Sparse Matrix Types

When solving linear equations involving sparse matrices Octave determines the appropriate algorithm from the properties of the matrix, as discussed in Section 20.2 [Sparse Linear Algebra], page 308. Octave probes the matrix type when the div (/) or ldiv (\) operators are first used and caches the type. The matrix_type function can also be used to determine the type of the sparse matrix prior to use of the div or ldiv operators. For example

      a = tril (sprandn(1024, 1024, 0.02), -1) ...
          + speye(1024);
      matrix_type (a);
      ans = Lower

The result shows that Octave correctly determines the matrix type, a lower triangular matrix. The function matrix_type can also be used to force the type of a matrix to a particular value. For example

# Chapter 20: Sparse Matrices

        a = matrix_type (tril (sprandn (1024, ...
            1024, 0.02), -1) + speye(1024), 'Lower');

This allows the cost of determining the matrix type to be avoided. However, incorrectly defining the matrix type will result in incorrect results from solutions of linear equations, and so it is entirely the responsibility of the user to correctly identify the matrix type.

## 20.1.5 Graphical Representations of Sparse Matrices

The functions described in this section display sparse matrices graphically. The spy command displays the structure of the non-zero elements of the matrix. More advanced graphical information can be obtained with the treeplot, etreeplot and gplot commands.

spy (x)                                                                Function File
spy (..., markersize)                                                  Function File
spy (..., LineSpec)                                                    Function File

Plot the sparsity pattern of the sparse matrix x. If the argument markersize is given as an scalar value, it is used to determine the point size in the plot. If the string LineSpec is given it is passed to plot and determines the appearance of the plot.

See also: plot.

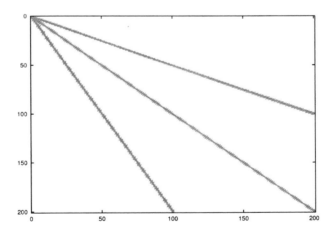

Figure 20.1: Structure of simple sparse matrix displayed with spy.

One use of sparse matrices is in graph theory, where the interconnections between nodes are represented as an adjacency matrix. That is, if the i-th node in a graph is connected to the j-th node. Then the ij-th node (and in the case of undirected graphs the ji-th node) of the sparse adjacency matrix is non-zero. If each node is then associated with a set of co-ordinates, then the gplot command can be used to graphically display the interconnections between nodes.

gplot (a, xy)                                                        Function File
gplot (a, xy, line_style)                       Function File
[x, y] = gplot (a, xy)                            Function File

Plot a graph defined by $A$ and $xy$ in the graph theory sense. $A$ is the adjacency matrix of the array to be plotted and $xy$ is an $n$-by-2 matrix containing the coordinates of the nodes of the graph.

The optional parameter *line_style* defines the output style for the plot. Called with no output arguments the graph is plotted directly. Otherwise, return the coordinates of the plot in $x$ and $y$.

**See also:** treeplot, etreeplot, spy.

As a trivial example of the use of gplot, consider the example

```
A = sparse([2,6,1,3,2,4,3,5,4,6,1,5],
    [1,1,2,2,3,3,4,4,5,5,6,6],1,6,6);
xy = [0,4,8,6,4,2;5,0,5,7,5,7]';
gplot(A,xy)
```

which creates an adjacency matrix A where node 1 is connected to nodes 2 and 6, node 2 with nodes 1 and 3, etc. The co-ordinates of the nodes are given in the n-by-2 matrix xy. See Figure 20.2.

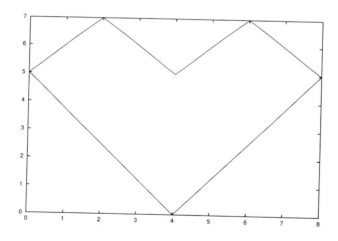

Figure 20.2: Simple use of the gplot command.

The dependencies between the nodes of a Cholesky factorization can be calculated in linear time without explicitly needing to calculate the Cholesky factorization. The etree command returns the elimination tree of the matrix. It can be displayed graphically with treeplot(etree(A)) if A is symmetric or treeplot(etree(A+A')) otherwise.

Chapter 20: Sparse Matrices                                                   297

p = etree (s)                                             Loadable Function
p = etree (s, typ)                                        Loadable Function
[p, q] = etree (s, typ)                                   Loadable Function
    Returns the elimination tree for the matrix s. By default s is assumed to be symmetric and the symmetric elimination tree is returned. The argument *typ* controls whether a symmetric or column elimination tree is returned. Valid values of *typ* are 'sym' or 'col', for symmetric or column elimination tree respectively

    Called with a second argument, *etree* also returns the postorder permutations on the tree.

etreeplot (*tree*)                                        Function File
etreeplot (*tree, node_style, edge_style*)                Function File
    Plot the elimination tree of the matrix *s* or *s+s* ' if *s* in non-symmetric. The optional parameters *line_style* and *edge_style* define the output style.

    **See also:** treeplot, gplot.

treeplot (*tree*)                                         Function File
treeplot (*tree, line_style, edge_style*)                 Function File
    Produces a graph of a tree (or forest). The first argument is a vector of predecessors, optional parameters *line_style* and *edge_style* define the output style. The vector *tree* is constructed by regarding each element of the vector as a node, and storing the index of the parent node in the element. For example, if node i is the parent of j, set tree(j) = i. The complexity of the algorithm is O(n) in terms of time and memory requirements.

    **See also:** etreeplot, gplot.

### 20.1.6 Basic Operators and Functions on Sparse Matrices

#### 20.1.6.1 Sparse Functions

    Many of the internal functions of Octave, such as `diag`, cannot accept sparse matrices as an input directly. The sparse implementation uses the `dispatch` function to overload these functions with equivalent functions that work with sparse matrices. The sparse matrix version of any function can be also used explicitly calling its function name.

    The table below lists all of the sparse functions of Octave. The names of the sparse forms of the functions are typically the same as the general versions with a sp prefix. In the table below, and the rest of this chapter the sparse versions of the function names are used.

Generate sparse matrices:
    spalloc, spdiags, speye, sprand, sprandn, sprandsym

Sparse matrix conversion:
    full, sparse, spconvert

Manipulate sparse matrices
    issparse, nnz, nonzeros, nzmax, spfun, spones, spy

Graph Theory:
: etree, etreeplot, gplot, treeplot

Sparse matrix reordering:
: amd, ccolamd, colamd, colperm, csymamd, dmperm, symamd, randperm, symrcm

Linear algebra:
: matrix_type, normest, condest, sprank spaugment

Iterative techniques:
: luinc, pcg, pcr

Miscellaneous:
: spparms, symbfact, spstats

Note that all of the standard Octave math functions that take a single argument, such as abs, etc can accept sparse matrices.

### 20.1.6.2 The Return Types of Operators and Functions

Octave attempts to return results of sparse matrix operations with an appropriate type, either sparse or full. When a matrix exceeds a certain density of non-zero elements it no longer makes sense to store it as a sparse matrix. For this reason, operators and functions that have a high probability of generating a full matrix result will always return a full matrix.

For example, adding a scalar constant to a sparse matrix returns a full matrix. This is true even if the constant is zero.

```
    speye(3) + 0
⇒   1 0 0
    0 1 0
    0 0 1
```

Additionally, if the optional setting sparse_auto_mutate is enabled, all sparse functions test the amount of memory occupied by their result to see if it is larger than the amount that would be used by the equivalent full matrix. Therefore speye (2) * 1 will return a full matrix, as the memory used for the full version is smaller than the sparse version.

As Octave provides a complete set of mixed operators and functions for full and sparse matrices, automatic conversion generally allows results to be computed in the most efficient way. However, a penalty can occur when a sparse matrix is promoted to a full matrix, and subsequent operations would resparsify the matrix. Such cases are rare, but can be artificially created. For example (fliplr(speye(3)) + speye(3)) - speye(3) gives a full matrix when it should give a sparse one. In general, where such cases occur, they impose only a small memory cost.

To avoid confusion, the sparse function, and other functions based on it (such as speye), always return a sparse matrix, even if the memory used will be larger than its full representation. This ensures that expressions such as diag (sparse([1,2,3]), -1) always return a sparse matrix.

val = sparse_auto_mutate ()                                    Built-in Function
old_val = sparse_auto_mutate (new_val)                         Built-in Function
> Query or set the internal variable that controls whether Octave will automatically mutate sparse matrices to real matrices to save memory. For example,
>
>     s = speye(3);
>     sparse_auto_mutate (false)
>     s (:, 1) = 1;
>     typeinfo (s)
>     ⇒ sparse matrix
>     sparse_auto_mutate (true)
>     s (1, :) = 1;
>     typeinfo (s)
>     ⇒ matrix

Note that the sparse_auto_mutate option is incompatible with MATLAB, and so it is off by default.

### 20.1.6.3 Mathematical Considerations

Octave's implementation of sparse matrices tries to behave in exactly the same manner as for full matrices. However, there are some additional considerations needed in the use of sparse matrices, and also differences from other sparse matrix implementations.

Firstly, the ./ and .^ operators must be used with care. In Octave their behavior is consistent with full matrices, and is different from sparse implementations in other products.

The following examples show how Octave's implementation behaves:

    s = speye (4);
    a1 = s .^ 2;      # sparse
    a2 = s .^ s;      # full
    a3 = s .^ -2;     # full, with Infs
    a4 = s ./ 2;      # sparse
    a5 = 2 ./ s;      # full, with Infs
    a6 = s ./ s;      # full, with NaNs

The first example of s raised to the power of 2 causes no problems. However s raised element-wise to itself involves a large number of terms 0 .^ 0 which is 1. Therefore s .^ s is a full matrix.

Likewise s .^ -2 involves terms like 0 .^ -2 which is infinity, and so s .^ -2 is equally a full matrix.

For the ./ operator s ./ 2 has no problems, but 2 ./ s involves a large number of infinity terms and is a full matrix. The case of s ./ s involves terms like 0 ./ 0 which is a NaN and so this is equally a full matrix with the zero elements of s filled with NaN values.

Calculations where the sign-bit of zero is important must not be performed using sparse matrices. The zeros of a sparse matrix are not stored, so their sign-bit is unknown. For example, the following calculation shows the lack of signed-zeros in the sparse matrix implementation:

```
a = 0 ./ [-1, 1; 1, -1];
b = 1 ./ a
⇒ -Inf           Inf
   Inf          -Inf
c = 1 ./ sparse (a)
⇒  Inf           Inf
   Inf           Inf
```

Correcting this behavior for sparse matrices would require the storage of zero elements with a negative sign-bit, and this is not implemented for reasons of efficiency.

### 20.1.7 Reordering

In general any function or operator used on a sparse matrix will result in a sparse matrix with the same or a larger number of non-zero elements than the original matrix. This is particularly true for the important case of sparse matrix factorizations. The usual way to address this is to reorder the matrix, such that its factorization is sparser than the factorization of the original matrix. That is the factorization of L * U = P * S * Q has sparser terms L and U than the equivalent factorization L * U = S.

Several functions are available to reorder depending on the type of the matrix to be factorized. If the matrix is symmetric positive-definite, then symamd or csymamd should be used. Otherwise amd, colamd or ccolamd should be used. For completeness the reordering functions colperm and randperm are also available.

See Figure 20.3 for an example of the structure of a simple positive definite matrix.

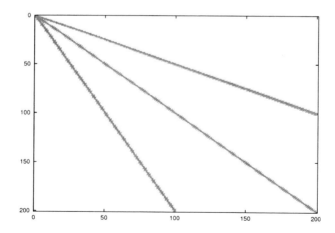

Figure 20.3: Structure of simple sparse matrix.

## Chapter 20: Sparse Matrices

The standard Cholesky factorization of this matrix can be obtained by the same command that would be used for a full matrix. This can be visualized with the command r = chol(A); spy(r);. See Figure 20.4. The original matrix had 598 non-zero terms, while this Cholesky factorization has 10200, with only half of the symmetric matrix being stored. This is a significant level of fill in, and although not an issue for such a small test case, can represents a large overhead in working with other sparse matrices.

The appropriate sparsity preserving permutation of the original matrix is given by symamd and the factorization using this reordering can be visualized using the command q = symamd(A); r = chol(A(q,q)); spy(r). This gives 399 non-zero terms which is a significant improvement.

The Cholesky factorization itself can be used to determine the appropriate sparsity preserving reordering of the matrix during the factorization, In that case this might be obtained with three return arguments as r[r, p, q] = chol(A); spy(r).

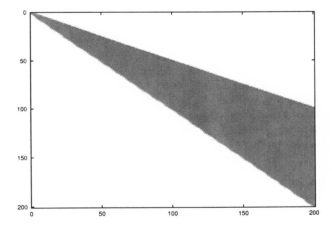

Figure 20.4: Structure of the un-permuted Cholesky factorization of the above matrix.

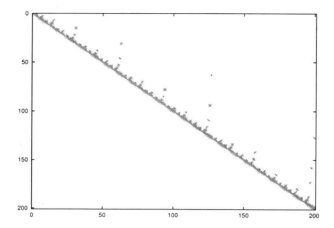

Figure 20.5: Structure of the permuted Cholesky factorization of the above matrix.

In the case of an asymmetric matrix, the appropriate sparsity preserving permutation is colamd and the factorization using this reordering can be visualized using the command q = colamd(A); [l, u, p] = lu(A(:,q)); spy(l+u).

Finally, Octave implicitly reorders the matrix when using the div (/) and ldiv (\) operators, and so no the user does not need to explicitly reorder the matrix to maximize performance.

p = ccolamd (s)                                    Loadable Function
p = ccolamd (s, knobs)                             Loadable Function
p = ccolamd (s, knobs, cmember)                    Loadable Function
[p, stats] = ccolamd (...)                         Loadable Function

Constrained column approximate minimum degree permutation. p = ccolamd (s) returns the column approximate minimum degree permutation vector for the sparse matrix s. For a non-symmetric matrix s, s (:, p) tends to have sparser LU factors than s. chol (s (:, p)' * s (:, p)) also tends to be sparser than chol (s' * s). p = ccolamd (s, 1) optimizes the ordering for lu (s (:, p)). The ordering is followed by a column elimination tree post-ordering.

knobs is an optional one- to five-element input vector, with a default value of [0 10 10 1 0] if not present or empty. Entries not present are set to their defaults.

knobs (1)
if nonzero, the ordering is optimized for lu (S (:, p)). It will be a poor ordering for chol (s (:, p)' * s (:, p)). This is the most important knob for ccolamd.

## Chapter 20: Sparse Matrices   303

knob(2)
: if s is m-by-n, rows with more than max (16, knobs (2) * sqrt (n)) entries are ignored.

knob(3)
: columns with more than max (16, knobs (3) * sqrt (min (m, n))) entries are ignored and ordered last in the output permutation (subject to the cmember constraints).

knob(4)
: if nonzero, aggressive absorption is performed.

knob(5)
: if nonzero, statistics and knobs are printed.

cmember is an optional vector of length n. It defines the constraints on the column ordering. If cmember (j) = c, then column j is in constraint set c (c must be in the range 1 to n). In the output permutation p, all columns in set 1 appear first, followed by all columns in set 2, and so on. cmember = ones(1,n) if not present or empty. ccolamd (s, [], 1 : n) returns 1 : n

p = ccolamd (s) is about the same as p = colamd (s). knobs and its default values differ. colamd always does aggressive absorption, and it finds an ordering suitable for both lu (s (:, p)) and chol (S (:, p)' * s (:, p)); it cannot optimize its ordering for lu (s (:, p)) to the extent that ccolamd (s, 1) can.

stats is an optional 20-element output vector that provides data about the ordering and the validity of the input matrix s. Ordering statistics are in stats (1 : 3). stats (1) and stats (2) are the number of dense or empty rows and columns ignored by CCOLAMD and stats (3) is the number of garbage collections performed on the internal data structure used by CCOLAMD (roughly of size 2.2 * nnz (s) + 4 * m + 7 * n integers).

stats (4 : 7) provide information if CCOLAMD was able to continue. The matrix is OK if stats (4) is zero, or 1 if invalid. stats (5) is the rightmost column index that is unsorted or contains duplicate entries, or zero if no such column exists. stats (6) is the last seen duplicate or out-of-order row index in the column index given by stats (5), or zero if no such row index exists. stats (7) is the number of duplicate or out-of-order row indices. stats (8 : 20) is always zero in the current version of CCOLAMD (reserved for future use).

The authors of the code itself are S. Larimore, T. Davis (Uni of Florida) and S. Rajamanickam in collaboration with J. Bilbert and E. Ng. Supported by the National Science Foundation (DMS-9504974, DMS-9803599, CCR-0203270), and a grant from Sandia National Lab.[5]

See also: colamd, csymamd.

---

[5] See http://www.cise.ufl.edu/research/sparse for ccolamd, csymamd, amd, colamd, symamd, and other related orderings.

p = colamd (s)                                              Loadable Function
p = colamd (s, knobs)                                       Loadable Function
[p, stats] = colamd (s)                                     Loadable Function
[p, stats] = colamd (s, knobs)                              Loadable Function

Column approximate minimum degree permutation. p = colamd (s) returns the column approximate minimum degree permutation vector for the sparse matrix s. For a non-symmetric matrix s, s (:,p) tends to have sparser LU factors than s. The Cholesky factorization of s (:,p)' * s (:,p) also tends to be sparser than that of s' * s.

knobs is an optional one- to three-element input vector. If s is m-by-n, then rows with more than max(16,knobs(1)*sqrt(n)) entries are ignored. Columns with more than max(16,knobs(2)*sqrt(min(m,n))) entries are removed prior to ordering, and ordered last in the output permutation p. Only completely dense rows or columns are removed if knobs (1) and knobs (2) are < 0, respectively. If knobs (3) is nonzero, stats and knobs are printed. The default is knobs = [10 10 0]. Note that knobs differs from earlier versions of colamd

stats is an optional 20-element output vector that provides data about the ordering and the validity of the input matrix s. Ordering statistics are in stats (1:3). stats (1) and stats (2) are the number of dense or empty rows and columns ignored by COLAMD and stats (3) is the number of garbage collections performed on the internal data structure used by COLAMD (roughly of size 2.2 * nnz(s) + 4 * m + 7 * n integers).

Octave built-in functions are intended to generate valid sparse matrices, with no duplicate entries, with ascending row indices of the nonzeros in each column, with a non-negative number of entries in each column (!) and so on. If a matrix is invalid, then COLAMD may or may not be able to continue. If there are duplicate entries (a row index appears two or more times in the same column) or if the row indices in a column are out of order, then COLAMD can correct these errors by ignoring the duplicate entries and sorting each column of its internal copy of the matrix s (the input matrix s is not repaired, however). If a matrix is invalid in other ways then COLAMD cannot continue, an error message is printed, and no output arguments (p or stats) are returned. COLAMD is thus a simple way to check a sparse matrix to see if it's valid.

stats (4:7) provide information if COLAMD was able to continue. The matrix is OK if stats (4) is zero, or 1 if invalid. stats (5) is the rightmost column index that is unsorted or contains duplicate entries, or zero if no such column exists. stats (6) is the last seen duplicate or out-of-order row index in the column index given by stats (5), or zero if no such row index exists. stats (7) is the number of duplicate or out-of-order row indices. stats (8:20) is always zero in the current version of COLAMD (reserved for future use).

The ordering is followed by a column elimination tree post-ordering.

The authors of the code itself are Stefan I. Larimore and Timothy A. Davis (davis@cise.ufl.edu), University of Florida. The algorithm was developed in

Chapter 20: Sparse Matrices                                                    305

collaboration with John Gilbert, Xerox PARC, and Esmond Ng, Oak Ridge
National Laboratory.[6]

See also: colperm, symamd.

p = colperm (s)                                                     Function File
Returns the column permutations such that the columns of s (:, p) are
ordered in terms of increase number of non-zero elements. If s is symmetric,
then p is chosen such that s (p, p) orders the rows and columns with
increasing number of non zeros elements.

p = csymamd (s)                                                Loadable Function
p = csymamd (s, knobs)                                         Loadable Function
p = csymamd (s, knobs, cmember)                                Loadable Function
[p, stats] = csymamd (...)                                     Loadable Function
For a symmetric positive definite matrix s, returns the permutation vector p
such that s (p,p) tends to have a sparser Cholesky factor than s. Sometimes
csymamd works well for symmetric indefinite matrices too. The matrix s is
assumed to be symmetric; only the strictly lower triangular part is referenced. s must be square. The ordering is followed by an elimination tree
post-ordering.

knobs is an optional one- to three-element input vector, with a default value
of [10 1 0] if present or empty. Entries not present are set to their defaults.

knobs (1)
   If s is n-by-n, then rows and columns with more than
   max(16,knobs(1)*sqrt(n)) entries are ignored, and ordered
   last in the output permutation (subject to the cmember constraints).

knobs (2)
   If nonzero, aggressive absorption is performed.

knobs (3)
   If nonzero, statistics and knobs are printed.

cmember is an optional vector of length n. It defines the constraints on the
ordering. If cmember(j) = s, then row/column j is in constraint set c (c
must be in the range 1 to n). In the output permutation p, rows/columns in
set 1 appear first, followed by all rows/columns in set 2, and so on. cmember
= ones(1,n) if not present or empty. csymamd(s,[],1:n) returns 1:n.

p = csymamd(s) is about the same as p = symamd(s). knobs and its default
values differ.

stats (4:7) provide information if CCOLAMD was able to continue. The
matrix is OK if stats (4) is zero, or 1 if invalid. stats (5) is the rightmost
column index that is unsorted or contains duplicate entries, or zero if no
such column exists. stats (6) is the last seen duplicate or out-of-order row
index in the column index given by stats (5), or zero if no such row index
exists. stats (7) is the number of duplicate or out-of-order row indices.

---

[6] See http://www.cise.ufl.edu/research/sparse/colamd.

*stats* (8:20) is always zero in the current version of CCOLAMD (reserved for future use).

The authors of the code itself are S. Larimore, T. Davis (Uni of Florida) and S. Rajamanickam in collaboration with J. Bilbert and E. Ng. Supported by the National Science Foundation (DMS-9504974, DMS-9803599, CCR-0203270), and a grant from Sandia National Lab.[7]

See also: symamd, ccolamd.

**p = dmperm (s)**     Loadable Function
**[p, q, r, s] = dmperm (s)**     Loadable Function

Perform a Dulmage-Mendelsohn permutation on the sparse matrix *s*. With a single output argument *dmperm* performs the row permutations *p* such that *s* (*p*,:) has no zero elements on the diagonal.

Called with two or more output arguments, returns the row and column permutations, such that *s* (*p*, *q*) is in block triangular form. The values of *r* and *s* define the boundaries of the blocks. If *s* is square then *r* == *s*.

The method used is described in: A. Pothen & C.-J. Fan. Computing the block triangular form of a sparse matrix. ACM Trans. Math. Software, 16(4):303-324, 1990.

See also: colamd, ccolamd.

**p = symamd (s)**     Loadable Function
**p = symamd (s, knobs)**     Loadable Function
**[p, stats] = symamd (s)**     Loadable Function
**[p, stats] = symamd (s, knobs)**     Loadable Function

For a symmetric positive definite matrix *s*, returns the permutation vector p such that *s* (*p*, *p*) tends to have a sparser Cholesky factor than *s*. Sometimes SYMAMD works well for symmetric indefinite matrices too. The matrix *s* is assumed to be symmetric; only the strictly lower triangular part is referenced. *s* must be square.

*knobs* is an optional one- to two-element input vector. If *s* is n-by-n, then rows and columns with more than max(16,*knobs*(1)*sqrt(n)) entries are removed prior to ordering, and ordered last in the output permutation *p*. No rows/columns are removed if *knobs* (1) < 0. If *knobs* (2) is nonzero, stats and *knobs* are printed. The default is *knobs* = [10 0]. Note that *knobs* differs from earlier versions of symamd.

*stats* is an optional 20-element output vector that provides data about the ordering and the validity of the input matrix *s*. Ordering statistics are in *stats* (1:3). *stats* (1) = *stats* (2) is the number of dense or empty rows and columns ignored by SYMAMD and *stats* (3) is the number of garbage collections performed on the internal data structure used by SYMAMD (roughly of size 8.4 * nnz (tril (*s*, -1)) + 9 * n integers).

---

[7] See http://www.cise.ufl.edu/research/sparse for ccolamd, csymamd, amd, colamd, symamd, and other related orderings.

Octave built-in functions are intended to generate valid sparse matrices, with no duplicate entries, with ascending row indices of the nonzeros in each column, with a non-negative number of entries in each column (!) and so on. If a matrix is invalid, then SYMAMD may or may not be able to continue. If there are duplicate entries (a row index appears two or more times in the same column) or if the row indices in a column are out of order, then SYMAMD can correct these errors by ignoring the duplicate entries and sorting each column of its internal copy of the matrix S (the input matrix S is not repaired, however). If a matrix is invalid in other ways then SYMAMD cannot continue, an error message is printed, and no output arguments (p or stats) are returned. SYMAMD is thus a simple way to check a sparse matrix to see if it's valid.

stats (4:7) provide information if SYMAMD was able to continue. The matrix is OK if stats (4) is zero, or 1 if invalid. stats (5) is the rightmost column index that is unsorted or contains duplicate entries, or zero if no such column exists. stats (6) is the last seen duplicate or out-of-order row index in the column index given by stats (5), or zero if no such row index exists. stats (7) is the number of duplicate or out-of-order row indices. stats (8:20) is always zero in the current version of SYMAMD (reserved for future use).

The ordering is followed by a column elimination tree post-ordering.

The authors of the code itself are Stefan I. Larimore and Timothy A. Davis (davis@cise.ufl.edu), University of Florida. The algorithm was developed in collaboration with John Gilbert, Xerox PARC, and Esmond Ng, Oak Ridge National Laboratory.[8]

See also: colperm, colamd.

*p* = symrcm (*S*) Loadable Function
Symmetric reverse Cuthill-McKee permutation of *S*. Return a permutation vector *p* such that *S* (*p*, *p*) tends to have its diagonal elements closer to the diagonal than *S*. This is a good preordering for LU or Cholesky factorization of matrices that come from 'long, skinny' problems. It works for both symmetric and asymmetric *S*.

The algorithm represents a heuristic approach to the NP-complete bandwidth minimization problem. The implementation is based in the descriptions found in

E. Cuthill, J. McKee: "Reducing the Bandwidth of Sparse Symmetric Matrices". Proceedings of the 24th ACM National Conference, 157-172 1969, Brandon Press, New Jersey.

Alan George, Joseph W. H. Liu: "Computer Solution of Large Sparse Positive Definite Systems", Prentice Hall Series in Computational Mathematics, ISBN 0-13-165274-5, 1981.

See also: colperm, colamd, symamd.

---

[8] See http://www.cise.ufl.edu/research/sparse/colamd.

## 20.2 Linear Algebra on Sparse Matrices

Octave includes a polymorphic solver for sparse matrices, where the exact solver used to factorize the matrix depends on the properties of the sparse matrix itself. Generally, the cost of determining the matrix type is small relative to the cost of factorizing the matrix, and the type is cached once it is calculated, so that it is not re-determined each time the matrix is used.

The selection tree for how the linear equation is solved is:

1. If the matrix is diagonal, solve directly and goto 8

2. If the matrix is a permuted diagonal, solve directly taking into account the permutations. Goto 8

3. If the matrix is square, banded and if the band density is less than that given by spparms ("bandden") continue, else goto 4.

    a. If the matrix is tridiagonal and the right-hand side is not sparse continue, else goto 3b.

        1. If the matrix is hermitian, with a positive real diagonal, attempt Cholesky factorization using LAPACK xPTSV.

        2. If the above failed or the matrix is not hermitian with a positive real diagonal use Gaussian elimination with pivoting using LAPACK xGTSV, and goto 8.

    b. If the matrix is hermitian with a positive real diagonal, attempt Cholesky factorization using LAPACK xPBTRF.

    c. if the above failed or the matrix is not hermitian with a positive real diagonal use Gaussian elimination with pivoting using LAPACK xGBTRF, and goto 8.

4. If the matrix is upper or lower triangular perform a sparse forward or backward substitution, and goto 8

5. If the matrix is a upper triangular matrix with column permutations or lower triangular matrix with row permutations, perform a sparse forward or backward substitution, and goto 8

6. If the matrix is square, hermitian with a real positive diagonal, attempt sparse Cholesky factorization using CHOLMOD.

7. If the sparse Cholesky factorization failed or the matrix is not hermitian with a real positive diagonal, and the matrix is square, factorize using UMFPACK.

8. If the matrix is not square, or any of the previous solvers flags a singular or near singular matrix, find a minimum norm solution using CXSPARSE[9].

---

[9] The CHOLMOD, UMFPACK and CXSPARSE packages were written by Tim Davis and are available at http://www.cise.ufl.edu/research/sparse/

# Chapter 20: Sparse Matrices

The band density is defined as the number of non-zero values in the matrix divided by the number of non-zero values in the matrix. The banded matrix solvers can be entirely disabled by using spparms to set bandden to 1 (i.e. spparms ("bandden", 1)).

The QR solver factorizes the problem with a Dulmage-Mendhelsohn decomposition, to separate the problem into blocks that can be treated as over-determined, multiple well determined blocks, and a final over-determined block. For matrices with blocks of strongly connected nodes this is a big win as LU decomposition can be used for many blocks. It also significantly improves the chance of finding a solution to over-determined problems rather than just returning a vector of NaN's.

All of the solvers above can calculate an estimate of the condition number. This can be used to detect numerical stability problems in the solution and force a minimum norm solution to be used. However, for narrow banded, triangular or diagonal matrices, the cost of calculating the condition number is significant, and can in fact exceed the cost of factoring the matrix. Therefore the condition number is not calculated in these cases, and Octave relies on simpler techniques to detect singular matrices or the underlying LAPACK code in the case of banded matrices.

The user can force the type of the matrix with the matrix_type function. This overcomes the cost of discovering the type of the matrix. However, it should be noted that identifying the type of the matrix incorrectly will lead to unpredictable results, and so matrix_type should be used with care.

[n, c] = normest (a, *tol*)   Function File

Estimate the 2-norm of the matrix a using a power series analysis. This is typically used for large matrices, where the cost of calculating the norm (a) is prohibitive and an approximation to the 2-norm is acceptable.

*tol* is the tolerance to which the 2-norm is calculated. By default *tol* is 1e-6.

c returns the number of iterations needed for normest to converge.

[est, v] = condest (a, t)   Function File
[est, v] = condest (a, *solve*, *solve_t*, t)   Function File
[est, v] = condest (*apply*, *apply_t*, *solve*, *solve_t*, n, t)   Function File

Estimate the 1-norm condition number of a matrix $A$ using $t$ test vectors using a randomized 1-norm estimator. If $t$ exceeds 5, then only 5 test vectors are used.

If the matrix is not explicit, e.g. when estimating the condition number of a given an LU factorization, condest uses the following functions:

*apply*
    A*x for a matrix x of size $n$ by $t$.

*apply_t*
    A'*x for a matrix x of size $n$ by $t$.

*solve*
    A \ b for a matrix b of size $n$ by $t$.

*solve_t*
: A' \ b for a matrix b of size *n* by *t*.

The implicit version requires an explicit dimension *n*.

condest uses a randomized algorithm to approximate the 1-norms.

condest returns the 1-norm condition estimate *est* and a vector *v* satisfying norm (A*v, 1) == norm (A, 1) * norm (v, 1) / *est*. When *est* is large, *v* is an approximate null vector.

References:

- Nicholas J. Higham and Françoise Tisseur, "A Block Algorithm for Matrix 1-Norm Estimation, with an Application to 1-Norm Pseudospectra." SIMAX vol 21, no 4, pp 1185-1201.[10]

See also: norm, cond, onenormest.

| | |
|---|---|
| spparms () | Loadable Function |
| *vals* = spparms () | Loadable Function |
| [*keys*, *vals*] = spparms () | Loadable Function |
| *val* = spparms (*key*) | Loadable Function |
| spparms (*vals*) | Loadable Function |
| spparms ('defaults') | Loadable Function |
| spparms ('tight') | Loadable Function |
| spparms (*key*, *val*) | Loadable Function |

Sets or displays the parameters used by the sparse solvers and factorization functions. The first four calls above get information about the current settings, while the others change the current settings. The parameters are stored as pairs of keys and values, where the values are all floats and the keys are one of the following strings:

spumoni
: Printing level of debugging information of the solvers (default 0)

ths_rel
: Included for compatibility. Not used. (default 1)

ths_abs
: Included for compatibility. Not used. (default 1)

exact_d
: Included for compatibility. Not used. (default 0)

supernd
: Included for compatibility. Not used. (default 3)

rreduce
: Included for compatibility. Not used. (default 3)

wh_frac
: Included for compatibility. Not used. (default 0.5)

---

[10] http://citeseer.ist.psu.edu/223007.html

**autommd**
> Flag whether the LU/QR and the \ and / operators will automatically use the sparsity preserving mmd functions (default 1)

**autoamd**
> Flag whether the LU and the \ and / operators will automatically use the sparsity preserving amd functions (default 1)

**piv_tol**
> The pivot tolerance of the UMFPACK solvers (default 0.1)

**bandden**
> Band density value for polymorphic solver (default 0.5)

**umfpack**
> Flag whether the UMFPACK or mmd solvers are used for the LU, \ and / operations (default 1)

The value of individual keys can be set with spparms (*key, val*). The default values can be restored with the special keyword 'defaults'. The special keyword 'tight' can be used to set the mmd solvers to attempt for a sparser solution at the potential cost of longer running time.

*p* = sprank (*s*)       Loadable Function
Calculates the structural rank of a sparse matrix *s*. Note that only the structure of the matrix is used in this calculation based on a Dulmage-Mendelsohn permutation to block triangular form. As such the numerical rank of the matrix *s* is bounded by sprank (*s*) >= rank (*s*). Ignoring floating point errors sprank (*s*) == rank (*s*).

**See also:** dmperm.

[*count, h, parent, post, r*] = symbfact (*s, typ, mode*)       Loadable Function
Performs a symbolic factorization analysis on the sparse matrix *s*. Where

*s*    *s* is a complex or real sparse matrix.

*typ*
> Is the type of the factorization and can be one of

> **sym**
>> Factorize *s*. This is the default.

> **col**
>> Factorize *s'* * *s*.

> **row**
>> Factorize *s* * *s'*.

> **lo**    Factorize *s'*

*mode*
: The default is to return the Cholesky factorization for *r*, and if *mode* is 'L', the conjugate transpose of the Cholesky factorization is returned. The conjugate transpose version is faster and uses less memory, but returns the same values for *count*, *h*, *parent* and *post* outputs.

The output variables are

*count*
: The row counts of the Cholesky factorization as determined by *typ*.

*h*  The height of the elimination tree.

*parent*
: The elimination tree itself.

*post*
: A sparse boolean matrix whose structure is that of the Cholesky factorization as determined by *typ*.

## 20.3 Iterative Techniques applied to sparse matrices

The left division \ and right division / operators, discussed in the previous section, use direct solvers to resolve a linear equation of the form x = A \ b or x = b / A. Octave equally includes a number of functions to solve sparse linear equations using iterative techniques.

x = pcg (a, b, tol, maxit, m1, m2, x0, ...)   Function File
[x, flag, relres, iter, resvec, eigest] = pcg (...)   Function File
: Solves the linear system of equations a * x = b by means of the Preconditioned Conjugate Gradient iterative method. The input arguments are

  - a can be either a square (preferably sparse) matrix or a function handle, inline function or string containing the name of a function which computes a * x. In principle a should be symmetric and positive definite; if pcg finds a to not be positive definite, you will get a warning message and the *flag* output parameter will be set.

  - b is the right hand side vector.

  - *tol* is the required relative tolerance for the residual error, b - a * x. The iteration stops if norm (b - a * x) <= *tol* * norm (b - a * x0). If *tol* is empty or is omitted, the function sets tol = 1e-6 by default.

  - maxit is the maximum allowable number of iterations; if [] is supplied for maxit, or pcg has less arguments, a default value equal to 20 is used.

  - m = m1 * m2 is the (left) preconditioning matrix, so that the iteration is (theoretically) equivalent to solving by pcg P * x = m \ b, with P = m \ a. Note that a proper choice of the preconditioner may dramatically improve the overall performance of the method. Instead of matrices m1 and m2, the user may pass two functions which return the results of applying the inverse of m1 and m2 to a vector (usually this is the preferred way of using the preconditioner). If [] is supplied for m1, or

# Chapter 20: Sparse Matrices

*m1* is omitted, no preconditioning is applied. If *m2* is omitted, $m = m1$ will be used as preconditioner.

- *x0* is the initial guess. If *x0* is empty or omitted, the function sets *x0* to a zero vector by default.

The arguments which follow *x0* are treated as parameters, and passed in a proper way to any of the functions (a or m) which are passed to pcg. See the examples below for further details. The output arguments are

- *x* is the computed approximation to the solution of a * x = b.
- *flag* reports on the convergence. `flag` = 0 means the solution converged and the tolerance criterion given by *tol* is satisfied. `flag` = 1 means that the *maxit* limit for the iteration count was reached. `flag` = 3 reports that the (preconditioned) matrix was found not positive definite.
- *relres* is the ratio of the final residual to its initial value, measured in the Euclidean norm.
- *iter* is the actual number of iterations performed.
- *resvec* describes the convergence history of the method. `resvec (i,1)` is the Euclidean norm of the residual, and `resvec (i,2)` is the preconditioned residual norm, after the (*i*-1)-th iteration, $i = 1, 2, \ldots,$ `iter`+1. The preconditioned residual norm is defined as norm $(r)$ ^ 2 = r' * (m \ r) where r = b - a * x, see also the description of *m*. If *eigest* is not required, only `resvec (:,1)` is returned.
- *eigest* returns the estimate for the smallest `eigest (1)` and largest `eigest (2)` eigenvalues of the preconditioned matrix $P = m \setminus a$. In particular, if no preconditioning is used, the estimates for the extreme eigenvalues of a are returned. `eigest (1)` is an overestimate and `eigest (2)` is an underestimate, so that `eigest (2)` / `eigest (1)` is a lower bound for cond (P, 2), which nevertheless in the limit should theoretically be equal to the actual value of the condition number. The method which computes *eigest* works only for symmetric positive definite a and m, and the user is responsible for verifying this assumption.

Let us consider a trivial problem with a diagonal matrix (we exploit the sparsity of A)

```
N = 10;
A = spdiag ([1:N]);
b = rand (N, 1);
[L, U, P, Q] = luinc (A,1.e-3);
```

EXAMPLE 1: Simplest use of pcg

```
x = pcg(A,b)
```

EXAMPLE 2: pcg with a function which computes a * x

```
function y = applyA (x)
  y = [1:N]'.*x;
endfunction

x = pcg ("applyA", b)
```

EXAMPLE 3: pcg with a preconditioner: $l * u$

```
x=pcg(A,b,1.e-6,500,L*U);
```

EXAMPLE 4: pcg with a preconditioner: $l * u$. Faster than EXAMPLE 3 since lower and upper triangular matrices are easier to invert

```
x=pcg(A,b,1.e-6,500,L,U);
```

EXAMPLE 5: Preconditioned iteration, with full diagnostics. The preconditioner (quite strange, because even the original matrix a is trivial) is defined as a function

```
function y = applyM(x)
  K = floor (length (x) - 2);
  y = x;
  y(1:K) = x(1:K)./[1:K]';
endfunction

[x, flag, relres, iter, resvec, eigest] = ...
              pcg (A, b, [], [], "applyM");
semilogy (1:iter+1, resvec);
```

EXAMPLE 6: Finally, a preconditioner which depends on a parameter $k$.

```
function y = applyM (x, varargin)
K = varargin{1};
y = x;
y(1:K) = x(1:K)./[1:K]';
endfunction

[x, flag, relres, iter, resvec, eigest] = ...
       pcg (A, b, [], [], "applyM", [], [], 3)
```

REFERENCES

[1] C.T.Kelley, "Iterative methods for linear and nonlinear equations", SIAM, 1995 (the base PCG algorithm)

[2] Y.Saad, "Iterative methods for sparse linear systems", PWS 1996 (condition number estimate from PCG). A revised version of this book is available online at http://www-users.cs.umn.edu/~saad/books.html.

See also: sparse, pcr.

x = pcr (a, b, tol, maxit, m, x0, ...)     Function File
[x, flag, relres, iter, resvec] = pcr (...)     Function File

Solves the linear system of equations $a * x = b$ by means of the Preconditioned Conjugate Residuals iterative method. The input arguments are

# Chapter 20: Sparse Matrices

- a can be either a square (preferably sparse) matrix or a function handle, inline function or string containing the name of a function which computes a * x. In principle a should be symmetric and non-singular; if pcr finds a to be numerically singular, you will get a warning message and the flag output parameter will be set.
- b is the right hand side vector.
- tol is the required relative tolerance for the residual error, b - a * x. The iteration stops if norm (b - a * x) <= tol * norm (b - a * x0). If tol is empty or is omitted, the function sets tol = 1e-6 by default.
- maxit is the maximum allowable number of iterations; if [] is supplied for maxit, or pcr has less arguments, a default value equal to 20 is used.
- m is the (left) preconditioning matrix, so that the iteration is (theoretically) equivalent to solving by pcr $P * x = m \setminus b$, with $P = m \setminus a$. Note that a proper choice of the preconditioner may dramatically improve the overall performance of the method. Instead of matrix m, the user may pass a function which returns the results of applying the inverse of m to a vector (usually this is the preferred way of using the preconditioner). If [] is supplied for m, or m is omitted, no preconditioning is applied.
- x0 is the initial guess. If x0 is empty or omitted, the function sets x0 to a zero vector by default.

The arguments which follow x0 are treated as parameters, and passed in a proper way to any of the functions (a or m) which are passed to pcr. See the examples below for further details. The output arguments are

- x is the computed approximation to the solution of a * x = b.
- flag reports on the convergence. flag = 0 means the solution converged and the tolerance criterion given by tol is satisfied. flag = 1 means that the maxit limit for the iteration count was reached. flag = 3 reports t pcr breakdown, see [1] for details.
- relres is the ratio of the final residual to its initial value, measured in the Euclidean norm.
- iter is the actual number of iterations performed.
- resvec describes the convergence history of the method, so that resvec (i) contains the Euclidean norms of the residual after the ($i$-1)-th iteration, $i$ = 1,2, ..., iter+1.

Let us consider a trivial problem with a diagonal matrix (we exploit the sparsity of A)

```
N = 10;
A = diag([1:N]); A = sparse(A);
b = rand(N,1);
```

EXAMPLE 1: Simplest use of pcr

```
x = pcr(A, b)
```

EXAMPLE 2: pcr with a function which computes a * x.

```
function y = applyA(x)
  y = [1:10]'.*x;
endfunction

x = pcr('applyA',b)
```

EXAMPLE 3: Preconditioned iteration, with full diagnostics. The preconditioner (quite strange, because even the original matrix a is trivial) is defined as a function

```
function y = applyM(x)
  K = floor(length(x)-2);
  y = x;
  y(1:K) = x(1:K)./[1:K]';
endfunction

[x, flag, relres, iter, resvec] = ...
                   pcr(A, b, [], [], 'applyM')
semilogy([1:iter+1], resvec);
```

EXAMPLE 4: Finally, a preconditioner which depends on a parameter k.

```
function y = applyM(x, varargin)
  K = varargin{1};
  y = x; y(1:K) = x(1:K)./[1:K]';
endfunction

[x, flag, relres, iter, resvec] = ...
                   pcr(A, b, [], [], 'applyM', [], 3)
```

REFERENCES

[1] W. Hackbusch, "Iterative Solution of Large Sparse Systems of Equations", section 9.5.4; Springer, 1994

**See also:** sparse, pcg.

The speed with which an iterative solver converges to a solution can be accelerated with the use of a pre-conditioning matrix $M$. In this case the linear equation $M^{-1} * x = M^{-1} * A \setminus b$ is solved instead. Typical pre-conditioning matrices are partial factorizations of the original matrix.

| | |
|---|---|
| [l, u, p, q] = luinc (a, '0') | Loadable Function |
| [l, u, p, q] = luinc (a, *droptol*) | Loadable Function |
| [l, u, p, q] = luinc (a, *opts*) | Loadable Function |

Produce the incomplete LU factorization of the sparse matrix a. Two types of incomplete factorization are possible, and the type is determined by the second argument to *luinc*.

Called with a second argument of '0', the zero-level incomplete LU factorization is produced. This creates a factorization of a where the position of the non-zero arguments correspond to the same positions as in the matrix a.

Alternatively, the fill-in of the incomplete LU factorization can be controlled through the variable *droptol* or the structure *opts*. The UMFPACK multifrontal factorization code by Tim A. Davis is used for the incomplete LU factorization.[11]

*droptol* determines the values below which the values in the LU factorization are dropped and replaced by zero. It must be a positive scalar, and any values in the factorization whose absolute value are less than this value are dropped, expect if leaving them increase the sparsity of the matrix. Setting *droptol* to zero results in a complete LU factorization which is the default.

*opts* is a structure containing one or more of the fields

droptol
: The drop tolerance as above. If *opts* only contains droptol then this is equivalent to using the variable *droptol*.

milu
: A logical variable flagging whether to use the modified incomplete LU factorization. In the case that milu is true, the dropped values are subtracted from the diagonal of the matrix U of the factorization. The default is false.

udiag
: A logical variable that flags whether zero elements on the diagonal of U should be replaced with *droptol* to attempt to avoid singular factors. The default is false.

thresh
: Defines the pivot threshold in the interval [0,1]. Values outside that range are ignored.

All other fields in *opts* are ignored. The outputs from *luinc* are the same as for *lu*.

See also: sparse, lu, cholinc.

---

[11] http://www.cise.ufl.edu/research/sparse/umfpack/

## 20.4 Real Life Example of the use of Sparse Matrices

A common application for sparse matrices is in the solution of Finite Element Models. Finite element models allow numerical solution of partial differential equations that do not have closed form solutions, typically because of the complex shape of the domain.

In order to motivate this application, we consider the boundary value Laplace equation. This system can model scalar potential fields, such as heat or electrical potential. Given a medium $\Omega$ with boundary $\partial\Omega$. At all points on the $\partial\Omega$ the boundary conditions are known, and we wish to calculate the potential in $\Omega$. Boundary conditions may specify the potential (Dirichlet boundary condition), its normal derivative across the boundary (Neumann boundary condition), or a weighted sum of the potential and its derivative (Cauchy boundary condition).

In a thermal model, we want to calculate the temperature in $\Omega$ and know the boundary temperature (Dirichlet condition) or heat flux (from which we can calculate the Neumann condition by dividing by the thermal conductivity at the boundary). Similarly, in an electrical model, we want to calculate the voltage in $\Omega$ and know the boundary voltage (Dirichlet) or current (Neumann condition after diving by the electrical conductivity). In an electrical model, it is common for much of the boundary to be electrically isolated; this is a Neumann boundary condition with the current equal to zero.

The simplest finite element models will divide $\Omega$ into simplexes (triangles in 2D, pyramids in 3D). We take as a 3D example a cylindrical liquid filled tank with a small non-conductive ball from the EIDORS project[12]. This is model is designed to reflect an application of electrical impedance tomography, where current patterns are applied to such a tank in order to image the internal conductivity distribution. In order to describe the FEM geometry, we have a matrix of vertices nodes and simplices elems.

The following example creates a simple rectangular 2D electrically conductive medium with 10 V and 20 V imposed on opposite sides (Dirichlet boundary conditions). All other edges are electrically isolated.

```
node_y= [1;1.2;1.5;1.8;2]*ones(1,11);
node_x= ones(5,1)*[1,1.05,1.1,1.2, ...
         1.3,1.5,1.7,1.8,1.9,1.95,2];
nodes= [node_x(:), node_y(:)];

[h,w]= size(node_x);
elems= [];
for idx= 1:w-1
  widx= (idx-1)*h;
  elems= [elems; ...
    widx+[(1:h-1);(2:h);h+(1:h-1)]'; ...
    widx+[(2:h);h+(2:h);h+(1:h-1)]' ];
endfor
```

---

[12] EIDORS - Electrical Impedance Tomography and Diffuse optical Tomography Reconstruction Software http://eidors3d.sourceforge.net

# Chapter 20: Sparse Matrices

```
E= size(elems,1); # No. of simplices
N= size(nodes,1); # No. of vertices
D= size(elems,2); # dimensions+1
```

This creates a N-by-2 matrix nodes and a E-by-3 matrix elems with values, which define finite element triangles:

```
nodes(1:7,:)'
   1.00  1.00  1.00  1.00  1.00  1.05  1.05 ...
   1.00  1.20  1.50  1.80  2.00  1.00  1.20 ...

elems(1:7,:)'
   1     2     3     4     2     3     4 ...
   2     3     4     5     7     8     9 ...
   6     7     8     9     6     7     8 ...
```

Using a first order FEM, we approximate the electrical conductivity distribution in $\Omega$ as constant on each simplex (represented by the vector conductivity). Based on the finite element geometry, we first calculate a system (or stiffness) matrix for each simplex (represented as 3-by-3 elements on the diagonal of the element-wise system matrix SE. Based on SE and a N-by-DE connectivity matrix C, representing the connections between simplices and vertices, the global connectivity matrix S is calculated.

```
# Element conductivity
conductivity= [1*ones(1,16), ...
     2*ones(1,48), 1*ones(1,16)];

# Connectivity matrix
C = sparse ((1:D*E), reshape (elems', ...
       D*E, 1), 1, D*E, N);

# Calculate system matrix
Siidx = floor ([0:D*E-1]'/D) * D * ...
       ones(1,D) + ones(D*E,1)*(1:D) ;
Sjidx = [1:D*E]'*ones(1,D);
Sdata = zeros(D*E,D);
dfact = factorial(D-1);
for j=1:E
   a = inv([ones(D,1), ...
       nodes(elems(j,:), :)]);
   const = conductivity(j) * 2 / ...
       dfact / abs(det(a));
   Sdata(D*(j-1)+(1:D),:) = const * ...
       a(2:D,:)' * a(2:D,:);
endfor
# Element-wise system matrix
SE= sparse(Siidx,Sjidx,Sdata);
# Global system matrix
S= C'* SE *C;
```

The system matrix acts like the conductivity $S$ in Ohm's law $SV = I$. Based on the Dirichlet and Neumann boundary conditions, we are able to solve for the voltages at each vertex V.

```
# Dirichlet boundary conditions
D_nodes=[1:5, 51:55];
D_value=[10*ones(1,5), 20*ones(1,5)];

V= zeros(N,1);
V(D_nodes) = D_value;
idx = 1:N; # vertices without Dirichlet
           # boundary condns
idx(D_nodes) = [];

# Neumann boundary conditions. Note that
# N_value must be normalized by the
# boundary length and element conductivity
N_nodes=[];
N_value=[];

Q = zeros(N,1);
Q(N_nodes) = N_value;

V(idx) = S(idx,idx) \ ( Q(idx) - ...
         S(idx,D_nodes) * V(D_nodes));
```

Finally, in order to display the solution, we show each solved voltage value in the z-axis for each simplex vertex. See Figure 20.6.

```
elemx = elems(:,[1,2,3,1])';
xelems = reshape (nodes(elemx, 1), 4, E);
yelems = reshape (nodes(elemx, 2), 4, E);
velems = reshape (V(elemx), 4, E);
plot3 (xelems,yelems,velems,'k');
print ('grid.eps');
```

Chapter 20: Sparse Matrices 321

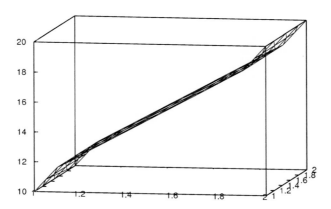

Figure 20.6: Example finite element model the showing triangular elements. The height of each vertex corresponds to the solution value.

# 21 Numerical Integration

Octave comes with several built-in functions for computing the integral of a function numerically. These functions all solve 1-dimensional integration problems.

## 21.1 Functions of One Variable

Octave supports three different algorithms for computing the integral

$$\int_a^b f(x)dx$$

of a function $f$ over the interval from $a$ to $b$. These are

quad
    Numerical integration based on Gaussian quadrature.

quadl
    Numerical integration using an adaptive Lobatto rule.

trapz
    Numerical integration using the trapezoidal method.

Besides these functions Octave also allows you to perform cumulative numerical integration using the trapezoidal method through the cumtrapz function.

[v, ier, nfun, err] = quad (f, a, b, tol, sing)        Loadable Function
    Integrate a nonlinear function of one variable using Quadpack. The first argument is the name of the function, the function handle or the inline function to call to compute the value of the integrand. It must have the form

        y = f (x)

    where y and x are scalars.

    The second and third arguments are limits of integration. Either or both may be infinite.

    The optional argument *tol* is a vector that specifies the desired accuracy of the result. The first element of the vector is the desired absolute tolerance, and the second element is the desired relative tolerance. To choose a relative test only, set the absolute tolerance to zero. To choose an absolute test only, set the relative tolerance to zero.

    The optional argument *sing* is a vector of values at which the integrand is known to be singular.

    The result of the integration is returned in *v* and *ier* contains an integer error code (0 indicates a successful integration). The value of *nfun* indicates how many function evaluations were required, and *err* contains an estimate of the error in the solution.

    You can use the function quad_options to set optional parameters for quad. It should be noted that since quad is written in Fortran it cannot be called recursively.

quad_options (*opt, val*)                                Loadable Function
When called with two arguments, this function allows you to set options parameters for the function quad. Given one argument, quad_options returns the value of the corresponding option. If no arguments are supplied, the names of all the available options and their current values are displayed.
Options include

"absolute tolerance"
: Absolute tolerance; may be zero for pure relative error test.

"relative tolerance"
: Nonnegative relative tolerance. If the absolute tolerance is zero, the relative tolerance must be greater than or equal to max (50*eps, 0.5e-28).

Here is an example of using quad to integrate the function

$$f(x) = x\sin(1/x)\sqrt{|1-x|}$$

from $x = 0$ to $x = 3$.

This is a fairly difficult integration (plot the function over the range of integration to see why).

The first step is to define the function:

```
function y = f (x)
  y = x .* sin (1 ./ x) .* sqrt (abs (1 - x));
endfunction
```

Note the use of the 'dot' forms of the operators. This is not necessary for the call to quad, but it makes it much easier to generate a set of points for plotting (because it makes it possible to call the function with a vector argument to produce a vector result).

Then we simply call quad:

```
[v, ier, nfun, err] = quad ("f", 0, 3)
    ⇒ 1.9819
    ⇒ 1
    ⇒ 5061
    ⇒ 1.1522e-07
```

Although quad returns a nonzero value for *ier*, the result is reasonably accurate (to see why, examine what happens to the result if you move the lower bound to 0.1, then 0.01, then 0.001, etc.).

q = quadl (f, a, b)                                                Function File
q = quadl (f, a, b, *tol*)                                         Function File
q = quadl (f, a, b, *tol, trace*)                                  Function File
q = quadl (f, a, b, *tol, trace, p1, p2, ...*)                     Function File
Numerically evaluate integral using adaptive Lobatto rule. quadl (f, a, b) approximates the integral of f(x) to machine precision. f is either a function handle, inline function or string containing the name of the function

to evaluate. The function *f* must return a vector of output values if given a vector of input values.

If defined, *tol* defines the relative tolerance to which to which to integrate f(x). While if *trace* is defined, displays the left end point of the current interval, the interval length, and the partial integral.

Additional arguments *p1*, etc, are passed directly to *f*. To use default values for *tol* and *trace*, one may pass empty matrices.

Reference: W. Gander and W. Gautschi, *Adaptive Quadrature—Revisited*, BIT Vol. 40, No. 1, March 2000, pp. 84–101.[1]

z = trapz (y)      Function File
z = trapz (x, y)      Function File
z = trapz (..., dim)      Function File

Numerical integration using trapezoidal method. trapz (y) computes the integral of the y along the first non-singleton dimension. If the argument x is omitted a equally spaced vector is assumed. trapz (x, y) evaluates the integral with respect to x.

See also: cumtrapz.

z = cumtrapz (y)      Function File
z = cumtrapz (x, y)      Function File
z = cumtrapz (..., dim)      Function File

Cumulative numerical integration using trapezoidal method. cumtrapz (y) computes the cumulative integral of the y along the first non-singleton dimension. If the argument x is omitted a equally spaced vector is assumed. cumtrapz (x, y) evaluates the cumulative integral with respect to x.

See also: trapz,cumsum.

## 21.2 Orthogonal Collocation

[r, amat, bmat, q] = colloc (n, "left", "right")      Loadable Function

Compute derivative and integral weight matrices for orthogonal collocation using the subroutines given in J. Villadsen and M. L. Michelsen, *Solution of Differential Equation Models by Polynomial Approximation*.

Here is an example of using colloc to generate weight matrices for solving the second order differential equation $u' - \alpha u'' = 0$ with the boundary conditions $u(0) = 0$ and $u(1) = 1$.

First, we can generate the weight matrices for *n* points (including the endpoints of the interval), and incorporate the boundary conditions in the right hand side (for a specific value of $\alpha$).

---

[1] http://www.inf.ethz.ch/personal/gander/

```
n = 7;
alpha = 0.1;
[r, a, b] = colloc (n-2, "left", "right");
at = a(2:n-1,2:n-1);
bt = b(2:n-1,2:n-1);
rhs = alpha * b(2:n-1,n) - a(2:n-1,n);
```
Then the solution at the roots r is
```
u = [ 0; (at - alpha * bt) \ rhs; 1]
  ⇒ [ 0.00; 0.004; 0.01 0.00; 0.12; 0.62; 1.00 ]
```

## 21.3 Functions of Multiple Variables

Octave does not have built-in functions for computing the integral of functions of multiple variables. It is however possible to compute the integral of a function of multiple variables using the functions for one-dimensional integrals.

To illustrate how the integration can be performed, we will integrate the function
$$f(x,y) = \sin(\pi x y)\sqrt{xy}$$
for $x$ and $y$ between 0 and 1.

The first approach creates a function that integrates $f$ with respect to $x$, and then integrates that function with respect to $y$. Since quad is written in Fortran it cannot be called recursively. This means that quad cannot integrate a function that calls quad, and hence cannot be used to perform the double integration. It is however possible with quadl, which is what the following code does.

```
function I = g(y)
  I = ones(1, length(y));
  for i = 1:length(y)
    f = @(x) sin(pi.*x.*y(i)).*sqrt(x.*y(i));
    I(i) = quadl(f, 0, 1);
  endfor
endfunction

I = quadl("g", 0, 1)
  ⇒ 0.30022
```

The above mentioned approach works but is fairly slow, and that problem increases exponentially with the dimensionality the problem. Another possible solution is to use Orthogonal Collocation as described in the previous section. The integral of a function $f(x,y)$ for $x$ and $y$ between 0 and 1 can be approximated using $n$ points by

$$\int_0^1 \int_0^1 f(x,y)dxdy \approx \sum_{i=1}^n \sum_{j=1}^n q_i q_j f(r_i, r_j),$$

where $q$ and $r$ is as returned by colloc(n). The generalisation to more than two variables is straight forward. The following code computes the studied integral using $n = 7$ points.

```
f = @(x,y) sin(pi*x*y').*sqrt(x*y');
n = 7;
[t, A, B, q] = colloc(n);
I = q'*f(t,t)*q;
    ⇒ 0.30022
```
It should be noted that the number of points determines the quality of the approximation. If the integration needs to be performed between $a$ and $b$ instead of 0 and 1, a change of variables is needed.

# 22 Differential Equations

Octave has built-in functions for solving ordinary differential equations, and differential-algebraic equations. All solvers are based on reliable ODE routines written in Fortran.

## 22.1 Ordinary Differential Equations

The function lsode can be used to solve ODEs of the form

$$\frac{dx}{dt} = f(x,t)$$

using Hindmarsh's ODE solver LSODE.

[x, istate, msg] = lsode (fcn, x_0, t, t_crit)                    Loadable Function
  Solve the set of differential equations

$$\frac{dx}{dt} = f(x,t)$$

with

$$x(t_0) = x_0$$

The solution is returned in the matrix x, with each row corresponding to an element of the vector t. The first element of t should be $t_0$ and should correspond to the initial state of the system x_0, so that the first row of the output is x_0.

The first argument, fcn, is a string, or cell array of strings, inline or function handles, that names the function to call to compute the vector of right hand sides for the set of equations. The function must have the form

    xdot = f (x, t)

in which xdot and x are vectors and t is a scalar.

If fcn is a two-element string array, the first element names the function f described above, and the second element names a function to compute the Jacobian of f. The Jacobian function must have the form

    jac = j (x, t)

in which jac is the matrix of partial derivatives

$$J = \frac{\partial f_i}{\partial x_j} = \begin{bmatrix} \frac{\partial f_1}{\partial x_1} & \frac{\partial f_1}{\partial x_2} & \cdots & \frac{\partial f_1}{\partial x_N} \\ \frac{\partial f_2}{\partial x_1} & \frac{\partial f_2}{\partial x_2} & \cdots & \frac{\partial f_2}{\partial x_N} \\ \vdots & \vdots & \ddots & \vdots \\ \frac{\partial f_M}{\partial x_1} & \frac{\partial f_M}{\partial x_2} & \cdots & \frac{\partial f_M}{\partial x_N} \end{bmatrix}$$

The second and third arguments specify the initial state of the system, $x_0$, and the initial value of the independent variable $t_0$.

The fourth argument is optional, and may be used to specify a set of times that the ODE solver should not integrate past. It is useful for avoiding difficulties with singularities and points where there is a discontinuity in the derivative.

After a successful computation, the value of *istate* will be 2 (consistent with the Fortran version of LSODE).

If the computation is not successful, *istate* will be something other than 2 and *msg* will contain additional information.

You can use the function lsode_options to set optional parameters for lsode.

**See also:** daspk, dassl, dasrt.

lsode_options (*opt, val*)                                                     Loadable Function

When called with two arguments, this function allows you to set options parameters for the function lsode. Given one argument, lsode_options returns the value of the corresponding option. If no arguments are supplied, the names of all the available options and their current values are displayed.

Options include

"absolute tolerance"
: Absolute tolerance. May be either vector or scalar. If a vector, it must match the dimension of the state vector.

"relative tolerance"
: Relative tolerance parameter. Unlike the absolute tolerance, this parameter may only be a scalar.

    The local error test applied at each integration step is

    ```
    abs (local error in x(i)) <= ...
         rtol * abs (y(i)) + atol(i)
    ```

"integration method"
: A string specifying the method of integration to use to solve the ODE system. Valid values are

    "adams"
    "non-stiff"
    : No Jacobian used (even if it is available).

    "bdf"
    "stiff"
    : Use stiff backward differentiation formula (BDF) method. If a function to compute the Jacobian is not supplied, lsode will compute a finite difference approximation of the Jacobian matrix.

"initial step size"
: The step size to be attempted on the first step (default is determined automatically).

Chapter 22: Differential Equations                                            331

"maximum order"
: Restrict the maximum order of the solution method. If using the Adams method, this option must be between 1 and 12. Otherwise, it must be between 1 and 5, inclusive.

"maximum step size"
: Setting the maximum stepsize will avoid passing over very large regions (default is not specified).

"minimum step size"
: The minimum absolute step size allowed (default is 0).

"step limit"
: Maximum number of steps allowed (default is 100000).

Here is an example of solving a set of three differential equations using lsode. Given the function

```
function xdot = f (x, t)

xdot = zeros (3,1);

xdot(1) = 77.27 * (x(2) - x(1)*x(2) + x(1) \
          - 8.375e-06*x(1)^2);
xdot(2) = (x(3) - x(1)*x(2) - x(2)) / 77.27;
xdot(3) = 0.161*(x(1) - x(3));

endfunction
```

and the initial condition x0 = [ 4; 1.1; 4 ], the set of equations can be integrated using the command

```
t = linspace (0, 500, 1000);
```

```
y = lsode ("f", x0, t);
```

If you try this, you will see that the value of the result changes dramatically between $t = 0$ and 5, and again around $t = 305$. A more efficient set of output points might be

```
t = [0, logspace (-1, log10(303), 150), \
     logspace (log10(304), log10(500), 150)];
```

See Alan C. Hindmarsh, *ODEPACK, A Systematized Collection of ODE Solvers*, in Scientific Computing, R. S. Stepleman, editor, (1983) for more information about the inner workings of lsode.

## 22.2 Differential-Algebraic Equations

The function daspk can be used to solve DAEs of the form

$$0 = f(\dot{x}, x, t), \qquad x(t=0) = x_0, \dot{x}(t=0) = \dot{x}_0$$

where $\dot{x} = \frac{dx}{dt}$ is the derivative of $x$. The equation is solved using Petzold's DAE solver DASPK.

[x, xdot, istate, msg] = daspk (fcn, x_0, xdot_0, t,     Loadable Function
    t_crit)
Solve the set of differential-algebraic equations

$$0 = f(x, \dot{x}, t)$$

with

$$x(t_0) = x_0, \dot{x}(t_0) = \dot{x}_0$$

The solution is returned in the matrices x and xdot, with each row in the result matrices corresponding to one of the elements in the vector t. The first element of t should be $t_0$ and correspond to the initial state of the system x_0 and its derivative xdot_0, so that the first row of the output x is x_0 and the first row of the output xdot is xdot_0.

The first argument, fcn, is a string or a two element cell array of strings, inline or function handle, that names the function, to call to compute the vector of residuals for the set of equations. It must have the form

    res = f (x, xdot, t)

in which x, xdot, and res are vectors, and t is a scalar.

If fcn is a two-element string array, the first element names the function $f$ described above, and the second element names a function to compute the modified Jacobian

$$J = \frac{\partial f}{\partial x} + c\frac{\partial f}{\partial \dot{x}}$$

The modified Jacobian function must have the form

    jac = j (x, xdot, t, c)

The second and third arguments to daspk specify the initial condition of the states and their derivatives, and the fourth argument specifies a vector of output times at which the solution is desired, including the time corresponding to the initial condition.

The set of initial states and derivatives are not strictly required to be consistent. If they are not consistent, you must use the daspk_options function to provide additional information so that daspk can compute a consistent starting point.

The fifth argument is optional, and may be used to specify a set of times that the DAE solver should not integrate past. It is useful for avoiding difficulties with singularities and points where there is a discontinuity in the derivative.

Chapter 22: Differential Equations                                              333

After a successful computation, the value of *istate* will be greater than zero (consistent with the Fortran version of DASPK).

If the computation is not successful, the value of *istate* will be less than zero and *msg* will contain additional information.

You can use the function daspk_options to set optional parameters for daspk.

**See also:** dassl.

daspk_options (*opt*, *val*)                                          Loadable Function

When called with two arguments, this function allows you to set options parameters for the function daspk. Given one argument, daspk_options returns the value of the corresponding option. If no arguments are supplied, the names of all the available options and their current values are displayed.

Options include

"absolute tolerance"
: Absolute tolerance. May be either vector or scalar. If a vector, it must match the dimension of the state vector, and the relative tolerance must also be a vector of the same length.

"relative tolerance"
: Relative tolerance. May be either vector or scalar. If a vector, it must match the dimension of the state vector, and the absolute tolerance must also be a vector of the same length.

    The local error test applied at each integration step is

    ```
    abs (local error in x(i))
         <= rtol(i) * abs (Y(i)) + atol(i)
    ```

"compute consistent initial condition"
: Denoting the differential variables in the state vector by Y_d and the algebraic variables by Y_a, ddaspk can solve one of two initialization problems:

    1. Given Y_d, calculate Y_a and Y'_d
    2. Given Y', calculate Y.

    In either case, initial values for the given components are input, and initial guesses for the unknown components must also be provided as input. Set this option to 1 to solve the first problem, or 2 to solve the second (the default is 0, so you must provide a set of initial conditions that are consistent).

    If this option is set to a nonzero value, you must also set the "algebraic variables" option to declare which variables in the problem are algebraic.

"use initial condition heuristics"
: Set to a nonzero value to use the initial condition heuristics options described below.

"initial condition heuristics"
: A vector of the following parameters that can be used to control the initial condition calculation.

    MXNIT
    : Maximum number of Newton iterations (default is 5).

    MXNJ
    : Maximum number of Jacobian evaluations (default is 6).

    MXNH
    : Maximum number of values of the artificial stepsize parameter to be tried if the "compute consistent initial condition" option has been set to 1 (default is 5).

        Note that the maximum total number of Newton iterations allowed is MXNIT*MXNJ*MXNH if the "compute consistent initial condition" option has been set to 1 and MXNIT*MXNJ if it is set to 2.

    LSOFF
    : Set to a nonzero value to disable the linesearch algorithm (default is 0).

    STPTOL
    : Minimum scaled step in linesearch algorithm (default is eps^(2/3)).

    EPINIT
    : Swing factor in the Newton iteration convergence test. The test is applied to the residual vector, premultiplied by the approximate Jacobian. For convergence, the weighted RMS norm of this vector (scaled by the error weights) must be less than EPINIT*EPCON, where EPCON = 0.33 is the analogous test constant used in the time steps. The default is EPINIT = 0.01.

"print initial condition info"
: Set this option to a nonzero value to display detailed information about the initial condition calculation (default is 0).

"exclude algebraic variables from error test"
: Set to a nonzero value to exclude algebraic variables from the error test. You must also set the "algebraic variables" option to declare which variables in the problem are algebraic (default is 0).

"algebraic variables"
: A vector of the same length as the state vector. A nonzero element indicates that the corresponding element of the state vector is an algebraic variable (i.e., its derivative does not appear explicitly in the equation set.

    This option is required by the compute consistent initial condition" and "exclude algebraic variables from error test" options.

Chapter 22: Differential Equations 335

"enforce inequality constraints"
Set to one of the following values to enforce the inequality constraints specified by the "inequality constraint types" option (default is 0).

1. To have constraint checking only in the initial condition calculation.
2. To enforce constraint checking during the integration.
3. To enforce both options 1 and 2.

"inequality constraint types"
A vector of the same length as the state specifying the type of inequality constraint. Each element of the vector corresponds to an element of the state and should be assigned one of the following codes

-2  Less than zero.

-1  Less than or equal to zero.

0  Not constrained.

1  Greater than or equal to zero.

2  Greater than zero.

This option only has an effect if the "enforce inequality constraints" option is nonzero.

"initial step size"
Differential-algebraic problems may occasionally suffer from severe scaling difficulties on the first step. If you know a great deal about the scaling of your problem, you can help to alleviate this problem by specifying an initial stepsize (default is computed automatically).

"maximum order"
Restrict the maximum order of the solution method. This option must be between 1 and 5, inclusive (default is 5).

"maximum step size"
Setting the maximum stepsize will avoid passing over very large regions (default is not specified).

Octave also includes DASSL, an earlier version of *Daspk*, and *dasrt*, which can be used to solve DAEs with constraints (stopping conditions).

[x, xdot, t_out, istat, msg] = dasrt (fcn [, g], x_0,     Loadable Function
          xdot_0, t [, t_crit])
Solve the set of differential-algebraic equations

$$0 = f(x, \dot{x}, t)$$

with

$$x(t_0) = x_0, \dot{x}(t_0) = \dot{x}_0$$

with functional stopping criteria (root solving).

The solution is returned in the matrices x and xdot, with each row in the result matrices corresponding to one of the elements in the vector t_out. The first element of t should be $t_0$ and correspond to the initial state of the system x_0 and its derivative xdot_0, so that the first row of the output x is x_0 and the first row of the output xdot is xdot_0.

The vector t provides an upper limit on the length of the integration. If the stopping condition is met, the vector t_out will be shorter than t, and the final element of t_out will be the point at which the stopping condition was met, and may not correspond to any element of the vector t.

The first argument, fcn, is a string, or cell array of strings or inline or function handles, that names the function to call to compute the vector of residuals for the set of equations. It must have the form

    res = f (x, xdot, t)

in which x, xdot, and res are vectors, and t is a scalar.

If fcn is a two-element string array, or two element cell array, the first element names the function $f$ described above, and the second element names a function to compute the modified Jacobian

$$J = \frac{\partial f}{\partial x} + c\frac{\partial f}{\partial \dot{x}}$$

The modified Jacobian function must have the form

    jac = j (x, xdot, t, c)

The optional second argument names a function that defines the constraint functions whose roots are desired during the integration. This function must have the form

    g_out = g (x, t)

and return a vector of the constraint function values. If the value of any of the constraint functions changes sign, DASRT will attempt to stop the integration at the point of the sign change.

If the name of the constraint function is omitted, dasrt solves the same problem as daspk or dassl.

Note that because of numerical errors in the constraint functions due to roundoff and integration error, DASRT may return false roots, or return the same root at two or more nearly equal values of $T$. If such false roots are suspected, the user should consider smaller error tolerances or higher precision in the evaluation of the constraint functions.

If a root of some constraint function defines the end of the problem, the input to DASRT should nevertheless allow integration to a point slightly past that root, so that DASRT can locate the root by interpolation.

The third and fourth arguments to dasrt specify the initial condition of the states and their derivatives, and the fourth argument specifies a vector of output times at which the solution is desired, including the time corresponding to the initial condition.

The set of initial states and derivatives are not strictly required to be consistent. In practice, however, DASSL is not very good at determining a

## Chapter 22: Differential Equations

consistent set for you, so it is best if you ensure that the initial values result in the function evaluating to zero.

The sixth argument is optional, and may be used to specify a set of times that the DAE solver should not integrate past. It is useful for avoiding difficulties with singularities and points where there is a discontinuity in the derivative.

After a successful computation, the value of *istate* will be greater than zero (consistent with the Fortran version of DASSL).

If the computation is not successful, the value of *istate* will be less than zero and *msg* will contain additional information.

You can use the function `dasrt_options` to set optional parameters for dasrt.

**See also:** daspk, dasrt, lsode.

dasrt_options (*opt, val*)　　　　　　　　　　　　　　Loadable Function
When called with two arguments, this function allows you to set options parameters for the function dasrt. Given one argument, dasrt_options returns the value of the corresponding option. If no arguments are supplied, the names of all the available options and their current values are displayed.

Options include

"absolute tolerance"
: Absolute tolerance. May be either vector or scalar. If a vector, it must match the dimension of the state vector, and the relative tolerance must also be a vector of the same length.

"relative tolerance"
: Relative tolerance. May be either vector or scalar. If a vector, it must match the dimension of the state vector, and the absolute tolerance must also be a vector of the same length.

    The local error test applied at each integration step is

    ```
    abs (local error in x(i)) <= ...
        rtol(i) * abs (Y(i)) + atol(i)
    ```

"initial step size"
: Differential-algebraic problems may occasionally suffer from severe scaling difficulties on the first step. If you know a great deal about the scaling of your problem, you can help to alleviate this problem by specifying an initial stepsize.

"maximum order"
: Restrict the maximum order of the solution method. This option must be between 1 and 5, inclusive.

"maximum step size"
: Setting the maximum stepsize will avoid passing over very large regions.

"step limit"
: Maximum number of integration steps to attempt on a single call to the underlying Fortran code.

See K. E. Brenan, et al., *Numerical Solution of Initial-Value Problems in Differential-Algebraic Equations*, North-Holland (1989) for more information about the implementation of DASSL.

# 23 Optimization

Octave comes with support for solving various kinds of optimization problems. Specifically Octave can solve problems in Linear Programming, Quadratic Programming, Nonlinear Programming, and Linear Least Squares Minimization.

## 23.1 Linear Programming

Octave can solve Linear Programming problems using the glpk function. That is, Octave can solve

$$\min_x c^T x$$

subject to the linear constraints $Ax = b$ where $x \geq 0$.

The glpk function also supports variations of this problem.

[xopt, fmin, status, extra] = glpk (c, a, b, lb, ub,  Function File
    ctype, vartype, sense, param)

Solve a linear program using the GNU GLPK library. Given three arguments, glpk solves the following standard LP:

$$\min_x C^T x$$

subject to

$$Ax = b \qquad x \geq 0$$

but may also solve problems of the form

$$[\min_x | \max_x] C^T x$$

subject to

$$Ax[= | \leq | \geq]b \qquad LB \leq x \leq UB$$

Input arguments:

c    A column array containing the objective function coefficients.

a    A matrix containing the constraints coefficients.

b    A column array containing the right-hand side value for each constraint in the constraint matrix.

lb    An array containing the lower bound on each of the variables. If *lb* is not supplied, the default lower bound for the variables is zero.

ub    An array containing the upper bound on each of the variables. If *ub* is not supplied, the default upper bound is assumed to be infinite.

*ctype*
: An array of characters containing the sense of each constraint in the constraint matrix. Each element of the array may be one of the following values

    "F"
    : A free (unbounded) constraint (the constraint is ignored).

    "U"
    : An inequality constraint with an upper bound (A(i,:)*x <= b(i)).

    "S"
    : An equality constraint (A(i,:)*x = b(i)).

    "L"
    : An inequality with a lower bound (A(i,:)*x >= b(i)).

    "D"
    : An inequality constraint with both upper and lower bounds (A(i,:)*x >= -b(i) *and* (A(i,:)*x <= b(i)).

*vartype*
: A column array containing the types of the variables.

    "C"
    : A continuous variable.

    "I"
    : An integer variable.

*sense*
: If *sense* is 1, the problem is a minimization. If *sense* is -1, the problem is a maximization. The default value is 1.

*param*
: A structure containing the following parameters used to define the behavior of solver. Missing elements in the structure take on default values, so you only need to set the elements that you wish to change from the default.

    Integer parameters:

    msglev (LPX_K_MSGLEV, default: 1)
    : Level of messages output by solver routines:

        0   No output.

        1   Error messages only.

        2   Normal output .

        3   Full output (includes informational messages).

    scale (LPX_K_SCALE, default: 1)
    : Scaling option:

Chapter 23: Optimization                                                    341

>  0    No scaling.
>
>  1    Equilibration scaling.
>
>  2    Geometric mean scaling, then equilibration scaling.

dual (LPX_K_DUAL, default: 0)
Dual simplex option:

>  0    Do not use the dual simplex.
>
>  1    If initial basic solution is dual feasible, use the dual simplex.

price (LPX_K_PRICE, default: 1)
Pricing option (for both primal and dual simplex):

>  0    Textbook pricing.
>
>  1    Steepest edge pricing.

round (LPX_K_ROUND, default: 0)
Solution rounding option:

>  0    Report all primal and dual values "as is".
>
>  1    Replace tiny primal and dual values by exact zero.

itlim (LPX_K_ITLIM, default: -1)
Simplex iterations limit. If this value is positive, it is decreased by one each time when one simplex iteration has been performed, and reaching zero value signals the solver to stop the search. Negative value means no iterations limit.

itcnt (LPX_K_OUTFRQ, default: 200)
Output frequency, in iterations. This parameter specifies how frequently the solver sends information about the solution to the standard output.

branch (LPX_K_BRANCH, default: 2)
Branching heuristic option (for MIP only):

>  0    Branch on the first variable.
>
>  1    Branch on the last variable.
>
>  2    Branch using a heuristic by Driebeck and Tomlin.

btrack (LPX_K_BTRACK, default: 2)
Backtracking heuristic option (for MIP only):

>  0    Depth first search.
>
>  1    Breadth first search.
>
>  2    Backtrack using the best projection heuristic.

presol (LPX_K_PRESOL, default: 1)
If this flag is set, the routine lpx_simplex solves the problem using the built-in LP presolver. Otherwise the LP presolver is not used.

lpsolver (default: 1)
: Select which solver to use. If the problem is a MIP problem this flag will be ignored.

    1   Revised simplex method.

    2   Interior point method.

save (default: 0)
: If this parameter is nonzero, save a copy of the problem in CPLEX LP format to the file '"outpb.lp"'. There is currently no way to change the name of the output file.

Real parameters:

relax (LPX_K_RELAX, default: 0.07)
: Relaxation parameter used in the ratio test. If it is zero, the textbook ratio test is used. If it is non-zero (should be positive), Harris' two-pass ratio test is used. In the latter case on the first pass of the ratio test basic variables (in the case of primal simplex) or reduced costs of non-basic variables (in the case of dual simplex) are allowed to slightly violate their bounds, but not more than relax*tolbnd or relax*toldj (thus, relax is a percentage of tolbnd or toldj.

tolbnd (LPX_K_TOLBND, default: 10e-7)
: Relative tolerance used to check if the current basic solution is primal feasible. It is not recommended that you change this parameter unless you have a detailed understanding of its purpose.

toldj (LPX_K_TOLDJ, default: 10e-7)
: Absolute tolerance used to check if the current basic solution is dual feasible. It is not recommended that you change this parameter unless you have a detailed understanding of its purpose.

tolpiv (LPX_K_TOLPIV, default: 10e-9)
: Relative tolerance used to choose eligible pivotal elements of the simplex table. It is not recommended that you change this parameter unless you have a detailed understanding of its purpose.

objll (LPX_K_OBJLL, default: -DBL_MAX)
: Lower limit of the objective function. If on the phase II the objective function reaches this limit and continues decreasing, the solver stops the search. This parameter is used in the dual simplex method only.

objul (LPX_K_OBJUL, default: +DBL_MAX)
: Upper limit of the objective function. If on the phase II the objective function reaches this limit and continues increasing, the solver stops the search. This parameter is used in the dual simplex only.

tmlim (LPX_K_TMLIM, default: -1.0)
: Searching time limit, in seconds. If this value is positive, it is decreased each time when one simplex iteration has been performed by the amount of time spent for the iteration, and reaching zero value signals the solver to stop the search. Negative value means no time limit.

outdly (LPX_K_OUTDLY, default: 0.0)
: Output delay, in seconds. This parameter specifies how long the solver should delay sending information about the solution to the standard output. Non-positive value means no delay.

tolint (LPX_K_TOLINT, default: 10e-5)
: Relative tolerance used to check if the current basic solution is integer feasible. It is not recommended that you change this parameter unless you have a detailed understanding of its purpose.

tolobj (LPX_K_TOLOBJ, default: 10e-7)
: Relative tolerance used to check if the value of the objective function is not better than in the best known integer feasible solution. It is not recommended that you change this parameter unless you have a detailed understanding of its purpose.

Output values:

*xopt*
: The optimizer (the value of the decision variables at the optimum).

*fopt*
: The optimum value of the objective function.

*status*
: Status of the optimization.

Simplex Method:

180 (LPX_OPT)
: Solution is optimal.

181 (LPX_FEAS)
: Solution is feasible.

182 (LPX_INFEAS)
: Solution is infeasible.

183 (LPX_NOFEAS)
: Problem has no feasible solution.

184 (LPX_UNBND)
: Problem has no unbounded solution.

185 (LPX_UNDEF)
: Solution status is undefined.

Interior Point Method:

150 (`LPX_T_UNDEF`)
: The interior point method is undefined.

151 (`LPX_T_OPT`)
: The interior point method is optimal.

Mixed Integer Method:

170 (`LPX_I_UNDEF`)
: The status is undefined.

171 (`LPX_I_OPT`)
: The solution is integer optimal.

172 (`LPX_I_FEAS`)
: Solution integer feasible but its optimality has not been proven

173 (`LPX_I_NOFEAS`)
: No integer feasible solution.

If an error occurs, *status* will contain one of the following codes:

204 (`LPX_E_FAULT`)
: Unable to start the search.

205 (`LPX_E_OBJLL`)
: Objective function lower limit reached.

206 (`LPX_E_OBJUL`)
: Objective function upper limit reached.

207 (`LPX_E_ITLIM`)
: Iterations limit exhausted.

208 (`LPX_E_TMLIM`)
: Time limit exhausted.

209 (`LPX_E_NOFEAS`)
: No feasible solution.

210 (`LPX_E_INSTAB`)
: Numerical instability.

211 (`LPX_E_SING`)
: Problems with basis matrix.

212 (`LPX_E_NOCONV`)
: No convergence (interior).

213 (`LPX_E_NOPFS`)
: No primal feasible solution (LP presolver).

214 (`LPX_E_NODFS`)
: No dual feasible solution (LP presolver).

*extra*
: A data structure containing the following fields:

lambda
: Dual variables.

redcosts
: Reduced Costs.

time
: Time (in seconds) used for solving LP/MIP problem.

mem
: Memory (in bytes) used for solving LP/MIP problem (this is not available if the version of GLPK is 4.15 or later).

Example:
```
c = [10, 6, 4]';
a = [ 1, 1, 1;
     10, 4, 5;
      2, 2, 6];
b = [100, 600, 300]';
lb = [0, 0, 0]';
ub = [];
ctype = "UUU";
vartype = "CCC";
s = -1;

param.msglev = 1;
param.itlim = 100;

[xmin, fmin, status, extra] = ...
    glpk (c, a, b, lb, ub, ctype, vartype, s, param);
```

## 23.2 Quadratic Programming

Octave can also solve Quadratic Programming problems, this is

$$\min_x \frac{1}{2} x^T H x + x^T q$$

subject to

$$Ax = b \qquad lb \leq x \leq ub \qquad A_{lb} \leq A_{in} \leq A_{ub}$$

[x, obj, info, lambda] = qp (x0, H, q, A, b, lb, ub, A_lb, A_in, A_ub)    *Function File*

Solve the quadratic program

$$\min_x \frac{1}{2} x^T H x + x^T q$$

subject to

$$Ax = b \qquad lb \leq x \leq ub \qquad A_{lb} \leq A_{in} \leq A_{ub}$$

using a null-space active-set method.

Any bound (A, b, lb, ub, A_lb, A_ub) may be set to the empty matrix ([]) if not present. If the initial guess is feasible the algorithm is faster.

The value *info* is a structure with the following fields:

**solveiter**
　　The number of iterations required to find the solution.

**info**
　　An integer indicating the status of the solution, as follows:

　　0　The problem is feasible and convex. Global solution found.

　　1　The problem is not convex. Local solution found.

　　2　The problem is not convex and unbounded.

　　3　Maximum number of iterations reached.

　　6　The problem is infeasible.

## 23.3 Nonlinear Programming

Octave can also perform general nonlinear minimization using a successive quadratic programming solver.

[x, obj, info, iter, nf, lambda] = sqp (x, phi, g, h)　　　　Function File
Solve the nonlinear program

$$\min_x \phi(x)$$

subject to

$$g(x) = 0 \quad h(x) \geq 0$$

using a successive quadratic programming method.

The first argument is the initial guess for the vector $x$.

The second argument is a function handle pointing to the objective function. The objective function must be of the form

　　y = phi (x)

in which $x$ is a vector and $y$ is a scalar.

The second argument may also be a 2- or 3-element cell array of function handles. The first element should point to the objective function, the second should point to a function that computes the gradient of the objective function, and the third should point to a function to compute the hessian of the objective function. If the gradient function is not supplied, the gradient is computed by finite differences. If the hessian function is not supplied, a BFGS update formula is used to approximate the hessian.

If supplied, the gradient function must be of the form

## Chapter 23: Optimization

```
        g = gradient (x)
```
in which x is a vector and g is a vector.

If supplied, the hessian function must be of the form
```
        h = hessian (x)
```
in which x is a vector and h is a matrix.

The third and fourth arguments are function handles pointing to functions that compute the equality constraints and the inequality constraints, respectively.

If your problem does not have equality (or inequality) constraints, you may pass an empty matrix for *cef* (or *cif*).

If supplied, the equality and inequality constraint functions must be of the form
```
        r = f (x)
```
in which x is a vector and r is a vector.

The third and fourth arguments may also be 2-element cell arrays of function handles. The first element should point to the constraint function and the second should point to a function that computes the gradient of the constraint function:

$$\left( \frac{\partial f(x)}{\partial x_1}, \frac{\partial f(x)}{\partial x_2}, \ldots, \frac{\partial f(x)}{\partial x_N} \right)^T$$

Here is an example of calling sqp:

```
function r = g (x)
  r = [ sumsq(x)-10;
        x(2)*x(3)-5*x(4)*x(5);
        x(1)^3+x(2)^3+1 ];
endfunction

function obj = phi (x)
  obj = exp(prod(x)) - 0.5*(x(1)^3+x(2)^3+1)^2;
endfunction

x0 = [-1.8; 1.7; 1.9; -0.8; -0.8];

[x, obj, info, iter, nf, lambda] = sqp (x0, @phi, @g, [])

x =

   -1.71714
    1.59571
    1.82725
   -0.76364
   -0.76364
```

```
obj = 0.053950
info = 101
iter = 8
nf = 10
lambda =

   -0.0401627
    0.0379578
   -0.0052227
```

The value returned in *info* may be one of the following:

101
> The algorithm terminated because the norm of the last step was less than tol * norm (x)) (the value of tol is currently fixed at sqrt (eps) in 'sqp.m').

102
> The BFGS update failed.

103
> The maximum number of iterations was reached (the maximum number of allowed iterations is currently fixed at 100 in 'sqp.m').

**See also:** qp.

## 23.4 Linear Least Squares

Octave also supports linear least squares minimization. That is, Octave can find the parameter $b$ such that the model $y = xb$ fits data $(x, y)$ as well as possible, assuming zero-mean Gaussian noise. If the noise is assumed to be isotropic the problem can be solved using the \ or / operators, or the ols function. In the general case where the noise is assumed to be anisotropic the gls is needed.

[beta, sigma, r] = ols (y, x)　　　　　　　　　　　　　　　　　Function File
> Ordinary least squares estimation for the multivariate model $y = xb + e$ with $\bar{e} = 0$, and $\text{cov}(\text{vec}(e)) = \text{kron}(s, I)$ where $y$ is a $t \times p$ matrix, $x$ is a $t \times k$ matrix, $b$ is a $k \times p$ matrix, and $e$ is a $t \times p$ matrix.
>
> Each row of $y$ and $x$ is an observation and each column a variable.
>
> The return values *beta*, *sigma*, and *r* are defined as follows.
>
> beta
>> The OLS estimator for $b$, beta = pinv (x) * y, where pinv (x) denotes the pseudoinverse of $x$.
>
> sigma
>> The OLS estimator for the matrix *s*,
>> ```
>> sigma = (y-x*beta)'
>>       * (y-x*beta)
>>       / (t-rank(x))
>> ```
>
> r　　The matrix of OLS residuals, r = y - x * beta.

[beta, v, r] = gls (y, x, o)  Function File
Generalized least squares estimation for the multivariate model $y = xb + e$ with $\bar{e} = 0$ and $\text{cov}(\text{vec}(e)) = (s^2)o$, where $y$ is a $t \times p$ matrix, $x$ is a $t \times k$ matrix, $b$ is a $k \times p$ matrix, $e$ is a $t \times p$ matrix, and $o$ is a $tp \times tp$ matrix.

Each row of $y$ and $x$ is an observation and each column a variable. The return values beta, v, and r are defined as follows.

beta
    The GLS estimator for $b$.

v    The GLS estimator for $s^2$.

r    The matrix of GLS residuals, $r = y - xbeta$.

# 24 Statistics

Octave has support for various statistical methods. This includes basic descriptive statistics, statistical tests, random number generation, and much more.

The functions that analyze data all assume that multidimensional data is arranged in a matrix where each row is an observation, and each column is a variable. So, the matrix defined by

```
a = [ 0.9, 0.7;
      0.1, 0.1;
      0.5, 0.4 ];
```

contains three observations from a two-dimensional distribution. While this is the default data arrangement, most functions support different arrangements.

It should be noted that the statistics functions don't test for data containing NaN, NA, or Inf. Such values need to be handled explicitly.

## 24.1 Descriptive Statistics

Octave can compute various statistics such as the moments of a data set.

mean (*x*, *dim*, *opt*)                                            Function File

If *x* is a vector, compute the mean of the elements of *x*

$$\mathrm{mean}(x) = \bar{x} = \frac{1}{N}\sum_{i=1}^{N} x_i$$

If *x* is a matrix, compute the mean for each column and return them in a row vector.

With the optional argument *opt*, the kind of mean computed can be selected. The following options are recognized:

"a"
    Compute the (ordinary) arithmetic mean. This is the default.

"g"
    Compute the geometric mean.

"h"
    Compute the harmonic mean.

If the optional argument *dim* is supplied, work along dimension *dim*.

Both *dim* and *opt* are optional. If both are supplied, either may appear first.

median (x, dim)                                                    Function File
   If x is a vector, compute the median value of the elements of x. If the
   elements of x are sorted, the median is defined as

$$\text{median}(x) = \begin{cases} x(\lceil N/2 \rceil), & N \text{ odd;} \\ (x(N/2) + x(N/2+1))/2, & N \text{ even.} \end{cases}$$

   If x is a matrix, compute the median value for each column and return them
   in a row vector. If the optional *dim* argument is given, operate along this
   dimension.
   See also: std, mean.

meansq (x)                                                         Function File
meansq (x, dim)                                                    Function File
   For vector arguments, return the mean square of the values. For matrix
   arguments, return a row vector containing the mean square of each column.
   With the optional *dim* argument, returns the mean squared of the values
   along this dimension.

std (x)                                                            Function File
std (x, opt)                                                       Function File
std (x, opt, dim)                                                  Function File
   If x is a vector, compute the standard deviation of the elements of x.

$$\text{std}(x) = \sigma(x) = \sqrt{\frac{\sum_{i=1}^{N}(x_i - \bar{x})^2}{N-1}}$$

   where $\bar{x}$ is the mean value of x. If x is a matrix, compute the standard
   deviation for each column and return them in a row vector.
   The argument *opt* determines the type of normalization to use. Valid values
   are

   0:   normalizes with $N-1$, provides the square root of best unbiased esti-
        mator of the variance [default]

   1:   normalizes with $N$, this provides the square root of the second moment
        around the mean

   The third argument *dim* determines the dimension along which the standard
   deviation is calculated.
   See also: mean, median.

# Chapter 24: Statistics

**var (x)** — Function File
For vector arguments, return the (real) variance of the values. For matrix arguments, return a row vector containing the variance for each column.
The argument *opt* determines the type of normalization to use. Valid values are

0: Normalizes with $N - 1$, provides the best unbiased estimator of the variance [default].

1: Normalizes with $N$, this provides the second moment around the mean.

The third argument *dim* determines the dimension along which the variance is calculated.

**[m, f, c] = mode (x, dim)** — Function File
Count the most frequently appearing value. mode counts the frequency along the first non-singleton dimension and if two or more values have te same frequency returns the smallest of the two in *m*. The dimension along which to count can be specified by the *dim* parameter.

The variable *f* counts the frequency of each of the most frequently occurring elements. The cell array *c* contains all of the elements with the maximum frequency.

**cov (x, y)** — Function File
Compute covariance.

If each row of *x* and *y* is an observation and each column is a variable, the $(i, j)$-th entry of cov (x, y) is the covariance between the $i$-th variable in *x* and the $j$-th variable in *y*.

$$\sigma_{ij} = \frac{1}{N-1} \sum_{i=1}^{N} (x_i - \bar{x})(y_i - \bar{y})$$

where $\bar{x}$ and $\bar{y}$ are the mean values of *x* and *y*. If called with one argument, compute cov (x, x).

**cor (x, y)** — Function File
Compute correlation.

The $(i, j)$-th entry of cor (x, y) is the correlation between the $i$-th variable in *x* and the $j$-th variable in *y*.

$$\text{corrcoef}(x, y) = \frac{\text{cov}(x, y)}{\text{std}(x)\text{std}(y)}$$

For matrices, each row is an observation and each column a variable; vectors are always observations and may be row or column vectors.

cor (x) is equivalent to cor (x, x).

Note that the corrcoef function does the same as cor.

**corrcoef (x, y)**  *Function File*
Compute correlation.

If each row of *x* and *y* is an observation and each column is a variable, the (*i*, *j*)-th entry of corrcoef (x, y) is the correlation between the *i*-th variable in *x* and the *j*-th variable in *y*.

$$\mathrm{corrcoef}(x, y) = \frac{\mathrm{cov}(x,y)}{\mathrm{std}(x)\mathrm{std}(y)}$$

If called with one argument, compute corrcoef (x, x).

**kurtosis (x, dim)**  *Function File*
If *x* is a vector of length $N$, return the kurtosis

$$\mathrm{kurtosis}(x) = \frac{1}{N\sigma(x)^4} \sum_{i=1}^{N} (x_i - \bar{x})^4 - 3$$

where $\bar{x}$ is the mean value of *x*.

If *x* is a matrix, return the kurtosis over the first non-singleton dimension. The optional argument *dim* can be given to force the kurtosis to be given over that dimension.

**skewness (x, dim)**  *Function File*
If *x* is a vector of length $n$, return the skewness

$$\mathrm{skewness}(x) = \frac{1}{N\sigma(x)^3} \sum_{i=1}^{N} (x_i - \bar{x})^3$$

where $\bar{x}$ is the mean value of *x*.

If *x* is a matrix, return the skewness along the first non-singleton dimension of the matrix. If the optional *dim* argument is given, operate along this dimension.

**statistics (x)**  *Function File*
If *x* is a matrix, return a matrix with the minimum, first quartile, median, third quartile, maximum, mean, standard deviation, skewness and kurtosis of the columns of *x* as its rows.

If *x* is a vector, treat it as a column vector.

**moment (x, p, opt, dim)**  *Function File*
If *x* is a vector, compute the *p*-th moment of *x*.

If *x* is a matrix, return the row vector containing the *p*-th moment of each column.

With the optional string *opt*, the kind of moment to be computed can be specified. If opt contains "c" or "a", central and/or absolute moments are returned. For example,

   moment (x, 3, "ac")

computes the third central absolute moment of *x*.

If the optional argument *dim* is supplied, work along dimension *dim*.

## 24.2 Basic Statistical Functions

Octave also supports various helpful statistical functions.

mahalanobis (*x*, *y*)     Function File
Return the Mahalanobis' D-square distance between the multivariate samples *x* and *y*, which must have the same number of components (columns), but may have a different number of observations (rows).

center (*x*)     Function File
center (*x*, *dim*)     Function File
If *x* is a vector, subtract its mean. If *x* is a matrix, do the above for each column. If the optional argument *dim* is given, perform the above operation along this dimension

studentize (*x*, *dim*)     Function File
If *x* is a vector, subtract its mean and divide by its standard deviation.
If *x* is a matrix, do the above along the first non-singleton dimension. If the optional argument *dim* is given then operate along this dimension.

c = nchoosek (*n*, *k*)     Function File
Compute the binomial coefficient or all combinations of *n*. If *n* is a scalar then, calculate the binomial coefficient of *n* and *k*, defined as

$$\binom{n}{k} = \frac{n(n-1)(n-2)\cdots(n-k+1)}{k!} = \frac{n!}{k!(n-k)!}$$

If *n* is a vector generate all combinations of the elements of *n*, taken *k* at a time, one row per combination. The resulting *c* has size [nchoosek (length (n), k), k].

See also: bincoeff.

perms (*v*)     Function File
Generate all permutations of *v*, one row per permutation. The result has size factorial (n) * n, where *n* is the length of *v*.

As an example, perms([1, 2, 3]) returns the matrix

```
1  2  3
2  1  3
1  3  2
2  3  1
3  1  2
3  2  1
```

values (*x*)     Function File
Return the different values in a column vector, arranged in ascending order.
As an example, values([1, 2, 3, 1]) returns the vector [1, 2, 3].

[t, l_x] = table (x)                                           Function File
[t, l_x, l_y] = table (x, y)                                   Function File
    Create a contingency table *t* from data vectors. The *l* vectors are the corresponding levels.

    Currently, only 1- and 2-dimensional tables are supported.

spearman (x, y)                                                Function File
    Compute Spearman's rank correlation coefficient *rho* for each of the variables specified by the input arguments.

    For matrices, each row is an observation and each column a variable; vectors are always observations and may be row or column vectors.

    spearman (x) is equivalent to spearman (x, x).

    For two data vectors *x* and *y*, Spearman's *rho* is the correlation of the ranks of *x* and *y*.

    If *x* and *y* are drawn from independent distributions, *rho* has zero mean and variance $1 / (n - 1)$, and is asymptotically normally distributed.

run_count (x, n)                                               Function File
    Count the upward runs along the first non-singleton dimension of *x* of length 1, 2, ..., $n - 1$ and greater than or equal to *n*. If the optional argument *dim* is given operate along this dimension

ranks (x, dim)                                                 Function File
    If *x* is a vector, return the (column) vector of ranks of *x* adjusted for ties.

    If *x* is a matrix, do the above for along the first non-singleton dimension. If the optional argument *dim* is given, operate along this dimension.

range (x)                                                      Function File
range (x, dim)                                                 Function File
    If *x* is a vector, return the range, i.e., the difference between the maximum and the minimum, of the input data.

    If *x* is a matrix, do the above for each column of *x*.

    If the optional argument *dim* is supplied, work along dimension *dim*.

probit (p)                                                     Function File
    For each component of *p*, return the probit (the quantile of the standard normal distribution) of *p*.

logit (p)                                                      Function File
    For each component of *p*, return the logit of *p* defined as

$$\mathrm{logit}(p) = \log\left(\frac{p}{1-p}\right)$$

## cloglog (x)   Function File
Return the complementary log-log function of x, defined as

$$\mathrm{cloglog}(x) = -\log(-\log(x))$$

## kendall (x, y)   Function File
Compute Kendall's *tau* for each of the variables specified by the input arguments.

For matrices, each row is an observation and each column a variable; vectors are always observations and may be row or column vectors.

kendall (x) is equivalent to kendall (x, x).

For two data vectors x, y of common length n, Kendall's *tau* is the correlation of the signs of all rank differences of x and y; i.e., if both x and y have distinct entries, then

$$\tau = \frac{1}{n(n-1)} \sum_{i,j} \mathrm{sign}(q_i - q_j)\mathrm{sign}(r_i - r_j)$$

in which the $q_i$ and $r_i$ are the ranks of x and y, respectively.

If x and y are drawn from independent distributions, Kendall's *tau* is asymptotically normal with mean 0 and variance $\frac{2(2n+5)}{9n(n-1)}$.

## iqr (x, dim)   Function File
If x is a vector, return the interquartile range, i.e., the difference between the upper and lower quartile, of the input data.

If x is a matrix, do the above for first non-singleton dimension of x. If the option *dim* argument is given, then operate along this dimension.

## cut (x, breaks)   Function File
Create categorical data out of numerical or continuous data by cutting into intervals.

If *breaks* is a scalar, the data is cut into that many equal-width intervals. If *breaks* is a vector of break points, the category has length (*breaks*) - 1 groups.

The returned value is a vector of the same size as x telling which group each point in x belongs to. Groups are labelled from 1 to the number of groups; points outside the range of *breaks* are labelled by NaN.

## 24.3 Statistical Plots

Octave can create Quantile Plots (QQ-Plots), and Probability Plots (PP-Plots). These are simple graphical tests for determining if a data set comes from a certain distribution.

Note that Octave can also show histograms of data using the hist function as described in Section 15.1.1 [Two-Dimensional Plots], page 193.

[q, s] = qqplot (x, dist, params)                                                         Function File
:   Perform a QQ-plot (quantile plot).

    If F is the CDF of the distribution *dist* with parameters *params* and G its inverse, and x a sample vector of length n, the QQ-plot graphs ordinate $s(i)$ = $i$-th largest element of x versus abscissa $q(if)$ = $G((i - 0.5)/n)$.

    If the sample comes from F except for a transformation of location and scale, the pairs will approximately follow a straight line.

    The default for *dist* is the standard normal distribution. The optional argument *params* contains a list of parameters of *dist*. For example, for a quantile plot of the uniform distribution on [2,4] and x, use

        qqplot (x, "uniform", 2, 4)

    *dist* can be any string for which a function *dist_inv* that calculates the inverse CDF of distribution *dist* exists.

    If no output arguments are given, the data are plotted directly.

[p, y] = ppplot (x, dist, params)                                                         Function File
:   Perform a PP-plot (probability plot).

    If F is the CDF of the distribution *dist* with parameters *params* and x a sample vector of length n, the PP-plot graphs ordinate $y(i)$ = F ($i$-th largest element of x) versus abscissa $p(i) = (i - 0.5)/n$. If the sample comes from F, the pairs will approximately follow a straight line.

    The default for *dist* is the standard normal distribution. The optional argument *params* contains a list of parameters of *dist*. For example, for a probability plot of the uniform distribution on [2,4] and x, use

        ppplot (x, "uniform", 2, 4)

    *dist* can be any string for which a function *dist_cdf* that calculates the CDF of distribution *dist* exists.

    If no output arguments are given, the data are plotted directly.

## 24.4 Tests

Octave can perform several different statistical tests. The following table summarizes the available tests.

| Hypothesis | Test Functions |
|---|---|
| Equal mean values | anova, hotelling_test2, t_test_2, welch_test, wilcoxon_test, z_test_2 |
| Equal medians | kruskal_wallis_test, sign_test |
| Equal variances | bartlett_test, manova, var_test |
| Equal distributions | chisquare_test_homogeneity, kolmogorov_smirnov_test_2, u_test |
| Equal marginal frequencies | mcnemar_test |
| Equal success probabilities | prop_test_2 |
| Independent observations | chisquare_test_independence, run_test |
| Uncorrelated observations | cor_test |
| Given mean value | hotelling_test, t_test, z_test |
| Observations from distribution | kolmogorov_smirnov_test |
| Regression | f_test_regression, t_test_regression |

The tests return a p-value that describes the outcome of the test. Assuming that the test hypothesis is true, the p-value is the probability of obtaining a worse result than the observed one. So large p-values corresponds to a successful test. Usually a test hypothesis is accepted if the p-value exceeds 0.05.

[pval, f, df_b, df_w] = anova (y, g)      Function File

Perform a one-way analysis of variance (ANOVA). The goal is to test whether the population means of data taken from $k$ different groups are all equal.

Data may be given in a single vector $y$ with groups specified by a corresponding vector of group labels $g$ (e.g. numbers from 1 to $k$). This is the general form which does not impose any restriction on the number of data in each group or the group labels.

If $y$ is a matrix and $g$ is omitted, each column of $y$ is treated as a group. This form is only appropriate for balanced ANOVA in which the numbers of samples from each group are all equal.

Under the null hypothesis of constant means, the statistic $f$ follows an F distribution with $df\_b$ and $df\_w$ degrees of freedom.

The p-value (1 minus the CDF of this distribution at $f$) is returned in pval.

If no output argument is given, the standard one-way ANOVA table is printed.

[pval, chisq, df] = bartlett_test (x1, ...)      Function File

Perform a Bartlett test for the homogeneity of variances in the data vectors $x1, x2, \ldots, xk$, where $k > 1$.

Under the null hypothesis of equal variances, the test statistic chisq approximately follows a chi-square distribution with $df$ degrees of freedom.

The p-value (1 minus the CDF of this distribution at chisq) is returned in pval.

If no output argument is given, the p-value is displayed.

[pval, chisq, df] = chisquare_test_homogeneity (x, y, c)    Function File
> Given two samples x and y, perform a chisquare test for homogeneity of the null hypothesis that x and y come from the same distribution, based on the partition induced by the (strictly increasing) entries of c.
>
> For large samples, the test statistic chisq approximately follows a chisquare distribution with df = length (c) degrees of freedom.
>
> The p-value (1 minus the CDF of this distribution at chisq) is returned in pval.
>
> If no output argument is given, the p-value is displayed.

[pval, chisq, df] = chisquare_test_independence (x)    Function File
> Perform a chi-square test for independence based on the contingency table x. Under the null hypothesis of independence, chisq approximately has a chi-square distribution with df degrees of freedom.
>
> The p-value (1 minus the CDF of this distribution at chisq) of the test is returned in pval.
>
> If no output argument is given, the p-value is displayed.

cor_test (x, y, alt, method)    Function File
> Test whether two samples x and y come from uncorrelated populations.
>
> The optional argument string alt describes the alternative hypothesis, and can be "!=" or "<>" (non-zero), ">" (greater than 0), or "<" (less than 0). The default is the two-sided case.
>
> The optional argument string method specifies on which correlation coefficient the test should be based. If method is "pearson" (default), the (usual) Pearson's product moment correlation coefficient is used. In this case, the data should come from a bivariate normal distribution. Otherwise, the other two methods offer nonparametric alternatives. If method is "kendall", then Kendall's rank correlation tau is used. If method is "spearman", then Spearman's rank correlation rho is used. Only the first character is necessary.
>
> The output is a structure with the following elements:
>
> pval
> > The p-value of the test.
>
> stat
> > The value of the test statistic.
>
> dist
> > The distribution of the test statistic.
>
> params
> > The parameters of the null distribution of the test statistic.

*alternative*
    The alternative hypothesis.

*method*
    The method used for testing.

If no output argument is given, the p-value is displayed.

[pval, f, df_num, df_den] = f_test_regression (y, x, rr,    Function File
    r)
Perform an F test for the null hypothesis rr * b = r in a classical normal regression model y = X * b + e.
Under the null hypothesis, the test statistic f follows an F distribution with *df_num* and *df_den* degrees of freedom.
The p-value (1 minus the CDF of this distribution at f) is returned in *pval*.
If not given explicitly, r = 0.
If no output argument is given, the p-value is displayed.

[pval, tsq] = hotelling_test (x, m)    Function File
For a sample x from a multivariate normal distribution with unknown mean and covariance matrix, test the null hypothesis that mean (x) == m.
Hotelling's $T^2$ is returned in *tsq*. Under the null hypothesis, $(n-p)T^2/(p(n-1))$ has an F distribution with $p$ and $n-p$ degrees of freedom, where $n$ and $p$ are the numbers of samples and variables, respectively.
The p-value of the test is returned in *pval*.
If no output argument is given, the p-value of the test is displayed.

[pval, tsq] = hotelling_test_2 (x, y)    Function File
For two samples x from multivariate normal distributions with the same number of variables (columns), unknown means and unknown equal covariance matrices, test the null hypothesis mean (x) == mean (y).
Hotelling's two-sample $T^2$ is returned in *tsq*. Under the null hypothesis,

$$\frac{n_x + n_y - p - 1)T^2}{p(n_x + n_y - 2)}$$

has an F distribution with $p$ and $n_x + n_y - p - 1$ degrees of freedom, where $n_x$ and $n_y$ are the sample sizes and $p$ is the number of variables.
The p-value of the test is returned in *pval*.
If no output argument is given, the p-value of the test is displayed.

[pval, ks] = kolmogorov_smirnov_test (x, dist, params,    Function File
    alt)
Perform a Kolmogorov-Smirnov test of the null hypothesis that the sample x comes from the (continuous) distribution dist, i.e. if $F$ and $G$ are the CDFs corresponding to the sample and distribution, respectively, then the null hypothesis is that $F = G$.

The optional argument params contains a list of parameters of dist. For example, to test whether a sample x comes from a uniform distribution on [2,4], use

    kolmogorov_smirnov_test(x, "uniform", 2, 4)

dist can be any string for which a function dist_cdf that calculates the CDF of distribution dist exists.

With the optional argument string alt, the alternative of interest can be selected. If alt is "!=" or "<>", the null hypothesis is tested against the two-sided alternative $F \neq G$. In this case, the test statistic ks follows a two-sided Kolmogorov-Smirnov distribution. If alt is ">", the one-sided alternative $F > G$ is considered. Similarly for "<", the one-sided alternative $F > G$ is considered. In this case, the test statistic ks has a one-sided Kolmogorov-Smirnov distribution. The default is the two-sided case.

The p-value of the test is returned in pval.

If no output argument is given, the p-value is displayed.

[pval, ks, d] = kolmogorov_smirnov_test_2 (x, y, alt)    Function File
Perform a 2-sample Kolmogorov-Smirnov test of the null hypothesis that the samples x and y come from the same (continuous) distribution, i.e. if $F$ and $G$ are the CDFs corresponding to the x and y samples, respectively, then the null hypothesis is that $F = G$.

With the optional argument string alt, the alternative of interest can be selected. If alt is "!=" or "<>", the null hypothesis is tested against the two-sided alternative $F \neq G$. In this case, the test statistic ks follows a two-sided Kolmogorov-Smirnov distribution. If alt is ">", the one-sided alternative $F > G$ is considered. Similarly for "<", the one-sided alternative $F < G$ is considered. In this case, the test statistic ks has a one-sided Kolmogorov-Smirnov distribution. The default is the two-sided case.

The p-value of the test is returned in pval.

The third returned value, d, is the test statistic, the maximum vertical distance between the two cumulative distribution functions.

If no output argument is given, the p-value is displayed.

[pval, k, df] = kruskal_wallis_test (x1, ...)    Function File
Perform a Kruskal-Wallis one-factor "analysis of variance".

Suppose a variable is observed for $k > 1$ different groups, and let $x1, \ldots, xk$ be the corresponding data vectors.

Under the null hypothesis that the ranks in the pooled sample are not affected by the group memberships, the test statistic $k$ is approximately chi-square with $df = k - 1$ degrees of freedom.

If the data contains ties (some value appears more than once) $k$ is divided by $1 - sumTies/(n^3 - n)$ where $sumTies$ is the sum of $t^2 - t$ over each group of ties where $t$ is the number of ties in the group and $n$ is the total number of values in the input data. For more information on this adjustment see "Use of Ranks in One-Criterion Variance Analysis" in Journal of the American Statistical Association, Vol. 47, No. 260 (Dec 1952) by William H. Kruskal and W. Allen Wallis.

The p-value (1 minus the CDF of this distribution at $k$) is returned in pval.

If no output argument is given, the p-value is displayed.

manova (y, g)                                                          Function File

Perform a one-way multivariate analysis of variance (MANOVA). The goal is to test whether the p-dimensional population means of data taken from $k$ different groups are all equal. All data are assumed drawn independently from p-dimensional normal distributions with the same covariance matrix.

The data matrix is given by $y$. As usual, rows are observations and columns are variables. The vector $g$ specifies the corresponding group labels (e.g. numbers from 1 to $k$).

The LR test statistic (Wilks' Lambda) and approximate p-values are computed and displayed.

[pval, chisq, df] = mcnemar_test (x)                          Function File

For a square contingency table $x$ of data cross-classified on the row and column variables, McNemar's test can be used for testing the null hypothesis of symmetry of the classification probabilities.

Under the null hypothesis, chisq is approximately distributed as chisquare with df degrees of freedom.

The p-value (1 minus the CDF of this distribution at chisq) is returned in pval.

If no output argument is given, the p-value of the test is displayed.

[pval, z] = prop_test_2 (x1, n1, x2, n2, alt)                Function File

If x1 and n1 are the counts of successes and trials in one sample, and x2 and n2 those in a second one, test the null hypothesis that the success probabilities p1 and p2 are the same. Under the null hypothesis, the test statistic z approximately follows a standard normal distribution.

With the optional argument string alt, the alternative of interest can be selected. If alt is "!=" or "<>", the null hypothesis is tested against the two-sided alternative p1 != p2. If alt is ">", the one-sided alternative p1 > p2 is used. Similarly for "<", the one-sided alternative p1 < p2 is used. The default is the two-sided case.

The p-value of the test is returned in pval.

If no output argument is given, the p-value of the test is displayed.

[pval, chisq] = run_test (x)  Function File
Perform a chi-square test with 6 degrees of freedom based on the upward runs in the columns of x. Can be used to test whether x contains independent data.

The p-value of the test is returned in pval.

If no output argument is given, the p-value is displayed.

[pval, b, n] = sign_test (x, y, alt)  Function File
For two matched-pair samples x and y, perform a sign test of the null hypothesis PROB (x > y) == PROB (x < y) == 1/2. Under the null hypothesis, the test statistic b roughly follows a binomial distribution with parameters n = sum (x != y) and $p = 1/2$.

With the optional argument alt, the alternative of interest can be selected. If alt is "!=" or "<>", the null hypothesis is tested against the two-sided alternative PROB (x < y) != 1/2. If alt is ">", the one-sided alternative PROB (x > y) > 1/2 ("x is stochastically greater than y") is considered. Similarly for "<", the one-sided alternative PROB (x > y) < 1/2 ("x is stochastically less than y") is considered. The default is the two-sided case.

The p-value of the test is returned in pval.

If no output argument is given, the p-value of the test is displayed.

[pval, t, df] = t_test (x, m, alt)  Function File
For a sample x from a normal distribution with unknown mean and variance, perform a t-test of the null hypothesis mean (x) == m. Under the null hypothesis, the test statistic t follows a Student distribution with df = length (x) - 1 degrees of freedom.

With the optional argument string alt, the alternative of interest can be selected. If alt is "!=" or "<>", the null hypothesis is tested against the two-sided alternative mean (x) != m. If alt is ">", the one-sided alternative mean (x) > m is considered. Similarly for "<", the one-sided alternative mean (x) < m is considered. The default is the two-sided case.

The p-value of the test is returned in pval.

If no output argument is given, the p-value of the test is displayed.

[pval, t, df] = t_test_2 (x, y, alt)  Function File
For two samples x and y from normal distributions with unknown means and unknown equal variances, perform a two-sample t-test of the null hypothesis of equal means. Under the null hypothesis, the test statistic t follows a Student distribution with df degrees of freedom.

With the optional argument string alt, the alternative of interest can be selected. If alt is "!=" or "<>", the null hypothesis is tested against the two-sided alternative mean (x) != mean (y). If alt is ">", the one-sided alternative mean (x) > mean (y) is used. Similarly for "<", the one-sided alternative mean (x) < mean (y) is used. The default is the two-sided case.

The p-value of the test is returned in pval.

If no output argument is given, the p-value of the test is displayed.

[pval, t, df] = t_test_regression (y, x, rr, r, alt)  Function File
Perform an t test for the null hypothesis rr * b = r in a classical normal regression model y = x * b + e. Under the null hypothesis, the test statistic t follows a t distribution with df degrees of freedom.

If r is omitted, a value of 0 is assumed.

With the optional argument string alt, the alternative of interest can be selected. If alt is "!=" or "<>", the null hypothesis is tested against the two-sided alternative rr * b != r. If alt is ">", the one-sided alternative rr * b > r is used. Similarly for "<", the one-sided alternative rr * b < r is used. The default is the two-sided case.

The p-value of the test is returned in pval.

If no output argument is given, the p-value of the test is displayed.

[pval, z] = u_test (x, y, alt)  Function File
For two samples x and y, perform a Mann-Whitney U-test of the null hypothesis PROB $(x > y)$ == $1/2$ == PROB $(x < y)$. Under the null hypothesis, the test statistic z approximately follows a standard normal distribution. Note that this test is equivalent to the Wilcoxon rank-sum test.

With the optional argument string alt, the alternative of interest can be selected. If alt is "!=" or "<>", the null hypothesis is tested against the two-sided alternative PROB $(x > y)$ != $1/2$. If alt is ">", the one-sided alternative PROB $(x > y) > 1/2$ is considered. Similarly for "<", the one-sided alternative PROB $(x > y) < 1/2$ is considered. The default is the two-sided case.

The p-value of the test is returned in pval.

If no output argument is given, the p-value of the test is displayed.

[pval, f, df_num, df_den] = var_test (x, y, alt)  Function File
For two samples x and y from normal distributions with unknown means and unknown variances, perform an F-test of the null hypothesis of equal variances. Under the null hypothesis, the test statistic f follows an F-distribution with df_num and df_den degrees of freedom.

With the optional argument string alt, the alternative of interest can be selected. If alt is "!=" or "<>", the null hypothesis is tested against the two-sided alternative var (x) != var (y). If alt is ">", the one-sided alternative var (x) > var (y) is used. Similarly for "<", the one-sided alternative var (x) > var (y) is used. The default is the two-sided case.

The p-value of the test is returned in pval.

If no output argument is given, the p-value of the test is displayed.

[*pval, t, df*] = welch_test (*x, y,* alt)  Function File
For two samples *x* and *y* from normal distributions with unknown means and unknown and not necessarily equal variances, perform a Welch test of the null hypothesis of equal means. Under the null hypothesis, the test statistic *t* approximately follows a Student distribution with *df* degrees of freedom.

With the optional argument string alt, the alternative of interest can be selected. If alt is "!=" or "<>", the null hypothesis is tested against the two-sided alternative mean (*x*) != *m*. If alt is ">", the one-sided alternative mean(*x*) > *m* is considered. Similarly for "<", the one-sided alternative mean(*x*) < *m* is considered. The default is the two-sided case.

The p-value of the test is returned in *pval*.

If no output argument is given, the p-value of the test is displayed.

[*pval, z*] = wilcoxon_test (*x, y,* alt)  Function File
For two matched-pair sample vectors *x* and *y*, perform a Wilcoxon signed-rank test of the null hypothesis PROB (*x* > *y*) == 1/2. Under the null hypothesis, the test statistic *z* approximately follows a standard normal distribution when *n* > 25.

**Warning**: This function assumes a normal distribution for *z* and thus is invalid for *n* <= 25.

With the optional argument string alt, the alternative of interest can be selected. If alt is "!=" or "<>", the null hypothesis is tested against the two-sided alternative PROB (*x* > *y*) != 1/2. If alt is ">", the one-sided alternative PROB (*x* > *y*) > 1/2 is considered. Similarly for "<", the one-sided alternative PROB (*x* > *y*) < 1/2 is considered. The default is the two-sided case.

The p-value of the test is returned in *pval*.

If no output argument is given, the p-value of the test is displayed.

[*pval, z*] = z_test (*x, m, v,* alt)  Function File
Perform a Z-test of the null hypothesis mean (*x*) == *m* for a sample *x* from a normal distribution with unknown mean and known variance *v*. Under the null hypothesis, the test statistic *z* follows a standard normal distribution.

With the optional argument string alt, the alternative of interest can be selected. If alt is "!=" or "<>", the null hypothesis is tested against the two-sided alternative mean (*x*) != *m*. If alt is ">", the one-sided alternative mean (*x*) > *m* is considered. Similarly for "<", the one-sided alternative mean (*x*) < *m* is considered. The default is the two-sided case.

The p-value of the test is returned in *pval*.

If no output argument is given, the p-value of the test is displayed along with some information.

[pval, z] = z_test_2 (x, y, v_x, v_y, alt)  Function File
For two samples x and y from normal distributions with unknown means and known variances v_x and v_y, perform a Z-test of the hypothesis of equal means. Under the null hypothesis, the test statistic z follows a standard normal distribution.

With the optional argument string alt, the alternative of interest can be selected. If alt is "!=" or "<>", the null hypothesis is tested against the two-sided alternative mean (x) != mean (y). If alt is ">", the one-sided alternative mean (x) > mean (y) is used. Similarly for "<", the one-sided alternative mean (x) < mean (y) is used. The default is the two-sided case.

The p-value of the test is returned in pval.

If no output argument is given, the p-value of the test is displayed along with some information.

## 24.5 Models

[theta, beta, dev, dl, d2l, p] = logistic_regression (y,  Function File
    x, print, theta, beta)
Perform ordinal logistic regression.

Suppose y takes values in k ordered categories, and let gamma_i (x) be the cumulative probability that y falls in one of the first i categories given the covariate x. Then
    [theta, beta] = logistic_regression (y, x)
fits the model
    logit (gamma_i (x)) = theta_i - beta' * x,    i = 1...k-1
The number of ordinal categories, k, is taken to be the number of distinct values of round (y). If k equals 2, y is binary and the model is ordinary logistic regression. The matrix x is assumed to have full column rank.

Given y only, theta = logistic_regression (y) fits the model with baseline logit odds only.

The full form is
    [theta, beta, dev, dl, d2l, gamma]
        = logistic_regression (y, x, print, theta, beta)
in which all output arguments and all input arguments except y are optional. Setting print to 1 requests summary information about the fitted model to be displayed. Setting print to 2 requests information about convergence at each iteration. Other values request no information to be displayed. The input arguments theta and beta give initial estimates for theta and beta.

The returned value dev holds minus twice the log-likelihood.

The returned values dl and d2l are the vector of first and the matrix of second derivatives of the log-likelihood with respect to theta and beta.

p holds estimates for the conditional distribution of y given x.

## 24.6 Distributions

Octave has functions for computing the Probability Density Function (PDF), the Cumulative Distribution function (CDF), and the quantile (the inverse of the CDF) of a large number of distributions.

The following table summarizes the supported distributions (in alphabetical order).

| Distribution | PDF | CDF | Quantile |
|---|---|---|---|
| Beta | betapdf | betacdf | betainv |
| Binomial | binopdf | binocdf | binoinv |
| Cauchy | cauchy_pdf | cauchy_cdf | cauchy_inv |
| Chi-Square | chi2pdf | chi2cdf | chi2inv |
| Univariate Discrete | discrete_pdf | discrete_cdf | discrete_inv |
| Empirical | empirical_pdf | empirical_cdf | empirical_inv |
| Exponential | exppdf | expcdf | expinv |
| F | fpdf | fcdf | finv |
| Gamma | gampdf | gamcdf | gaminv |
| Geometric | geopdf | geocdf | geoinv |
| Hypergeometric | hygepdf | hygecdf | hygeinv |
| Kolmogorov Smirnov | *Not Available* | kolmogorov_smirnov_cdf | *Not Available* |
| Laplace | laplace_pdf | laplace_cdf | laplace_inv |
| Logistic | logistic_pdf | logistic_cdf | logistic_inv |
| Log-Normal | lognpdf | logncdf | logninv |
| Pascal | nbinpdf | nbincdf | nbininv |
| Univariate Normal | normpdf | normcdf | norminv |
| Poisson | poisspdf | poisscdf | poissinv |
| t (Student) | tpdf | tcdf | tinv |
| Univariate Discrete | unidpdf | unidcdf | unidinv |
| Uniform | unifpdf | unifcdf | unifinv |
| Weibull | wblpdf | wblcdf | wblinv |

betacdf (*x*, *a*, *b*)                                                                             Function File
    For each element of *x*, returns the CDF at *x* of the beta distribution with parameters *a* and *b*, i.e., PROB (beta (a, b) <= x).

betainv (*x*, *a*, *b*)                                                                              Function File
    For each component of *x*, compute the quantile (the inverse of the CDF) at *x* of the Beta distribution with parameters *a* and *b*.

betapdf (*x*, *a*, *b*)                                                                              Function File
    For each element of *x*, returns the PDF at *x* of the beta distribution with parameters *a* and *b*.

Chapter 24: Statistics                                                         369

binocdf (x, n, p)                                                    Function File
    For each element of x, compute the CDF at x of the binomial distribution
    with parameters n and p.

binoinv (x, n, p)                                                    Function File
    For each element of x, compute the quantile at x of the binomial distribution
    with parameters n and p.

binopdf (x, n, p)                                                    Function File
    For each element of x, compute the probability density function (PDF) at x
    of the binomial distribution with parameters n and p.

cauchy_cdf (x, lambda, sigma)                                        Function File
    For each element of x, compute the cumulative distribution function (CDF)
    at x of the Cauchy distribution with location parameter lambda and scale
    parameter sigma. Default values are lambda = 0, sigma = 1.

cauchy_inv (x, lambda, sigma)                                        Function File
    For each element of x, compute the quantile (the inverse of the CDF) at
    x of the Cauchy distribution with location parameter lambda and scale
    parameter sigma. Default values are lambda = 0, sigma = 1.

cauchy_pdf (x, lambda, sigma)                                        Function File
    For each element of x, compute the probability density function (PDF) at
    x of the Cauchy distribution with location parameter lambda and scale
    parameter sigma > 0. Default values are lambda = 0, sigma = 1.

chi2cdf (x, n)                                                       Function File
    For each element of x, compute the cumulative distribution function (CDF)
    at x of the chisquare distribution with n degrees of freedom.

chi2inv (x, n)                                                       Function File
    For each element of x, compute the quantile (the inverse of the CDF) at x
    of the chisquare distribution with n degrees of freedom.

chisquare_pdf (x, n)                                                 Function File
    For each element of x, compute the probability density function (PDF) at x
    of the chisquare distribution with n degrees of freedom.

discrete_cdf (x, v, p)                                               Function File
    For each element of x, compute the cumulative distribution function (CDF)
    at x of a univariate discrete distribution which assumes the values in v with
    probabilities p.

discrete_inv (x, v, p)          Function File
: For each component of x, compute the quantile (the inverse of the CDF) at x of the univariate distribution which assumes the values in v with probabilities p.

discrete_pdf (x, v, p)          Function File
: For each element of x, compute the probability density function (PDF) at x of a univariate discrete distribution which assumes the values in v with probabilities p.

empirical_cdf (x, data)          Function File
: For each element of x, compute the cumulative distribution function (CDF) at x of the empirical distribution obtained from the univariate sample data.

empirical_inv (x, data)          Function File
: For each element of x, compute the quantile (the inverse of the CDF) at x of the empirical distribution obtained from the univariate sample data.

empirical_pdf (x, data)          Function File
: For each element of x, compute the probability density function (PDF) at x of the empirical distribution obtained from the univariate sample data.

expcdf (x, lambda)          Function File
: For each element of x, compute the cumulative distribution function (CDF) at x of the exponential distribution with mean lambda.

  The arguments can be of common size or scalar.

expinv (x, lambda)          Function File
: For each element of x, compute the quantile (the inverse of the CDF) at x of the exponential distribution with mean lambda.

exppdf (x, lambda)          Function File
: For each element of x, compute the probability density function (PDF) of the exponential distribution with mean lambda.

fcdf (x, m, n)          Function File
: For each element of x, compute the CDF at x of the F distribution with m and n degrees of freedom, i.e., PROB (F $(m, n)$ <= x).

finv (x, m, n)          Function File
: For each component of x, compute the quantile (the inverse of the CDF) at x of the F distribution with parameters m and n.

fpdf (x, m, n)          Function File
: For each element of x, compute the probability density function (PDF) at x of the F distribution with m and n degrees of freedom.

Chapter 24: Statistics                                                                 371

gamcdf (x, a, b)                                                          Function File
    For each element of x, compute the cumulative distribution function (CDF)
    at x of the Gamma distribution with parameters a and b.

    **See also:** gamma, gammaln, gammainc, gampdf, gaminv, gamrnd.

gaminv (x, a, b)                                                          Function File
    For each component of x, compute the quantile (the inverse of the CDF) at
    x of the Gamma distribution with parameters a and b.

    **See also:** gamma, gammaln, gammainc, gampdf, gamcdf, gamrnd.

gampdf (x, a, b)                                                          Function File
    For each element of x, return the probability density function (PDF) at x
    of the Gamma distribution with parameters a and b.

    **See also:** gamma, gammaln, gammainc, gamcdf, gaminv, gamrnd.

geocdf (x, p)                                                             Function File
    For each element of x, compute the CDF at x of the geometric distribution
    with parameter p.

geoinv (x, p)                                                             Function File
    For each element of x, compute the quantile at x of the geometric distribution
    with parameter p.

geopdf (x, p)                                                             Function File
    For each element of x, compute the probability density function (PDF) at x
    of the geometric distribution with parameter p.

hygecdf (x, t, m, n)                                                      Function File
    Compute the cumulative distribution function (CDF) at x of the hypergeo-
    metric distribution with parameters t, m, and n. This is the probability of
    obtaining not more than x marked items when randomly drawing a sample
    of size n without replacement from a population of total size t containing m
    marked items.

    The parameters t, m, and n must positive integers with m and n not greater
    than t.

hygeinv (x, t, m, n)                                                      Function File
    For each element of x, compute the quantile at x of the hypergeometric
    distribution with parameters t, m, and n.

    The parameters t, m, and n must positive integers with m and n not greater
    than t.

**hygepdf** (*x*, *t*, *m*, *n*)  Function File
Compute the probability density function (PDF) at *x* of the hypergeometric distribution with parameters *t*, *m*, and *n*. This is the probability of obtaining *x* marked items when randomly drawing a sample of size *n* without replacement from a population of total size *t* containing *m* marked items.
The arguments must be of common size or scalar.

**kolmogorov_smirnov_cdf** (*x*, *tol*)  Function File
Return the CDF at *x* of the Kolmogorov-Smirnov distribution,

$$Q(x) = \sum_{k=-\infty}^{\infty} (-1)^k \exp(-2k^2 x^2)$$

for *x* > 0.
The optional parameter *tol* specifies the precision up to which the series should be evaluated; the default is *tol* = eps.

**laplace_cdf** (*x*)  Function File
For each element of *x*, compute the cumulative distribution function (CDF) at *x* of the Laplace distribution.

**laplace_inv** (*x*)  Function File
For each element of *x*, compute the quantile (the inverse of the CDF) at *x* of the Laplace distribution.

**laplace_pdf** (*x*)  Function File
For each element of *x*, compute the probability density function (PDF) at *x* of the Laplace distribution.

**logistic_cdf** (*x*)  Function File
For each component of *x*, compute the CDF at *x* of the logistic distribution.

**logistic_inv** (*x*)  Function File
For each component of *x*, compute the quantile (the inverse of the CDF) at *x* of the logistic distribution.

**logistic_pdf** (*x*)  Function File
For each component of *x*, compute the PDF at *x* of the logistic distribution.

**logncdf** (*x*, *mu*, *sigma*)  Function File
For each element of *x*, compute the cumulative distribution function (CDF) at *x* of the lognormal distribution with parameters *mu* and *sigma*. If a random variable follows this distribution, its logarithm is normally distributed with mean *mu* and standard deviation *sigma*.
Default values are *mu* = 1, *sigma* = 1.

## Chapter 24: Statistics

**logninv (x, mu, sigma)**      Function File
For each element of x, compute the quantile (the inverse of the CDF) at x of the lognormal distribution with parameters *mu* and *sigma*. If a random variable follows this distribution, its logarithm is normally distributed with mean log (*mu*) and variance *sigma*.
Default values are *mu* = 1, *sigma* = 1.

**lognpdf (x, mu, sigma)**      Function File
For each element of x, compute the probability density function (PDF) at x of the lognormal distribution with parameters *mu* and *sigma*. If a random variable follows this distribution, its logarithm is normally distributed with mean *mu* and standard deviation *sigma*.
Default values are *mu* = 1, *sigma* = 1.

**nbincdf (x, n, p)**      Function File
For each element of x, compute the CDF at x of the Pascal (negative binomial) distribution with parameters n and p.
The number of failures in a Bernoulli experiment with success probability p before the n-th success follows this distribution.

**nbininv (x, n, p)**      Function File
For each element of x, compute the quantile at x of the Pascal (negative binomial) distribution with parameters n and p.
The number of failures in a Bernoulli experiment with success probability p before the n-th success follows this distribution.

**nbinpdf (x, n, p)**      Function File
For each element of x, compute the probability density function (PDF) at x of the Pascal (negative binomial) distribution with parameters n and p.
The number of failures in a Bernoulli experiment with success probability p before the n-th success follows this distribution.

**normcdf (x, m, s)**      Function File
For each element of x, compute the cumulative distribution function (CDF) at x of the normal distribution with mean m and standard deviation s.
Default values are m = 0, s = 1.

**norminv (x, m, s)**      Function File
For each element of x, compute the quantile (the inverse of the CDF) at x of the normal distribution with mean m and standard deviation s.
Default values are m = 0, s = 1.

**normpdf (x, m, s)**      Function File
For each element of x, compute the probability density function (PDF) at x of the normal distribution with mean m and standard deviation s.
Default values are m = 0, s = 1.

poisscdf (x, lambda)  Function File
   For each element of x, compute the cumulative distribution function (CDF) at x of the Poisson distribution with parameter lambda.

poissinv (x, lambda)  Function File
   For each component of x, compute the quantile (the inverse of the CDF) at x of the Poisson distribution with parameter lambda.

poisspdf (x, lambda)  Function File
   For each element of x, compute the probability density function (PDF) at x of the poisson distribution with parameter lambda.

tcdf (x, n)  Function File
   For each element of x, compute the cumulative distribution function (CDF) at x of the t (Student) distribution with n degrees of freedom, i.e., PROB (t($n$) <= x).

tinv (x, n)  Function File
   For each probability value x, compute the inverse of the cumulative distribution function (CDF) of the t (Student) distribution with degrees of freedom n. This function is analagous to looking in a table for the t-value of a single-tailed distribution.

tpdf (x, n)  Function File
   For each element of x, compute the probability density function (PDF) at x of the t (Student) distribution with n degrees of freedom.

unidcdf (x, v)  Function File
   For each element of x, compute the cumulative distribution function (CDF) at x of a univariate discrete distribution which assumes the values in v with equal probability.

unidinv (x, v)  Function File
   For each component of x, compute the quantile (the inverse of the CDF) at x of the univariate discrete distribution which assumes the values in v with equal probability

unidpdf (x, v)  Function File
   For each element of x, compute the probability density function (PDF) at x of a univariate discrete distribution which assumes the values in v with equal probability.

unifcdf (x, a, b)  Function File
   Return the CDF at x of the uniform distribution on [a, b], i.e., PROB (uniform (a, b) <= x).
   Default values are $a = 0$, $b = 1$.

Chapter 24: Statistics

**unifinv (x, a, b)**      Function File
For each element of x, compute the quantile (the inverse of the CDF) at x of the uniform distribution on [a, b].
Default values are $a = 0$, $b = 1$.

**unifpdf (x, a, b)**      Function File
For each element of x, compute the PDF at x of the uniform distribution on [a, b].
Default values are $a = 0$, $b = 1$.

**wblcdf (x, scale, shape)**      Function File
Compute the cumulative distribution function (CDF) at x of the Weibull distribution with shape parameter scale and scale parameter shape, which is

$$1 - \exp(-(x/shape)^{scale})$$

for $x \geq 0$.

**wblinv (x, scale, shape)**      Function File
Compute the quantile (the inverse of the CDF) at x of the Weibull distribution with shape parameter scale and scale parameter shape.

**wblpdf (x, scale, shape)**      Function File
Compute the probability density function (PDF) at x of the Weibull distribution with shape parameter scale and scale parameter shape which is given by

$$scale \cdot shape^{-scale} x^{scale-1} \exp(-(x/shape)^{scale})$$

for x > 0.

## 24.7 Random Number Generation

Octave can generate random numbers from a large number of distributions. The samples are computed from the basic random number generators described in Section 16.5 [Random Matrices], page 246.

The following table summarizes the available random number generators (in alphabetical order).

| Distribution | Function |
|---|---|
| Beta Distribution | betarnd |
| Binomial Distribution | binornd |
| Cauchy Distribution | cauchy_rnd |
| Chi-Square Distribution | chi2rnd |
| Univariate Discrete Distribution | discrete_rnd |
| Empirical Distribution | empirical_rnd |
| Exponential Distribution | exprnd |
| F Distribution | frnd |
| Gamma Distribution | gamrnd |
| Geometric Distribution | geornd |
| Hypergeometric Distribution | hygernd |
| Laplace Distribution | laplace_rnd |
| Logistic Distribution | logistic_rnd |
| Log-Normal Distribution | lognrnd |
| Pascal Distribution | nbinrnd |
| Univariate Normal Distribution | normrnd |
| Poisson Distribution | poissrnd |
| t (Student) Distribution | trnd |
| Univariate Discrete Distribution | unidrnd |
| Uniform Distribution | unifrnd |
| Weibull Distribution | wblrnd |
| Wiener Process | wienrnd |

betarnd (*a*, *b*, *r*, *c*)　　　　　　　　　　　　　　　　　　　　Function File
betarnd (*a*, *b*, *sz*)　　　　　　　　　　　　　　　　　　　　　Function File

　　Return an *r* by *c* or size (*sz*) matrix of random samples from the Beta distribution with parameters *a* and *b*. Both *a* and *b* must be scalar or of size *r* by *c*.

　　If *r* and *c* are omitted, the size of the result matrix is the common size of *a* and *b*.

binornd (*n*, *p*, *r*, *c*)　　　　　　　　　　　　　　　　　　　　Function File
binornd (*n*, *p*, *sz*)　　　　　　　　　　　　　　　　　　　　　Function File

　　Return an *r* by *c* or a size (*sz*) matrix of random samples from the binomial distribution with parameters *n* and *p*. Both *n* and *p* must be scalar or of size *r* by *c*.

　　If *r* and *c* are omitted, the size of the result matrix is the common size of *n* and *p*.

Chapter 24: Statistics                                                             377

cauchy_rnd (lambda, sigma, r, c)                                      Function File
cauchy_rnd (lambda, sigma, sz)                                        Function File
  Return an r by c or a size (sz) matrix of random samples from the Cauchy distribution with parameters lambda and sigma which must both be scalar or of size r by c.

  If r and c are omitted, the size of the result matrix is the common size of lambda and sigma.

chi2rnd (n, r, c)                                                     Function File
chi2rnd (n, sz)                                                       Function File
  Return an r by c or a size (sz) matrix of random samples from the chisquare distribution with n degrees of freedom. n must be a scalar or of size r by c.

  If r and c are omitted, the size of the result matrix is the size of n.

discrete_rnd (n, v, p)                                                Function File
discrete_rnd (v, p, r, c)                                             Function File
discrete_rnd (v, p, sz)                                               Function File
  Generate a row vector containing a random sample of size n from the univariate distribution which assumes the values in v with probabilities p. n must be a scalar.

  If r and c are given create a matrix with r rows and c columns. Or if sz is a vector, create a matrix of size sz.

empirical_rnd (n, data)                                               Function File
empirical_rnd (data, r, c)                                            Function File
empirical_rnd (data, sz)                                              Function File
  Generate a bootstrap sample of size n from the empirical distribution obtained from the univariate sample data.

  If r and c are given create a matrix with r rows and c columns. Or if sz is a vector, create a matrix of size sz.

exprnd (lambda, r, c)                                                 Function File
exprnd (lambda, sz)                                                   Function File
  Return an r by c matrix of random samples from the exponential distribution with mean lambda, which must be a scalar or of size r by c. Or if sz is a vector, create a matrix of size sz.

  If r and c are omitted, the size of the result matrix is the size of lambda.

frnd (m, n, r, c)                                                     Function File
frnd (m, n, sz)                                                       Function File
  Return an r by c matrix of random samples from the F distribution with m and n degrees of freedom. Both m and n must be scalar or of size r by c. If sz is a vector the random samples are in a matrix of size sz.

  If r and c are omitted, the size of the result matrix is the common size of m and n.

gamrnd (a, b, r, c)                                       Function File
gamrnd (a, b, sz)                                         Function File
   Return an r by c or a size (sz) matrix of random samples from the Gamma
   distribution with parameters a and b. Both a and b must be scalar or of
   size r by c.

   If r and c are omitted, the size of the result matrix is the common size of a
   and b.

   See also: gamma, gammaln, gammainc, gampdf, gamcdf, gaminv.

geornd (p, r, c)                                          Function File
geornd (p, sz)                                            Function File
   Return an r by c matrix of random samples from the geometric distribution
   with parameter p, which must be a scalar or of size r by c.

   If r and c are given create a matrix with r rows and c columns. Or if sz is
   a vector, create a matrix of size sz.

hygernd (t, m, n, r, c)                                   Function File
hygernd (t, m, n, sz)                                     Function File
hygernd (t, m, n)                                         Function File
   Return an r by c matrix of random samples from the hypergeometric distribution with parameters t, m, and n.

   The parameters t, m, and n must positive integers with m and n not greater
   than t.

   The parameter sz must be scalar or a vector of matrix dimensions. If sz is
   scalar, then a sz by sz matrix of random samples is generated.

laplace_rnd (r, c)                                        Function File
laplace_rnd (sz);                                         Function File
   Return an r by c matrix of random numbers from the Laplace distribution.
   Or if sz is a vector, create a matrix of sz.

logistic_rnd (r, c)                                       Function File
logistic_rnd (sz)                                         Function File
   Return an r by c matrix of random numbers from the logistic distribution.
   Or if sz is a vector, create a matrix of sz.

lognrnd (mu, sigma, r, c)                                 Function File
lognrnd (mu, sigma, sz)                                   Function File
   Return an r by c matrix of random samples from the lognormal distribution
   with parameters mu and sigma. Both mu and sigma must be scalar or of
   size r by c. Or if sz is a vector, create a matrix of size sz.

   If r and c are omitted, the size of the result matrix is the common size of
   mu and sigma.

Chapter 24: Statistics                                                              379

nbinrnd (n, p, r, c)                                                      Function File
nbinrnd (n, p, sz)                                                        Function File
> Return an r by c matrix of random samples from the Pascal (negative binomial) distribution with parameters n and p. Both n and p must be scalar or of size r by c.
>
> If r and c are omitted, the size of the result matrix is the common size of n and p. Or if sz is a vector, create a matrix of size sz.

normrnd (m, s, r, c)                                                      Function File
normrnd (m, s, sz)                                                        Function File
> Return an r by c or size (sz) matrix of random samples from the normal distribution with parameters mean m and standard deviation s. Both m and s must be scalar or of size r by c.
>
> If r and c are omitted, the size of the result matrix is the common size of m and s.

poissrnd (lambda, r, c)                                                   Function File
> Return an r by c matrix of random samples from the Poisson distribution with parameter lambda, which must be a scalar or of size r by c.
>
> If r and c are omitted, the size of the result matrix is the size of lambda.

trnd (n, r, c)                                                            Function File
trnd (n, sz)                                                              Function File
> Return an r by c matrix of random samples from the t (Student) distribution with n degrees of freedom. n must be a scalar or of size r by c. Or if sz is a vector create a matrix of size sz.
>
> If r and c are omitted, the size of the result matrix is the size of n.

unidrnd (mx);                                                             Function File
unidrnd (mx, v);                                                          Function File
unidrnd (mx, m, n, ...);                                                  Function File
> Return random values from discrete uniform distribution, with maximum value(s) given by the integer mx, which may be a scalar or multidimensional array.
>
> If mx is a scalar, the size of the result is specified by the vector v, or by the optional arguments m, n, .... Otherwise, the size of the result is the same as the size of mx.

unifrnd (a, b, r, c)                                                      Function File
unifrnd (a, b, sz)                                                        Function File
> Return an r by c or a size (sz) matrix of random samples from the uniform distribution on [a, b]. Both a and b must be scalar or of size r by c.
>
> If r and c are omitted, the size of the result matrix is the common size of a and b.

wblrnd (*scale, shape, r, c*)  Function File
wblrnd (*scale, shape, sz*)  Function File
 Return an *r* by *c* matrix of random samples from the Weibull distribution with parameters *scale* and *shape* which must be scalar or of size *r* by *c*. Or if *sz* is a vector return a matrix of size *sz*.

 If *r* and *c* are omitted, the size of the result matrix is the common size of *alpha* and *sigma*.

wienrnd (*t, d, n*)  Function File
 Return a simulated realization of the *d*-dimensional Wiener Process on the interval $[0, t]$. If *d* is omitted, $d = 1$ is used. The first column of the return matrix contains time, the remaining columns contain the Wiener process.

 The optional parameter *n* gives the number of summands used for simulating the process over an interval of length 1. If *n* is omitted, $n = 1000$ is used.

# 25 Sets

Octave has a limited number of functions for managing sets of data, where a set is defined as a collection of unique elements. In Octave a set is represented as a vector of numbers.

create_set (x)                                                    Function File
    Return a row vector containing the unique values in x, sorted in ascending order. For example,

        create_set ([ 1, 2; 3, 4; 4, 2 ])
        ⇒ [ 1, 2, 3, 4 ]

    See also: union, intersection, complement.

unique (x)                                                        Function File
    Return the unique elements of x, sorted in ascending order. If x is a row vector, return a row vector, but if x is a column vector or a matrix return a column vector.

unique (A, 'rows')                                                Function File
    Return the unique rows of A, sorted in ascending order.

[y, i, j] = unique (x)                                            Function File
    Return index vectors i and j such that x(i)==y and y(j)==x.

    See also: union, intersect, setdiff, setxor, ismember.

## 25.1 Set Operations

Octave supports the basic set operations. That is, Octave can compute the union, intersection, complement, and difference of two sets. Octave also supports the *Exclusive Or* set operation, and membership determination. The functions for set operations all work in much the same way. As an example, assume that x and y contains two sets, then

    union(x, y)

computes the union of the two sets.

[tf, a_idx] = ismember (A, S)                                     Function File
[tf, a_idx] = ismember (A, S, "rows")                             Function File
    Return a matrix *tf* the same shape as A which has 1 if A(i,j) is in S or 0 if it isn't. If a second output argument is requested, the indexes into S of the matching elements are also returned.

        a = [3, 10, 1];
        s = [0:9];
        [tf, a_idx] = residue (a, s);
        ⇒ tf = [1, 0, 1]
        ⇒ a_idx = [4, 0, 2]

    The inputs, A and S, may also be cell arrays.

```
a = {'abc'};
s = {'abc', 'def'};
[tf, a_idx] = residue (a, s);
⇒ tf = [1, 0]
⇒ a_idx = [1, 0]
```
With the optional third argument "rows", and matrices $A$ and $S$ with the same number of columns, compare rows in $A$ with the rows in $S$.
```
a = [1:3; 5:7; 4:6];
s = [0:2; 1:3; 2:4; 3:5; 4:6];
[tf, a_idx] = ismember(a, s, 'rows');
⇒ tf = logical ([1; 0; 1])
⇒ a_idx = [2; 0; 5];
```
See also: unique, union, intersection, setxor, setdiff.

union (*x*, *y*)         Function File

Return the set of elements that are in either of the sets *x* and *y*. For example,
```
union ([ 1, 2, 4 ], [ 2, 3, 5 ])
⇒ [ 1, 2, 3, 4, 5 ]
```
See also: create_set, intersection, complement.

intersect (*a*, *b*)         Function File
[*c*, *ia*, *ib*] = intersect (*a*, *b*)         Function File

Return the elements in both *a* and *b*, sorted in ascending order. If *a* and *b* are both column vectors return a column vector, otherwise return a row vector.

Return index vectors *ia* and *ib* such that a(ia)==c and b(ib)==c.

See also: unique, union, setxor, setdiff, ismember.

complement (*x*, *y*)         Function File

Return the elements of set *y* that are not in set *x*. For example,
```
complement ([ 1, 2, 3 ], [ 2, 3, 5 ])
⇒ 5
```
See also: create_set, union, intersection.

setdiff (*a*, *b*)         Function File
setdiff (*a*, *b*, "rows")         Function File

Return the elements in *a* that are not in *b*, sorted in ascending order. If *a* and *b* are both column vectors return a column vector, otherwise return a row vector.

Given the optional third argument "rows", return the rows in *a* that are not in *b*, sorted in ascending order by rows.

See also: unique, union, intersect, setxor, ismember.

setxor (a, b)                                                      Function File
　　Return the elements exclusive to a or b, sorted in ascending order. If a and
　　b are both column vectors return a column vector, otherwise return a row
　　vector.

　　**See also:** unique, union, intersect, setdiff, ismember.

# 26 Polynomial Manipulations

In Octave, a polynomial is represented by its coefficients (arranged in descending order). For example, a vector $c$ of length $N+1$ corresponds to the following polynomial of order $N$

$$p(x) = c_1 x^N + \ldots + c_N x + c_{N+1}.$$

## 26.1 Evaluating Polynomials

The value of a polynomial represented by the vector $c$ can be evaluated at the point $x$ very easily, as the following example shows:

```
N = length(c)-1;
val = dot( x.^(N:-1:0), c );
```

While the above example shows how easy it is to compute the value of a polynomial, it isn't the most stable algorithm. With larger polynomials you should use more elegant algorithms, such as Horner's Method, which is exactly what the Octave function polyval does.

In the case where $x$ is a square matrix, the polynomial given by $c$ is still well-defined. As when $x$ is a scalar the obvious implementation is easily expressed in Octave, but also in this case more elegant algorithms perform better. The polyvalm function provides such an algorithm.

polyval (c, x)     Function File
    Evaluate a polynomial.

    polyval (c, x) will evaluate the polynomial at the specified value of $x$.

    If $x$ is a vector or matrix, the polynomial is evaluated at each of the elements of $x$.

    See also: polyvalm, poly, roots, conv, deconv, residue, filter, polyderiv, polyinteg.

polyvalm (c, x)     Function File
    Evaluate a polynomial in the matrix sense.

    polyvalm (c, x) will evaluate the polynomial in the matrix sense, i.e. matrix multiplication is used instead of element by element multiplication as is used in polyval.

    The argument $x$ must be a square matrix.

    See also: polyval, poly, roots, conv, deconv, residue, filter, polyderiv, and polyinteg.

## 26.2 Finding Roots

Octave can find the roots of a given polynomial. This is done by computing the companion matrix of the polynomial (see the compan function for a definition), and then finding its eigenvalues.

**roots** (*v*)                                Function File
    For a vector $v$ with $N$ components, return the roots of the polynomial

$$v_1 z^{N-1} + \cdots + v_{N-1} z + v_N.$$

As an example, the following code finds the roots of the quadratic polynomial

$$p(x) = x^2 - 5.$$

```
c = [1, 0, -5];
roots(c)
⇒   2.2361
⇒  -2.2361
```

Note that the true result is $\pm\sqrt{5}$ which is roughly $\pm 2.2361$.

**See also:** compan.

**compan** (*c*)                               Function File
    Compute the companion matrix corresponding to polynomial coefficient vector $c$.

The companion matrix is

$$A = \begin{bmatrix} -c_2/c_1 & -c_3/c_1 & \cdots & -c_N/c_1 & -c_{N+1}/c_1 \\ 1 & 0 & \cdots & 0 & 0 \\ 0 & 1 & \cdots & 0 & 0 \\ \vdots & \vdots & \ddots & \vdots & \vdots \\ 0 & 0 & \cdots & 1 & 0 \end{bmatrix}.$$

The eigenvalues of the companion matrix are equal to the roots of the polynomial.

**See also:** poly, roots, residue, conv, deconv, polyval, polyderiv, polyinteg.

## 26.3 Products of Polynomials

conv (a, b)                  Function File
    Convolve two vectors.

    y = conv (a, b) returns a vector of length equal to length (a) + length (b) - 1. If a and b are polynomial coefficient vectors, conv returns the coefficients of the product polynomial.

    See also: deconv, poly, roots, residue, polyval, polyderiv, polyinteg.

deconv (y, a)                Function File
    Deconvolve two vectors.

    [b, r] = deconv (y, a) solves for b and r such that y = conv (a, b) + r.

    If y and a are polynomial coefficient vectors, b will contain the coefficients of the polynomial quotient and r will be a remainder polynomial of lowest order.

    See also: conv, poly, roots, residue, polyval, polyderiv, polyinteg.

y = conv2 (a, b, shape)                Loadable Function
y = conv2 (v1, v2, M, shape)              Loadable Function
    Returns 2D convolution of a and b where the size of c is given by

    *shape*= 'full'
        returns full 2-D convolution

    *shape*= 'same'
        same size as a. 'central' part of convolution

    *shape*= 'valid'
        only parts which do not include zero-padded edges

    By default *shape* is 'full'. When the third argument is a matrix returns the convolution of the matrix *M* by the vector *v1* in the column direction and by vector *v2* in the row direction

q = polygcd (b, a, tol)                Function File
    Find greatest common divisor of two polynomials. This is equivalent to the polynomial found by multiplying together all the common roots. Together with deconv, you can reduce a ratio of two polynomials. Tolerance defaults to

        sqrt(eps).

    Note that this is an unstable algorithm, so don't try it on large polynomials.
    Example

```
polygcd (poly(1:8), poly(3:12)) - poly(3:8)
⇒ [ 0, 0, 0, 0, 0, 0, 0 ]
deconv (poly(1:8), polygcd (poly(1:8), poly(3:12))) ...
   - poly(1:2)
⇒ [ 0, 0, 0 ]
```

**See also:** poly, polyinteg, polyderiv, polyreduce, roots, conv, deconv, residue, filter, polyval, and polyvalm.

[r, p, k, e] = residue (b, a)    Function File
Compute the partial fraction expansion for the quotient of the polynomials, b and a.

$$\frac{B(s)}{A(s)} = \sum_{m=1}^{M} \frac{r_m}{(s-p_m)_m^e} + \sum_{i=1}^{N} k_i s^{N-i}.$$

where $M$ is the number of poles (the length of the r, p, and e), the k vector is a polynomial of order $N-1$ representing the direct contribution, and the e vector specifies the multiplicity of the m-th residue's pole.

For example,
```
    b = [1, 1, 1];
    a = [1, -5, 8, -4];
    [r, p, k, e] = residue (b, a);
⇒ r = [-2; 7; 3]
⇒ p = [2; 2; 1]
⇒ k = [](0x0)
⇒ e = [1; 2; 1]
```
which represents the following partial fraction expansion

$$\frac{s^2+s+1}{s^3-5s^2+8s-4} = \frac{-2}{s-2} + \frac{7}{(s-2)^2} + \frac{3}{s-1}$$

[b, a] = residue (r, p, k)    Function File
[b, a] = residue (r, p, k, e)    Function File
Compute the reconstituted quotient of polynomials, b(s)/a(s), from the partial fraction expansion; represented by the residues, poles, and a direct polynomial specified by r, p and k, and the pole multiplicity e.

If the multiplicity, e, is not explicitly specified the multiplicity is determined by the script mpoles.m.

For example,

```
r = [-2; 7; 3];
p = [2; 2; 1];
k = [1, 0];
[b, a] = residue (r, p, k);
⇒ b = [1, -5, 9, -3, 1]
⇒ a = [1, -5, 8, -4]
```

where mpoles.m is used to determine e = [1; 2; 1]

Alternatively the multiplicity may be defined explicitly, for example,

```
r = [7; 3; -2];
p = [2; 1; 2];
k = [1, 0];
e = [2; 1; 1];
[b, a] = residue (r, p, k, e);
⇒ b = [1, -5, 9, -3, 1]
⇒ a = [1, -5, 8, -4]
```

which represents the following partial fraction expansion

$$\frac{-2}{s-2} + \frac{7}{(s-2)^2} + \frac{3}{s-1} + s = \frac{s^4 - 5s^3 + 9s^2 - 3s + 1}{s^3 - 5s^2 + 8s - 4}$$

**See also:** poly, roots, conv, deconv, mpoles, polyval, polyderiv, polyinteg.

## 26.4 Derivatives and Integrals

Octave comes with functions for computing the derivative and the integral of a polynomial. The functions polyderiv and polyint both return new polynomials describing the result. As an example we'll compute the definite integral of $p(x) = x^2 + 1$ from 0 to 3.

```
c = [1, 0, 1];
integral = polyint(c);
area = polyval(integral, 3) - polyval(integral, 0)
⇒ 12
```

**polyderiv** (*c*)                                                                  Function File
**[*q*] = polyderiv** (*b*, *a*)                                                     Function File
**[*q*, *r*] = polyderiv** (*b*, *a*)                                                Function File

Return the coefficients of the derivative of the polynomial whose coefficients are given by vector *c*. If a pair of polynomials is given *b* and *a*, the derivative of the product is returned in *q*, or the quotient numerator in *q* and the quotient denominator in *r*.

**See also:** poly, polyinteg, polyreduce, roots, conv, deconv, residue, filter, polygcd, polyval, polyvalm.

polyder (c)                                                    Function File
[q] = polyder (b, a)                                           Function File
[q, r] = polyder (b, a)                                        Function File
   See polyderiv.

polyint (c, k)                                                 Function File
   Return the coefficients of the integral of the polynomial whose coefficients are represented by the vector c. The variable k is the constant of integration, which by default is set to zero.

   **See also:** poly, polyderiv, polyreduce, roots, conv, deconv, residue, filter, polyval, and polyvalm.

## 26.5 Polynomial Interpolation

   Octave comes with good support for various kinds of interpolation, most of which are described in Chapter 27 [Interpolation], page 393. One simple alternative to the functions described in the aforementioned chapter, is to fit a single polynomial to some given data points. To avoid a highly fluctuating polynomial, one most often wants to fit a low-order polynomial to data. This usually means that it is necessary to fit the polynomial in a least-squares sense, which is what the polyfit function does.

[p, s] = polyfit (x, y, n)                                     Function File
   Return the coefficients of a polynomial $p(x)$ of degree $n$ that minimizes

$$\sum_{i=1}^{N}(p(x_i) - y_i)^2$$

   to best fit the data in the least squares sense.

   The polynomial coefficients are returned in a row vector.

   If two output arguments are requested, the second is a structure containing the following fields:

R    The Cholesky factor of the Vandermonde matrix used to compute the polynomial coefficients.

X    The Vandermonde matrix used to compute the polynomial coefficients.

df   The degrees of freedom.

normr
     The norm of the residuals.

yf   The values of the polynomial for each value of x.

Chapter 26: Polynomial Manipulations    391

In situations where a single polynomial isn't good enough, a solution is to use several polynomials pieced together. The function mkpp creates a piece-wise polynomial, ppval evaluates the function created by mkpp, and unmkpp returns detailed information about the function.

The following example shows how to combine two linear functions and a quadratic into one function. Each of these functions is expressed on adjoined intervals.

```
x = [-2, -1, 1, 2];
p = [ 0,  1, 0;
      1, -2, 1;
      0, -1, 1 ];
pp = mkpp(x, p);
xi = linspace(-2, 2, 50);
yi = ppval(pp, xi);
plot(xi, yi);
```

*yi* = ppval (*pp*, *xi*)                                                    Function File
Evaluate piece-wise polynomial *pp* at the points *xi*. If *pp*.d is a scalar greater than 1, or an array, then the returned value *yi* will be an array that is d1, d1, ..., dk, length (*xi*)].

See also: mkpp, unmkpp, spline.

*pp* = mkpp (*x*, *p*)                                                       Function File
*pp* = mkpp (*x*, *p*, *d*)                                                  Function File
Construct a piece-wise polynomial structure from sample points *x* and coefficients *p*. The i-th row of *p*, *p* (i,:), contains the coefficients for the polynomial over the *i*-th interval, ordered from highest to lowest. There must be one row for each interval in *x*, so rows (*p*) == length (*x*) - 1.

You can concatenate multiple polynomials of the same order over the same set of intervals using *p* = [ *p1* ; *p2* ; ... ; *pd* ]. In this case, rows (*p*) == *d* * (length (*x*) - 1).

*d* specifies the shape of the matrix *p* for all except the last dimension. If *d* is not specified it will be computed as round (rows (*p*) / (length (*x*) - 1)) instead.

See also: unmkpp, ppval, spline.

[*x*, *p*, *n*, *k*, *d*] = unmkpp (*pp*)                                    Function File
Extract the components of a piece-wise polynomial structure *pp*. These are as follows:

*x*      Sample points.

*p*      Polynomial coefficients for points in sample interval. *p* (i, :) contains the coefficients for the polynomial over interval *i* ordered from highest to lowest. If d > 1, *p* (r, i, :) contains the coefficients for the r-th polynomial defined on interval *i*. However, this is stored as a 2-D array such that c = reshape (*p* (:, j), *n*, *d*) gives c (i, r) is the j-th coefficient of the r-th polynomial over the i-th interval.

n     Number of polynomial pieces.

k     Order of the polynomial plus 1.

d     Number of polynomials defined for each interval.

**See also:** mkpp, ppval, spline.

## 26.6 Miscellaneous Functions

poly (a)                                                                                                      Function File

If a is a square $N$-by-$N$ matrix, poly (a) is the row vector of the coefficients of det (z * eye (N) - a), the characteristic polynomial of a. As an example we can use this to find the eigenvalues of a as the roots of poly (a).

    roots(poly(eye(3)))
⇒ 1.00000 + 0.00000i
⇒ 1.00000 - 0.00000i
⇒ 1.00000 + 0.00000i

In real-life examples you should, however, use the eig function for computing eigenvalues.

If x is a vector, poly (x) is a vector of coefficients of the polynomial whose roots are the elements of x. That is, of c is a polynomial, then the elements of d = roots (poly (c)) are contained in c. The vectors c and d are, however, not equal due to sorting and numerical errors.

**See also:** eig, roots.

polyout (c, x)                                                                           Function File

Write formatted polynomial

$$c(x) = c_1 x^n + \ldots + c_n x + c_{n+1}$$

and return it as a string or write it to the screen (if *nargout* is zero). x defaults to the string "s".

**See also:** polyval, polyvalm, poly, roots, conv, deconv, residue, filter, polyderiv, and polyinteg.

polyreduce (c)                                                                    Function File

Reduces a polynomial coefficient vector to a minimum number of terms by stripping off any leading zeros.

**See also:** poly, roots, conv, deconv, residue, filter, polyval, polyvalm, polyderiv, polyinteg.

# 27 Interpolation

## 27.1 One-dimensional Interpolation

Octave supports several methods for one-dimensional interpolation, most of which are described in this section. Section 26.5 [Polynomial Interpolation], page 390 and Section 28.4 [Interpolation on Scattered Data], page 413 describe further methods.

| | |
|---|---|
| `yi = interp1 (x, y, xi)` | Function File |
| `yi = interp1 (..., method)` | Function File |
| `yi = interp1 (..., extrap)` | Function File |
| `pp = interp1 (..., 'pp')` | Function File |

One-dimensional interpolation. Interpolate *y*, defined at the points *x*, at the points *xi*. The sample points *x* must be strictly monotonic. If *y* is an array, treat the columns of *y* separately.

Method is one of:

'nearest'
: Return the nearest neighbour.

'linear'
: Linear interpolation from nearest neighbours

'pchip'
: Piece-wise cubic hermite interpolating polynomial

'cubic'
: Cubic interpolation from four nearest neighbours

'spline'
: Cubic spline interpolation–smooth first and second derivatives throughout the curve

Appending '*' to the start of the above method forces `interp1` to assume that *x* is uniformly spaced, and only *x* (1) and *x* (2) are referenced. This is usually faster, and is never slower. The default method is 'linear'.

If *extrap* is the string 'extrap', then extrapolate values beyond the endpoints. If *extrap* is a number, replace values beyond the endpoints with that number. If *extrap* is missing, assume NA.

If the string argument 'pp' is specified, then *xi* should not be supplied and `interp1` returns the piece-wise polynomial that can later be used with ppval to evaluate the interpolation. There is an equivalence, such that  ppval (interp1 (*x*, *y*, *method*, 'pp'), *xi*) == interp1 (*x*, *y*, *xi*, *method*, 'extrap').

An example of the use of `interp1` is

```
xf=[0:0.05:10]; yf = sin(2*pi*xf/5);
xp=[0:10];      yp = sin(2*pi*xp/5);
lin=interp1(xp,yp,xf);
spl=interp1(xp,yp,xf,'spline');
cub=interp1(xp,yp,xf,'cubic');
near=interp1(xp,yp,xf,'nearest');
plot(xf,yf,"r",xf,lin,"g",xf,spl,"b", ...
     xf,cub,"c",xf,near,"m",xp,yp,"r*");
legend ("original","linear","spline","cubic","nearest")
```

See also: interpft.

There are some important differences between the various interpolation methods. The 'spline' method enforces that both the first and second derivatives of the interpolated values have a continuous derivative, whereas the other methods do not. This means that the results of the 'spline' method are generally smoother. If the function to be interpolated is in fact smooth, then 'spline' will give excellent results. However, if the function to be evaluated is in some manner discontinuous, then 'pchip' interpolation might give better results.

This can be demonstrated by the code

```
t = -2:2;
dt = 1;
ti =-2:0.025:2;
dti = 0.025;
y = sign(t);
ys = interp1(t,y,ti,'spline');
yp = interp1(t,y,ti,'pchip');
ddys = diff(diff(ys)./dti)./dti;
ddyp = diff(diff(yp)./dti)./dti;
figure(1);
plot (ti, ys,'r-', ti, yp,'g-');
legend('spline','pchip',4);
figure(2);
plot (ti, ddys,'r+', ti, ddyp,'g*');
legend('spline','pchip');
```

The result of which can be seen in Figure 27.1 and Figure 27.2.

Chapter 27: Interpolation    395

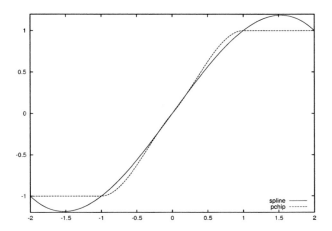

Figure 27.1: Comparison of 'phcip' and 'spline' interpolation methods for a step function

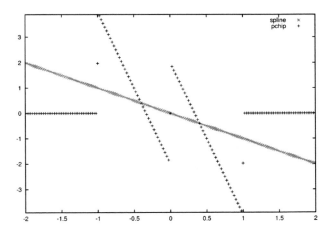

Figure 27.2: Comparison of the second derivative of the 'phcip' and 'spline' interpolation methods for a step function

Fourier interpolation, is a resampling technique where a signal is converted to the frequency domain, padded with zeros and then reconverted to the time domain.

interpft (x, n)                                                    Function File
interpft (x, n, dim)                                               Function File
   Fourier interpolation. If x is a vector, then x is resampled with n points. The
   data in x is assumed to be equispaced. If x is an array, then operate along
   each column of the array separately. If dim is specified, then interpolate
   along the dimension dim.
   interpft assumes that the interpolated function is periodic, and so assumptions are made about the end points of the interpolation.
   See also: interp1.

There are two significant limitations on Fourier interpolation. Firstly, the
function signal is assumed to be periodic, and so non-periodic signals will be
poorly represented at the edges. Secondly, both the signal and its interpolation
are required to be sampled at equispaced points. An example of the use of
interpft is

```
t = 0 : 0.3 : pi; dt = t(2)-t(1);
n = length (t); k = 100;
ti = t(1) + [0 : k-1]*dt*n/k;
y = sin (4*t + 0.3) .* cos (3*t - 0.1);
yp = sin (4*ti + 0.3) .* cos (3*ti - 0.1);
plot (ti, yp, 'g', ti, interp1(t, y, ti, 'spline'), 'b', ...
      ti, interpft (y, k), 'c', t, y, 'r+');
legend ('sin(4t+0.3)cos(3t-0.1','spline','interpft','data');
```

which demonstrates the poor behavior of Fourier interpolation for non-periodic
functions, as can be seen in Figure 27.3.

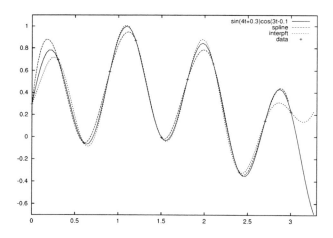

Figure 27.3: Comparison of interp1 and interpft for non-periodic data

The support functions spline and lookup that underlie the interp1 function
can be called directly.

## Chapter 27: Interpolation

*pp* = spline (*x, y*)                                                    Function File
*yi* = spline (*x, y, xi*)                                    Function File

Returns the cubic spline interpolation of *y* at the point *x*. Called with two arguments the piece-wise polynomial *pp* that may later be used with ppval to evaluate the polynomial at specific points.

The variable *x* must be a vector of length *n*, and *y* can be either a vector or array. In the case where *y* is a vector, it can have a length of either *n* or *n* + 2. If the length of *y* is *n*, then the 'not-a-knot' end condition is used. If the length of *y* is *n* + 2, then the first and last values of the vector *y* are the first derivative of the cubic spline at the end-points.

If *y* is an array, then the size of *y* must have the form

$$[s_1, s_2, \cdots, s_k, n]$$

or

$$[s_1, s_2, \cdots, s_k, n+2].$$

The array is then reshaped internally to a matrix where to leading dimension is given by

$$s_1 s_2 \cdots s_k$$

and each row this matrix is then treated separately. Note that this is exactly the opposite treatment than interp1 and is done for compatibility.

Called with a third input argument, spline evaluates the piece-wise spline at the points *xi*. There is an equivalence between ppval (spline (*x, y*), *xi*) and spline (*x, y, xi*).

See also: ppval, mkpp, unmkpp.

The lookup function is used by other interpolation functions to identify the points of the original data that are closest to the current point of interest.

*idx* = lookup (*table, y*)                                 Function File

Lookup values in a sorted table. Usually used as a prelude to interpolation. If table is strictly increasing and idx = lookup (table, y), then table(idx(i)) <= y(i) < table(idx(i+1)) for all y(i) within the table. If y(i) is before the table, then idx(i) is 0. If y(i) is after the table then idx(i) is table(n).

If the table is strictly decreasing, then the tests are reversed. There are no guarantees for tables which are non-monotonic or are not strictly monotonic.

To get an index value which lies within an interval of the table, use:

    idx = lookup (table(2:length(table)-1), y) + 1

This expression puts values before the table into the first interval, and values after the table into the last interval.

## 27.2 Multi-dimensional Interpolation

There are three multi-dimensional interpolation functions in Octave, with similar capabilities. Methods using Delaunay tessellation are described in Section 28.4 [Interpolation on Scattered Data], page 413.

| | |
|---|---|
| zi= interp2 (x, y, z, xi, yi) | Function File |
| zi= interp2 (Z, xi, yi) | Function File |
| zi= interp2 (Z, n) | Function File |
| zi= interp2 (..., method) | Function File |
| zi= interp2 (..., method, extrapval) | Function File |

Two-dimensional interpolation. $x$, $y$ and $z$ describe a surface function. If $x$ and $y$ are vectors their length must correspondent to the size of $z$. $x$ and $y$ must be monotonic. If they are matrices they must have the meshgrid format.

interp2 (x, y, Z, xi, yi, ...)
: Returns a matrix corresponding to the points described by the matrices $xi$, $yi$.

If the last argument is a string, the interpolation method can be specified. The method can be 'linear', 'nearest' or 'cubic'. If it is omitted 'linear' interpolation is assumed.

interp2 (z, xi, yi)
: Assumes $x$ = 1:rows ($z$) and $y$ = 1:columns ($z$)

interp2 (z, n)
: Interleaves the matrix $z$ n-times. If $n$ is omitted a value of $n$ = 1 is assumed.

The variable *method* defines the method to use for the interpolation. It can take one of the following values

'nearest'
: Return the nearest neighbor.

'linear'
: Linear interpolation from nearest neighbors.

'pchip'
: Piece-wise cubic hermite interpolating polynomial (not implemented yet).

'cubic'
: Cubic interpolation from four nearest neighbors.

'spline'
: Cubic spline interpolation–smooth first and second derivatives throughout the curve.

If a scalar value *extrapval* is defined as the final value, then values outside the mesh as set to this value. Note that in this case *method* must be defined as well. If *extrapval* is not defined then NA is assumed.

See also: interp1.

Chapter 27: Interpolation                                              399

vi = interp3 (x, y,z, v, xi, yi, zi)                         Function File
vi = interp3 (v, xi, yi, zi)                                 Function File
vi = interp3 (v, m)                                          Function File
vi = interp3 (v)                                             Function File
vi = interp3 (..., method)                                   Function File
vi = interp3 (..., method, extrapval)                        Function File

Perform 3-dimensional interpolation. Each element of the 3-dimensional array v represents a value at a location given by the parameters x, y, and z. The parameters x, x, and z are either 3-dimensional arrays of the same size as the array v in the 'meshgrid' format or vectors. The parameters xi, etc respect a similar format to x, etc, and they represent the points at which the array vi is interpolated.

If x, y, z are omitted, they are assumed to be x = 1 : size (v, 2), y = 1 : size (v, 1) and z = 1 : size (v, 3). If m is specified, then the interpolation adds a point half way between each of the interpolation points. This process is performed m times. If only v is specified, then m is assumed to be 1.

Method is one of:

'nearest'
    Return the nearest neighbour.

'linear'
    Linear interpolation from nearest neighbours.

'cubic'
    Cubic interpolation from four nearest neighbours (not implemented yet).

'spline'
    Cubic spline interpolation–smooth first and second derivatives throughout the curve.

The default method is 'linear'.

If extrap is the string 'extrap', then extrapolate values beyond the endpoints. If extrap is a number, replace values beyond the endpoints with that number. If extrap is missing, assume NA.

See also: interp1, interp2, spline, meshgrid.

vi = interpn (x1, x2, ..., v, y1, y2, ...)                   Function File
vi = interpn (v, y1, y2, ...)                                Function File
vi = interpn (v, m)                                          Function File
vi = interpn (v)                                             Function File
vi = interpn (..., method)                                   Function File
vi = interpn (..., method, extrapval)                        Function File

Perform n-dimensional interpolation, where n is at least two. Each element of the n-dimensional array v represents a value at a location given by the parameters x1, x2, ..., xn. The parameters x1, x2, ..., xn are either n-dimensional arrays of the same size as the array v in the 'ndgrid' format or vectors. The parameters *y1*, etc respect a similar format to *x1*, etc, and they represent the points at which the array *vi* is interpolated.

If *x1*, ..., *xn* are omitted, they are assumed to be x1 = 1 : size (v, 1), etc. If *m* is specified, then the interpolation adds a point half way between each of the interpolation points. This process is performed *m* times. If only *v* is specified, then *m* is assumed to be 1.

Method is one of:

'nearest'
> Return the nearest neighbour.

'linear'
> Linear interpolation from nearest neighbours.

'cubic'
> Cubic interpolation from four nearest neighbours (not implemented yet).

'spline'
> Cubic spline interpolation–smooth first and second derivatives throughout the curve.

The default method is 'linear'.

If *extrap* is the string 'extrap', then extrapolate values beyond the endpoints.
If *extrap* is a number, replace values beyond the endpoints with that number.
If *extrap* is missing, assume NA.

See also: interp1, interp2, spline, ndgrid.

A significant difference between `interpn` and the other two multidimensional interpolation functions is the fashion in which the dimensions are treated. For `interp2` and `interp3`, the 'y' axis is considered to be the columns of the matrix, whereas the 'x' axis corresponds to the rows of the array. As Octave indexes arrays in column major order, the first dimension of any array is the columns, and so `interpn` effectively reverses the 'x' and 'y' dimensions. Consider the example

```
x = y = z = -1:1;
f = @(x,y,z) x.^2 - y - z.^2;
[xx, yy, zz] = meshgrid (x, y, z);
v = f (xx,yy,zz);
xi = yi = zi = -1:0.1:1;
[xxi, yyi, zzi] = meshgrid (xi, yi, zi);
vi = interp3(x, y, z, v, xxi, yyi, zzi, 'spline');
[xxi, yyi, zzi] = ndgrid (xi, yi, zi);
vi2 = interpn(x, y, z, v, xxi, yyi, zzi, 'spline');
mesh (zi, yi, squeeze (vi2(1,:,:)));
```

where vi and vi2 are identical. The reversal of the dimensions is treated in the `meshgrid` and `ndgrid` functions respectively. The result of this code can be seen in Figure 27.4.

# Chapter 27: Interpolation

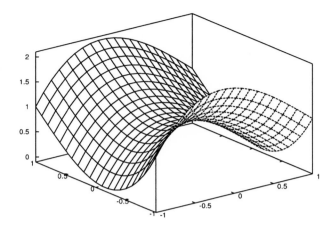

Figure 27.4: Demonstration of the use of interpn

The support function bicubic that underlies the cubic interpolation of interp2 function can be called directly.

*zi* = bicubic (*x, y, z, xi, yi, extrapval*)  Function File

Return a matrix *zi* corresponding to the bicubic interpolations at *xi* and *yi* of the data supplied as *x*, *y* and *z*. Points outside the grid are set to *extrapval*.

See http://wiki.woodpecker.org.cn/moin/Octave/Bicubic for further information.

**See also:** interp2.

# 28 Geometry

Much of geometry code in Octave is based on the QHull library[1]. Some of the documentation for Qhull, particularly for the options that can be passed to delaunay, voronoi and convhull, etc, is relevant to Octave users.

## 28.1 Delaunay Triangulation

The Delaunay triangulation of a set of points is constructed from a set of *circum-circles*. These are circles which are chosen so that at least three points in the set lie on each circumference and no point in the set falls inside any circum-circle.

In general there are only three points on the circumference of any circum-circle. However, in the some cases, and in particular for the case of a regular grid, 4 or more points can be on a single circum-circle. In this case the Delaunay triangulation is not unique.

*tri* = delaunay (*x*, *y*)  *Function File*
*tri* = delaunay (*x*, *y*, *opt*)  *Function File*

The return matrix of size [n, 3] contains a set triangles which are described by the indices to the data point x and y vector. The triangulation satisfies the Delaunay circumcircle criterion. No other data point is in the circumcircle of the defining triangle.

A third optional argument, which must be a string, contains extra options passed to the underlying qhull command. See the documentation for the Qhull library for details.

```
x = rand (1, 10);
y = rand (size (x));
T = delaunay (x, y);
X = [x(T(:,1)); x(T(:,2)); x(T(:,3)); x(T(:,1))];
Y = [y(T(:,1)); y(T(:,2)); y(T(:,3)); y(T(:,1))];
axis ([0,1,0,1]);
plot (X, Y, "b", x, y, "r*");
```

See also: voronoi, delaunay3, delaunayn.

The 3- and N-dimensional extension of the Delaunay triangulation are given by delaunay3 and delaunayn respectively. delaunay3 returns a set of tetrahedra that satisfy the Delaunay circum-circle criteria. Similarly, delaunayn returns the N-dimensional simplex satisfying the Delaunay circum-circle criteria. The N-dimensional extension of a triangulation is called a tessellation.

---

[1] Barber, C.B., Dobkin, D.P., and Huhdanpaa, H.T., "The Quickhull algorithm for convex hulls," ACM Trans. on Mathematical Software, 22(4):469-483, Dec 1996, http://www.qhull.org

T = delaunay3 (x, y, z)                                          Function File
T = delaunay3 (x, y, z, opt)                                     Function File
  A matrix of size [n, 4] is returned. Each row contains a set of tetrahedron
  which are described by the indices to the data point vectors (x,y,z).

  A fourth optional argument, which must be a string or cell array of strings,
  contains extra options passed to the underlying qhull command. See the
  documentation for the Qhull library for details.

  See also: delaunay,delaunayn.

T = delaunayn (P)                                                Function File
T = delaunayn (P, opt)                                           Function File
  Form the Delaunay triangulation for a set of points. The Delaunay triangulation is a tessellation of the convex hull of the points such that no n-sphere
  defined by the n-triangles contains any other points from the set. The input
  matrix P of size [n, dim] contains n points in a space of dimension dim.
  The return matrix T has the size [m, dim+1]. It contains for each row a set
  of indices to the points, which describes a simplex of dimension dim. For
  example, a 2d simplex is a triangle and 3d simplex is a tetrahedron.

  Extra options for the underlying Qhull command can be specified by the
  second argument. This argument is a cell array of strings. The default
  options depend on the dimension of the input:

  - 2D and 3D: $opt = $ {"Qt", "Qbb", "Qc"}
  - 4D and higher: $opt = $ {"Qt", "Qbb", "Qc", "Qz"}

  If opt is [], then the default arguments are used. If opt is {""}, then none
  of the default arguments are used by Qhull. See the Qhull documentation
  for the available options.

  All options can also be specified as single string, for example "Qt Qbb Qc
  Qz".

  An example of a Delaunay triangulation of a set of points is
```
      rand ("state", 2);
      x = rand (10, 1);
      y = rand (10, 1);
      T = delaunay (x, y);
      X = [ x(T(:,1)); x(T(:,2)); x(T(:,3)); x(T(:,1)) ];
      Y = [ y(T(:,1)); y(T(:,2)); y(T(:,3)); y(T(:,1)) ];
      axis ([0, 1, 0, 1]);
      plot(X, Y, "b", x, y, "r*");
```
The result of which can be seen in Figure 28.1.

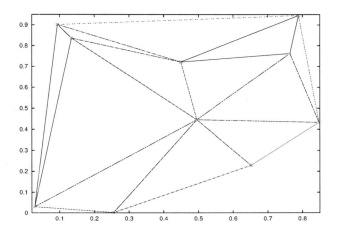

Figure 28.1: Delaunay triangulation of a random set of points

### 28.1.1 Plotting the Triangulation

Octave has the functions `triplot` and `trimesh` to plot the Delaunay triangulation of a 2-dimensional set of points.

| | |
|---|---:|
| `triplot` (*tri*, *x*, *y*) | Function File |
| `triplot` (*tri*, *x*, *y*, *linespec*) | Function File |
| *h* = `triplot` (...) | Function File |

Plot a triangular mesh in 2D. The variable *tri* is the triangular meshing of the points (*x*, *y*) which is returned from delaunay. If given, the *linespec* determines the properties to use for the lines. The output argument *h* is the graphic handle to the plot.

**See also:** plot, trimesh, delaunay.

| | |
|---|---:|
| `trimesh` (*tri*, *x*, *y*, *z*) | Function File |
| *h* = `trimesh` (...) | Function File |

Plot a triangular mesh in 3D. The variable *tri* is the triangular meshing of the points (*x*, *y*) which is returned from delaunay. The variable *z* is value at the point (*x*, *y*). The output argument *h* is the graphic handle to the plot.

**See also:** triplot, delaunay3.

The difference between `triplot` and `trimesh` is that the former only plots the 2-dimensional triangulation itself, whereas the second plots the value of some function f (*x*, *y*). An example of the use of the `triplot` function is

```
rand ("state", 2)
x = rand (20, 1);
y = rand (20, 1);
tri = delaunay (x, y);
triplot (tri, x, y);
```
that plot the Delaunay triangulation of a set of random points in 2-dimensions. The output of the above can be seen in Figure 28.2.

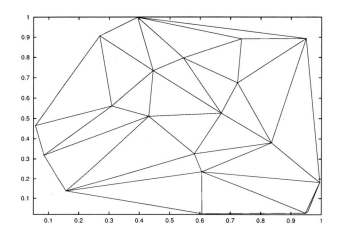

Figure 28.2: Delaunay triangulation of a random set of points

### 28.1.2 Identifying points in Triangulation

It is often necessary to identify whether a particular point in the N-dimensional space is within the Delaunay tessellation of a set of points in this N-dimensional space, and if so which N-simplex contains the point and which point in the tessellation is closest to the desired point. The functions tsearch and dsearch perform this function in a triangulation, and tsearchn and dsearchn in an N-dimensional tessellation.

To identify whether a particular point represented by a vector p falls within one of the simplices of an N-simplex, we can write the Cartesian coordinates of the point in a parametric form with respect to the N-simplex. This parametric form is called the barycentric coordinates of the point. If the points defining the N-simplex are given by $N + 1$ vectors $t(i,:)$, then the barycentric coordinates defining the point p are given by

p = sum (beta(1:N+1) * t(1:N+1),:)

where there are $N + 1$ values beta(i) that together as a vector represent the barycentric coordinates of the point p. To ensure a unique solution for the values of beta(i) an additional criteria of sum (beta(1:N+1)) == 1

is imposed, and we can therefore write the above as

p - t(end, :) = beta(1:end-1) * (t(1:end-1, :)
    - ones(N, 1) * t(end, :)

Solving for beta we can then write

beta(1:end-1) = (p - t(end, :)) / (t(1:end-1, :)
    - ones(N, 1) * t(end, :))
beta(end) = sum(beta(1:end-1))

which gives the formula for the conversion of the Cartesian coordinates of the point $p$ to the barycentric coordinates beta. An important property of the barycentric coordinates is that for all points in the N-simplex

0 <= beta(i) <= 1

Therefore, the test in tsearch and tsearchn essentially only needs to express each point in terms of the barycentric coordinates of each of the simplices of the N-simplex and test the values of beta. This is exactly the implementation used in tsearchn. tsearch is optimized for 2-dimensions and the barycentric coordinates are not explicitly formed.

idx = tsearch (x, y, t, xi, yi)                                Loadable Function
    Searches for the enclosing Delaunay convex hull. For t = delaunay (x, y), finds the index in t containing the points (xi, yi). For points outside the convex hull, idx is NaN.

    See also: delaunay, delaunayn.

[idx, p] = tsearchn (x, t, xi)                                     Function File
    Searches for the enclosing Delaunay convex hull. For t = delaunayn (x), finds the index in t containing the points xi. For points outside the convex hull, idx is NaN. If requested tsearchn also returns the barycentric coordinates p of the enclosing triangles.

    See also: delaunay, delaunayn.

An example of the use of tsearch can be seen with the simple triangulation

    x = [-1; -1; 1; 1];
    y = [-1; 1; -1; 1];
    tri = [1, 2, 3; 2, 3, 1];

consisting of two triangles defined by tri. We can then identify which triangle a point falls in like

    tsearch (x, y, tri, -0.5, -0.5)
    ⇒ 1
    tsearch (x, y, tri, 0.5, 0.5)
    ⇒ 2

and we can confirm that a point doesn't lie within one of the triangles like

```
tsearch (x, y, tri, 2, 2)
⇒ NaN
```

The dsearch and dsearchn find the closest point in a tessellation to the desired point. The desired point does not necessarily have to be in the tessellation, and even if it the returned point of the tessellation does not have to be one of the vertexes of the N-simplex within which the desired point is found.

*idx* = dsearch (*x*, *y*, *tri*, *xi*, *yi*)                         *Function File*
*idx* = dsearch (*x*, *y*, *tri*, *xi*, *yi*, *s*)                      *Function File*
    Returns the index *idx* or the closest point in *x*, *y* to the elements [*xi*(:), *yi*(:)]. The variable *s* is accepted but ignored for compatibility.

    **See also:** dsearchn, tsearch.

*idx* = dsearchn (*x*, *tri*, *xi*)                                     *Function File*
*idx* = dsearchn (*x*, *tri*, *xi*, *outval*)                         *Function File*
*idx* = dsearchn (*x*, *xi*)                                          *Function File*
[*idx*, *d*] = dsearchn (...)                                     *Function File*
    Returns the index *idx* or the closest point in *x* to the elements *xi*. If *outval* is supplied, then the values of *xi* that are not contained within one of the simplices *tri* are set to *outval*. Generally, *tri* is returned from delaunayn (*x*).

    **See also:** dsearch, tsearch.

An example of the use of dsearch, using the above values of *x*, *y* and *tri* is
```
dsearch (x, y, tri, -2, -2)
⇒ 1
```
If you wish the points that are outside the tessellation to be flagged, then dsearchn can be used as
```
dsearchn ([x, y], tri, [-2, -2], NaN)
⇒ NaN
dsearchn ([x, y], tri, [-0.5, -0.5], NaN)
⇒ 1
```
where the point outside the tessellation are then flagged with NaN.

## 28.2 Voronoi Diagrams

A Voronoi diagram or Voronoi tessellation of a set of points *s* in N-dimensional space divides the space into volumes v(p) around each point *p* such that everywhere in v(p) is closer to *p* than any other point in *s*. The Voronoi diagram is related to the Delaunay triangulation of a set of points, in that the vertexes of the Voronoi tessellation are the centers of the circum-circles of the simplices of the Delaunay tessellation.

voronoi (x, y)                                      Function File
voronoi (x, y, "plotstyle")                         Function File
voronoi (x, y, "plotstyle", *options*)              Function File
[vx, vy] = voronoi (...)                            Function File

plots voronoi diagram of points (x, y). The voronoi facets with points at infinity are not drawn. [vx, vy] = voronoi(...) returns the vertices instead of plotting the diagram. plot (vx, vy) shows the voronoi diagram.

A fourth optional argument, which must be a string, contains extra options passed to the underlying qhull command. See the documentation for the Qhull library for details.

```
x = rand (10, 1);
y = rand (size (x));
h = convhull (x, y);
[vx, vy] = voronoi (x, y);
plot (vx, vy, "-b", x, y, "o", x(h), y(h), "-g")
legend ("", "points", "hull");
```

See also: voronoin, delaunay, convhull.

[C, F] = voronoin (*pts*)                           Function File
[C, F] = voronoin (*pts*, *options*)                Function File

computes n- dimensional voronoi facets. The input matrix *pts* of size [n, dim] contains n points of dimension dim. *C* contains the points of the voronoi facets. The list *F* contains for each facet the indices of the voronoi points.

A second optional argument, which must be a string, contains extra options passed to the underlying qhull command. See the documentation for the Qhull library for details.

See also: voronoin, delaunay, convhull.

An example of the use of voronoi is
```
rand("state",9);
x = rand(10,1);
y = rand(10,1);
tri = delaunay (x, y);
[vx, vy] = voronoi (x, y, tri);
triplot (tri, x, y, "b");
hold on;
plot (vx, vy, "r");
```

The result of which can be seen in Figure 28.3. Note that the circum-circle of one of the triangles has been added to this figure, to make the relationship between the Delaunay tessellation and the Voronoi diagram clearer.

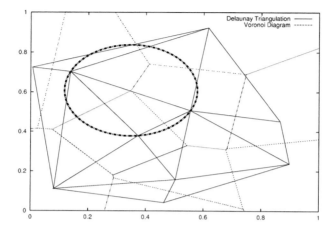

Figure 28.3: Delaunay triangulation and Voronoi diagram of a random set of points

Additional information about the size of the facets of a Voronoi diagram, and which points of a set of points is in a polygon can be had with the polyarea and inpolygon functions respectively.

polyarea (*x*, *y*)                                                           Function File
polyarea (*x*, *y*, *dim*)                                       Function File

Determines area of a polygon by triangle method. The variables *x* and *y* define the vertex pairs, and must therefore have the same shape. They can be either vectors or arrays. If they are arrays then the columns of *x* and *y* are treated separately and an area returned for each.

If the optional *dim* argument is given, then polyarea works along this dimension of the arrays *x* and *y*.

An example of the use of polyarea might be

```
rand ("state", 2);
x = rand (10, 1);
y = rand (10, 1);
[c, f] = voronoin ([x, y]);
af = zeros (size(f));
for i = 1 : length (f)
   af(i) = polyarea (c (f {i, :}, 1), c (f {i, :}, 2));
endfor
```

Facets of the Voronoi diagram with a vertex at infinity have infinity area.

Chapter 28: Geometry                                                  411

[in, on] = inpolygon (x, y, xv, xy)                        Function File
   For a polygon defined by (xv, yv) points, determine if the points (x, y)
   are inside or outside the polygon. The variables x, y, must have the same
   dimension. The optional output on gives the points that are on the polygon.

   An example of the use of inpolygon might be
      randn ("state", 2);
      x = randn (100, 1);
      y = randn (100, 1);
      vx = cos (pi * [-1 : 0.1: 1]);
      vy = sin (pi * [-1 : 0.1 : 1]);
      in = inpolygon (x, y, vx, vy);
      plot(vx, vy, x(in), y(in), "r+", x(!in), y(!in), "bo");
      axis ([-2, 2, -2, 2]);
The result of which can be seen in Figure 28.4.

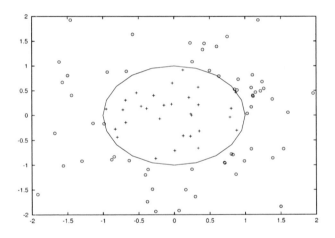

Figure 28.4: Demonstration of the inpolygon function to determine the points inside a polygon

## 28.3 Convex Hull

The convex hull of a set of points is the minimum convex envelope containing all of the points. Octave has the functions convhull and convhulln to calculate the convex hull of 2-dimensional and N-dimensional sets of points.

H = convhull (*x*, *y*)                                                    Function File
H = convhull (*x*, *y*, *opt*)                                      Function File

    Returns the index vector to the points of the enclosing convex hull. The data points are defined by the x and y vectors.

    A third optional argument, which must be a string, contains extra options passed to the underlying qhull command. See the documentation for the Qhull library for details.

    **See also:** delaunay, convhulln.

H = convhulln (*p*)                                                    Loadable Function
H = convhulln (*p*, *opt*)                                       Loadable Function

    Returns an index vector to the points of the enclosing convex hull. The input matrix of size [n, dim] contains n points of dimension dim.

    If a second optional argument is given, it must be a string or cell array of strings containing options for the underlying qhull command. (See the Qhull documentation for the available options.) The default options are "s Qci Tcv".

    **See also:** convhull, delaunayn.

An example of the use of convhull is

```
x = -3:0.05:3;
y = abs (sin (x));
k = convhull (x, y);
plot (x(k), y(k), "r-", x, y, "b+");
axis ([-3.05, 3.05, -0.05, 1.05]);
```

The output of the above can be seen in Figure 28.5.

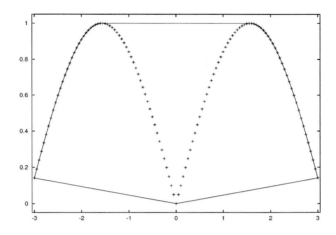

Figure 28.5: The convex hull of a simple set of points

## 28.4 Interpolation on Scattered Data

An important use of the Delaunay tessellation is that it can be used to interpolate from scattered data to an arbitrary set of points. To do this the N-simplex of the known set of points is calculated with delaunay, delaunay3 or delaunayn. Then the simplices in which the desired points are found are identified. Finally the vertices of the simplices are used to interpolate to the desired points. The functions that perform this interpolation are griddata, griddata3 and griddatan.

zi = griddata (x, y, z, xi, yi, method)        Function File
[xi, yi, zi] = griddata (x, y, z, xi, yi, method)    Function File
 Generate a regular mesh from irregular data using interpolation. The function is defined by $z = f(x, y)$. The interpolation points are all $(xi, yi)$. If xi, yi are vectors then they are made into a 2D mesh.

 The interpolation method can be "nearest", "cubic" or "linear". If method is omitted it defaults to "linear".

 See also: delaunay.

vi = griddata3 (x, y, z, v xi, yi, zi, method, options)    Function File
 Generate a regular mesh from irregular data using interpolation. The function is defined by $y = f(x,y,z)$. The interpolation points are all xi.

 The interpolation method can be "nearest" or "linear". If method is omitted it defaults to "linear".

 See also: griddata, delaunayn.

*yi* = griddatan (*x*, *y*, *xi*, *method*, *options*)　　　　Function File
Generate a regular mesh from irregular data using interpolation. The function is defined by y = f (x). The interpolation points are all *xi*.

The interpolation method can be "nearest" or "linear". If method is omitted it defaults to "linear".

**See also:** griddata, delaunayn.

An example of the use of the griddata function is
```
rand("state",1);
x=2*rand(1000,1)-1;
y=2*rand(size(x))-1;
z=sin(2*(x.^2+y.^2));
[xx,yy]=meshgrid(linspace(-1,1,32));
griddata(x,y,z,xx,yy);
```
that interpolates from a random scattering of points, to a uniform grid. The output of the above can be seen in Figure 28.6.

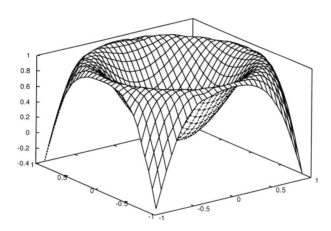

Figure 28.6: Interpolation from a scattered data to a regular grid

# 29 Signal Processing

This chapter describes the signal processing and fast fourier transform functions available in Octave. Fast fourier transforms are computed with the FFTW or FFTPACK libraries depending on how Octave is built.

## 29.1 Fast Fourier Transforms

fft (a, n, dim)                                                    Loadable Function
Compute the FFT of a. The FFT is calculated along the first non-singleton dimension of the array. Thus if a is a matrix, fft (a) computes the FFT for each column of a.

If called with two arguments, n is expected to be an integer specifying the number of elements of a to use, or an empty matrix to specify that its value should be ignored. If n is larger than the dimension along which the FFT is calculated, then a is resized and padded with zeros. Otherwise, if n is smaller than the dimension along which the FFT is calculated, then a is truncated.

If called with three arguments, dim is an integer specifying the dimension of the matrix along which the FFT is performed

See also: ifft, fft2, fftn, fftw.

ifft (a, n, dim)                                                   Loadable Function
Compute the inverse FFT of a. The inverse FFT is calculated along the first non-singleton dimension of the array. Thus if a is a matrix, fft (a) computes the inverse FFT for each column of a.

If called with two arguments, n is expected to be an integer specifying the number of elements of a to use, or an empty matrix to specify that its value should be ignored. If n is larger than the dimension along which the inverse FFT is calculated, then a is resized and padded with zeros. Otherwise, if n is smaller than the dimension along which the inverse FFT is calculated, then a is truncated.

If called with three arguments, dim is an integer specifying the dimension of the matrix along which the inverse FFT is performed

See also: fft, ifft2, ifftn, fftw.

fft2 (a, n, m)                                                     Loadable Function
Compute the two-dimensional FFT of a. The optional arguments n and m may be used specify the number of rows and columns of a to use. If either of these is larger than the size of a, a is resized and padded with zeros.

If a is a multi-dimensional matrix, each two-dimensional sub-matrix of a is treated separately

See also: ifft2, fft, fftn, fftw.

fft2 (a, n, m)                                              Loadable Function
   Compute the inverse two-dimensional FFT of a. The optional arguments n
   and m may be used specify the number of rows and columns of a to use.
   If either of these is larger than the size of a, a is resized and padded with
   zeros.

   If a is a multi-dimensional matrix, each two-dimensional sub-matrix of a is
   treated separately

   See also: fft2, ifft, ifftn, fftw.

fftn (a, size)                                              Loadable Function
   Compute the N-dimensional FFT of a. The optional vector argument size
   may be used specify the dimensions of the array to be used. If an element
   of size is smaller than the corresponding dimension, then the dimension is
   truncated prior to performing the FFT. Otherwise if an element of size is
   larger than the corresponding dimension a is resized and padded with zeros.

   See also: ifftn, fft, fft2, fftw.

ifftn (a, size)                                             Loadable Function
   Compute the inverse N-dimensional FFT of a. The optional vector argument
   size may be used specify the dimensions of the array to be used. If an element
   of size is smaller than the corresponding dimension, then the dimension is
   truncated prior to performing the inverse FFT. Otherwise if an element of
   size is larger than the corresponding dimension a is resized and padded with
   zeros.

   See also: fftn, ifft, ifft2, fftw.

fftconv (a, b, n)                                                Function File
   Return the convolution of the vectors a and b, as a vector with length equal
   to the length (a) + length (b) - 1. If a and b are the coefficient vectors of
   two polynomials, the returned value is the coefficient vector of the product
   polynomial.

   The computation uses the FFT by calling the function fftfilt. If the
   optional argument n is specified, an N-point FFT is used.

fftfilt (b, x, n)                                                Function File
   With two arguments, fftfilt filters x with the FIR filter b using the FFT.

   Given the optional third argument, n, fftfilt uses the overlap-add method
   to filter x with b using an N-point FFT.

   If x is a matrix, filter each column of the matrix.

# Chapter 29: Signal Processing

`method = fftw ('planner')`                     Loadable Function
`fftw ('planner', method)`                      Loadable Function
`wisdom = fftw ('dwisdom')`                     Loadable Function
`wisdom = fftw ('dwisdom', wisdom)`             Loadable Function

This function is specific to FFTW. It is used to manage FFTW's precomputed tables known as *wisdom*, which impose an initial cost but significantly increase the speed of FFTW. The wisdom used by Octave can be manipulated with the function fftw. For example

   `wisdom = fftw ('dwisdom')`

will save the existing wisdom used by Octave to the string *wisdom*. This string can then be saved in the usual manner. This existing wisdom can be reimported as follows

   `fftw ('dwisdom', wisdom)`

If *wisdom* is an empty matrix, then the wisdom used is cleared.

During the calculation of fourier transforms further wisdom is generated. The fashion in which this wisdom is generated is equally controlled by the *fftw* function. There are five different manners in which the wisdom can be treated, these being

'estimate'
: This specifies that no run-time measurement of the optimal means of calculating a particular is performed, and a simple heuristic is used to pick a (probably sub-optimal) plan. The advantage of this method is that there is little or no overhead in the generation of the plan, which is appropriate for a fourier transform that will be calculated once.

'measure'
: In this case a range of algorithms to perform the transform is considered and the best is selected based on their execution time.

'patient'
: This is like 'measure', but a wider range of algorithms is considered.

'exhaustive'
: This is like 'measure', but all possible algorithms that may be used to treat the transform are considered.

'hybrid'
: As run-time measurement of the algorithm can be expensive, this is a compromise where 'measure' is used for transforms up to the size of 8192 and beyond that the 'estimate' method is used.

The default method is 'estimate', and the method currently being used can be probed with

   `method = fftw ('planner')`

and the method used can be set using

    fftw ('planner', method)
    Note that calculated wisdom will be lost when restarting Octave. However, the wisdom data can be reloaded if it is saved to a file as described above. Also, any system-wide wisdom file that has been found will also be used. Saved wisdom files should not be used on different platforms since they will not be efficient and the point of calculating the wisdom is lost.

    See also: fft, ifft, fft2, ifft2, fftn, ifftn.

## 29.2 Filters and Windowing Functions

detrend (x, p)  Function File
   If x is a vector, detrend (x, p) removes the best fit of a polynomial of order p from the data x.

   If x is a matrix, detrend (x, p) does the same for each column in x.

   The second argument is optional. If it is not specified, a value of 1 is assumed. This corresponds to removing a linear trend.

y = filter (b, a, x)  Loadable Function
[y, sf] = filter (b, a, x, si)  Loadable Function
[y, sf] = filter (b, a, x, [], dim)  Loadable Function
[y, sf] = filter (b, a, x, si, dim)  Loadable Function
   Return the solution to the following linear, time-invariant difference equation:

$$\sum_{k=0}^{N} a_{k+1} y_{n-k} = \sum_{k=0}^{M} b_{k+1} x_{n-k}, \quad 1 \leq n \leq P$$

where $a \in \Re^{N-1}$, $b \in \Re^{M-1}$, and $x \in \Re^{P}$. over the first non-singleton dimension of x or over dim if supplied. An equivalent form of this equation is:

$$y_n = -\sum_{k=1}^{N} c_{k+1} y_{n-k} + \sum_{k=0}^{M} d_{k+1} x_{n-k}, \quad 1 \leq n \leq P$$

where $c = a/a_1$ and $d = b/a_1$.

If the fourth argument si is provided, it is taken as the initial state of the system and the final state is returned as sf. The state vector is a column vector whose length is equal to the length of the longest coefficient vector minus one. If si is not supplied, the initial state vector is set to all zeros.

In terms of the z-transform, y is the result of passing the discrete- time signal x through a system characterized by the following rational system function:

$$H(z) = \frac{\sum_{k=0}^{M} d_{k+1} z^{-k}}{1 + \sum_{k+1}^{N} c_{k+1} z^{-k}}$$

y = filter2 (b, x)                                           Function File
y = filter2 (b, x, shape)                                    Function File
    Apply the 2-D FIR filter b to x. If the argument shape is specified, return an array of the desired shape. Possible values are:

'full'
    pad x with zeros on all sides before filtering.

'same'
    unpadded x (default)

'valid'
    trim x after filtering so edge effects are no included.

Note this is just a variation on convolution, with the parameters reversed and b rotated 180 degrees.

See also: conv2.

[h, w] = freqz (b, a, n, "whole")                            Function File
    Return the complex frequency response h of the rational IIR filter whose numerator and denominator coefficients are b and a, respectively. The response is evaluated at n angular frequencies between 0 and $2\pi$.

The output value w is a vector of the frequencies.

If the fourth argument is omitted, the response is evaluated at frequencies between 0 and $\pi$.

If n is omitted, a value of 512 is assumed.

If a is omitted, the denominator is assumed to be 1 (this corresponds to a simple FIR filter).

For fastest computation, n should factor into a small number of small primes.

h = freqz (b, a, w)                                          Function File
    Evaluate the response at the specific frequencies in the vector w. The values for w are measured in radians.

[...] = freqz (..., Fs)                                      Function File
    Return frequencies in Hz instead of radians assuming a sampling rate Fs. If you are evaluating the response at specific frequencies w, those frequencies should be requested in Hz rather than radians.

freqz (...)                                                  Function File
    Plot the pass band, stop band and phase response of h rather than returning them.

freqz_plot (w, h)                                            Function File
    Plot the pass band, stop band and phase response of h.

sinc (x)                                                     Function File
    Return $\sin(\pi x)/(\pi x)$.

`b = unwrap (a, tol, dim)`     Function File
Unwrap radian phases by adding multiples of 2*pi as appropriate to remove jumps greater than *tol*. *tol* defaults to pi.

Unwrap will unwrap along the first non-singleton dimension of *a*, unless the optional argument *dim* is given, in which case the data will be unwrapped along this dimension

`[a, b] = arch_fit (y, x, p, iter, gamma, a0, b0)`     Function File
Fit an ARCH regression model to the time series *y* using the scoring algorithm in Engle's original ARCH paper. The model is
```
y(t) = b(1) * x(t,1) + ... + b(k) * x(t,k) + e(t),
h(t) = a(1) + a(2) * e(t-1)^2 + ... + a(p+1) * e(t-p)^2
```
in which $e(t)$ is $N(0, h(t))$, given a time-series vector *y* up to time $t-1$ and a matrix of (ordinary) regressors *x* up to *t*. The order of the regression of the residual variance is specified by *p*.

If invoked as `arch_fit (y, k, p)` with a positive integer *k*, fit an ARCH(*k*, *p*) process, i.e., do the above with the *t*-th row of *x* given by
```
[1, y(t-1), ..., y(t-k)]
```
Optionally, one can specify the number of iterations *iter*, the updating factor *gamma*, and initial values *a0* and *b0* for the scoring algorithm.

`arch_rnd (a, b, t)`     Function File
Simulate an ARCH sequence of length *t* with AR coefficients *b* and CH coefficients *a*. I.e., the result $y(t)$ follows the model
```
y(t) = b(1) + b(2) * y(t-1) + ... + b(lb) * y(t-lb+1) + e(t),
```
where $e(t)$, given *y* up to time $t-1$, is $N(0, h(t))$, with
```
h(t) = a(1) + a(2) * e(t-1)^2 + ... + a(la) * e(t-la+1)^2
```

`[pval, lm] = arch_test (y, x, p)`     Function File
For a linear regression model
```
y = x * b + e
```
perform a Lagrange Multiplier (LM) test of the null hypothesis of no conditional heteroscedascity against the alternative of CH(*p*).
I.e., the model is
```
y(t) = b(1) * x(t,1) + ... + b(k) * x(t,k) + e(t),
```
given *y* up to $t-1$ and *x* up to *t*, $e(t)$ is $N(0, h(t))$ with
```
h(t) = v + a(1) * e(t-1)^2 + ... + a(p) * e(t-p)^2,
```
and the null is $a(1) = \ldots = a(p) = 0$.

If the second argument is a scalar integer, *k*, perform the same test in a linear autoregression model of order *k*, i.e., with

Chapter 29: Signal Processing                                              421

>        [1, y(t-1), ..., y(t-k)]

as the $t$-th row of $x$.

Under the null hypothesis, LM approximately has a chisquare distribution with $p$ degrees of freedom and *pval* is the $p$-value (1 minus the CDF of this distribution at LM) of the test.

If no output argument is given, the $p$-value is displayed.

**arma_rnd (a, b, v, t, n)**                                  *Function File*
>   Return a simulation of the ARMA model
>
>        x(n) = a(1) * x(n-1) + ... + a(k) * x(n-k)
>             + e(n) + b(1) * e(n-1) + ... + b(l) * e(n-l)
>
>   in which $k$ is the length of vector $a$, $l$ is the length of vector $b$ and $e$ is gaussian white noise with variance $v$. The function returns a vector of length $t$.
>
>   The optional parameter $n$ gives the number of dummy $x(i)$ used for initialization, i.e., a sequence of length $t+n$ is generated and $x(n+1:t+n)$ is returned. If $n$ is omitted, $n = 100$ is used.

**autocor (x, h)**                                            *Function File*
>   Return the autocorrelations from lag 0 to $h$ of vector $x$. If $h$ is omitted, all autocorrelations are computed. If $x$ is a matrix, the autocorrelations of each column are computed.

**autocov (x, h)**                                            *Function File*
>   Return the autocovariances from lag 0 to $h$ of vector $x$. If $h$ is omitted, all autocovariances are computed. If $x$ is a matrix, the autocovariances of each column are computed.

**autoreg_matrix (y, k)**                                     *Function File*
>   Given a time series (vector) $y$, return a matrix with ones in the first column and the first $k$ lagged values of $y$ in the other columns. I.e., for $t > k$, $[1, y(t-1), ..., y(t-k)]$ is the t-th row of the result. The resulting matrix may be used as a regressor matrix in autoregressions.

**bartlett (m)**                                              *Function File*
>   Return the filter coefficients of a Bartlett (triangular) window of length $m$.
>
>   For a definition of the Bartlett window, see e.g. A. V. Oppenheim & R. W. Schafer, *Discrete-Time Signal Processing*.

**blackman (m)**                                              *Function File*
>   Return the filter coefficients of a Blackman window of length $m$.
>
>   For a definition of the Blackman window, see e.g. A. V. Oppenheim & R. W. Schafer, *Discrete-Time Signal Processing*.

[d, dd] = diffpara (x, a, b)  Function File
Return the estimator d for the differencing parameter of an integrated time series.

The frequencies from $[2*pi*a/t, 2*pi*b/T]$ are used for the estimation. If b is omitted, the interval $[2*pi/T, 2*pi*a/T]$ is used. If both b and a are omitted then $a = 0.5 * sqrt(T)$ and $b = 1.5 * sqrt(T)$ is used, where $T$ is the sample size. If x is a matrix, the differencing parameter of each column is estimated.

The estimators for all frequencies in the intervals described above are returned in dd. The value of d is simply the mean of dd.

Reference: Brockwell, Peter J. & Davis, Richard A. Time Series: Theory and Methods Springer 1987.

durbinlevinson (c, oldphi, oldv)  Function File
Perform one step of the Durbin-Levinson algorithm.

The vector c specifies the autocovariances [gamma_0, ..., gamma_t] from lag 0 to t, oldphi specifies the coefficients based on c(t-1) and oldv specifies the corresponding error.

If oldphi and oldv are omitted, all steps from 1 to t of the algorithm are performed.

fftshift (v)  Function File
fftshift (v, dim)  Function File
Perform a shift of the vector v, for use with the fft and ifft functions, in order the move the frequency 0 to the center of the vector or matrix.

If v is a vector of $N$ elements corresponding to $N$ time samples spaced of $Dt$ each, then fftshift (fft (v)) corresponds to frequencies

```
f = ((1:N) - ceil(N/2)) / N / Dt
```

If v is a matrix, the same holds for rows and columns. If v is an array, then the same holds along each dimension.

The optional dim argument can be used to limit the dimension along which the permutation occurs.

ifftshift (v)  Function File
ifftshift (v, dim)  Function File
Undo the action of the fftshift function. For even length v, fftshift is its own inverse, but odd lengths differ slightly.

fractdiff (x, d)  Function File
Compute the fractional differences $(1 - L)^d x$ where $L$ denotes the lag-operator and d is greater than -1.

hamming (m)  Function File
Return the filter coefficients of a Hamming window of length m.

For a definition of the Hamming window, see e.g. A. V. Oppenheim & R. W. Schafer, *Discrete-Time Signal Processing*.

hanning (m)                                                                Function File
Return the filter coefficients of a Hanning window of length $m$.
For a definition of this window type, see e.g. A. V. Oppenheim & R. W. Schafer, *Discrete-Time Signal Processing*.

hurst (x)                                                                     Function File
Estimate the Hurst parameter of sample $x$ via the rescaled range statistic. If $x$ is a matrix, the parameter is estimated for every single column.

pp = pchip (x, y)                                                Function File
yi = pchip (x, y, xi)                                            Function File
Piecewise Cubic Hermite interpolating polynomial. Called with two arguments, the piece-wise polynomial *pp* is returned, that may later be used with ppval to evaluate the polynomial at specific points.

The variable $x$ must be a strictly monotonic vector (either increasing or decreasing). While $y$ can be either a vector or array. In the case where $y$ is a vector, it must have a length of $n$. If $y$ is an array, then the size of $y$ must have the form

$$[s_1, s_2, \cdots, s_k, n]$$

The array is then reshaped internally to a matrix where the leading dimension is given by

$$s_1 s_2 \cdots s_k$$

and each row in this matrix is then treated separately. Note that this is exactly the opposite treatment than interp1 and is done for compatibility.

Called with a third input argument, pchip evaluates the piece-wise polynomial at the points *xi*. There is an equivalence between ppval (pchip (x, y), xi) and pchip (x, y, xi).

**See also:** spline, ppval, mkpp, unmkpp.

periodogram (x)                                             Function File
For a data matrix $x$ from a sample of size $n$, return the periodogram.

rectangle_lw (n, b)                                      Function File
Rectangular lag window. Subfunction used for spectral density estimation.

rectangle_sw (n, b)                                      Function File
Rectangular spectral window. Subfunction used for spectral density estimation.

sinetone (freq, rate, sec, ampl)                      Function File
Return a sinetone of frequency *freq* with length of *sec* seconds at sampling rate *rate* and with amplitude *ampl*. The arguments *freq* and *ampl* may be vectors of common size.
Defaults are $rate = 8000$, $sec = 1$ and $ampl = 64$.

sinewave (m, n, d)                                         Function File
    Return an m-element vector with i-th element given by sin (2 * pi * (i+d-1) / n).
    The default value for d is 0 and the default value for n is m.

spectral_adf (c, win, b)                                   Function File
    Return the spectral density estimator given a vector of autocovariances c, window name win, and bandwidth, b.
    The window name, e.g., "triangle" or "rectangle" is used to search for a function called win_sw.
    If win is omitted, the triangle window is used. If b is omitted, 1 / sqrt (length (x)) is used.

spectral_xdf (x, win, b)                                   Function File
    Return the spectral density estimator given a data vector x, window name win, and bandwidth, b.
    The window name, e.g., "triangle" or "rectangle" is used to search for a function called win_sw.
    If win is omitted, the triangle window is used. If b is omitted, 1 / sqrt (length (x)) is used.

spencer (x)                                                Function File
    Return Spencer's 15 point moving average of every single column of x.

[y, c] = stft (x, win_size, inc, num_coef, w_type)         Function File
    Compute the short-term Fourier transform of the vector x with num_coef coefficients by applying a window of win_size data points and an increment of inc points.
    Before computing the Fourier transform, one of the following windows is applied:
    hanning
        w_type = 1
    hamming
        w_type = 2
    rectangle
        w_type = 3
    The window names can be passed as strings or by the w_type number.
    If not all arguments are specified, the following defaults are used: win_size = 80, inc = 24, num_coef = 64, and w_type = 1.
    y = stft (x, ...) returns the absolute values of the Fourier coefficients according to the num_coef positive frequencies.
    [y, c] = stft (x, ...) returns the entire STFT-matrix y and a 3-element vector c containing the window size, increment, and window type, which is needed by the synthesis function.

synthesis (y, c)                                                    Function File
　　Compute a signal from its short-time Fourier transform y and a 3-element
　　vector c specifying window size, increment, and window type.
　　The values y and c can be derived by
　　　　[y, c] = stft (x , ...)

triangle_lw (n, b)                                                  Function File
　　Triangular lag window. Subfunction used for spectral density estimation.

triangle_sw (n, b)                                                  Function File
　　Triangular spectral window. Subfunction used for spectral density estimation.

[a, v] = yulewalker (c)                                             Function File
　　Fit an AR (p)-model with Yule-Walker estimates given a vector c of autocovariances [gamma_0, ..., gamma_p].
　　Returns the AR coefficients, a, and the variance of white noise, v.

# 30 Image Processing

Since an image basically is a matrix Octave is a very powerful environment for processing and analysing images. To illustrate how easy it is to do image processing in Octave, the following example will load an image, smooth it by a 5-by-5 averaging filter, and compute the gradient of the smoothed image.

```
I = loadimage ("default.img");
S = conv2 (I, ones (5, 5) / 25, "same");
[Dx, Dy] = gradient (S);
```

In this example S contains the smoothed image, and Dx and Dy contains the partial spatial derivatives of the image.

## 30.1 Loading and Saving Images

The first step in most image processing tasks is to load an image into Octave. Currently Octave only support saving images in the Portable Pixmap Format (PPM), PostScript, and Octave's own format, and loading images in Octave's format. Most image processing code will follow the structure of this code

```
I = loadimage ("my_input_image.img");
J = process_my_image (I);
saveimage ("my_output_image.img", J);
```

[x, map] = loadimage (*file*)  *Function File*
Load an image file and its associated color map from the specified *file*. The image must be stored in Octave's image format.

See also: saveimage, load, save.

saveimage (*file, x, fmt, map*)  *Function File*
Save the matrix *x* to *file* in image format *fmt*. Valid values for *fmt* are

"img"
    Octave's image format. The current colormap is also saved in the file.

"ppm"
    Portable pixmap format.

"ps"
    PostScript format. Note that images saved in PostScript format cannot be read back into Octave with loadimage.

If the fourth argument is supplied, the specified colormap will also be saved along with the image.

Note: if the colormap contains only two entries and these entries are black and white, the bitmap ppm and PostScript formats are used. If the image is a gray scale image (the entries within each row of the colormap are equal) the gray scale ppm and PostScript image formats are used, otherwise the full color formats are used.

See also: loadimage, save, load, colormap.

*val* = IMAGE_PATH ()     Built-in Function
*old_val* = IMAGE_PATH (*new_val*)     Built-in Function
    Query or set the internal variable that specifies a colon separated list of directories in which to search for image files.

## 30.2 Displaying Images

A natural part of image processing is visualization of an image. The most basic function for this is the imshow function that shows the image given in the first input argument. This function uses an external program to show the image. If gnuplot 4.2 or later is available it will be used to display the image, otherwise the display, xv, or xloadimage program is used. The actual program can be selected with the image_viewer function.

imshow (*im*)     Function File
imshow (*im, limits*)     Function File
imshow (*im, map*)     Function File
imshow (*rgb, ...*)     Function File
imshow (*filename*)     Function File
imshow (*..., string_param1, value1, ...*)     Function File
    Display the image *im*, where *im* can be a 2-dimensional (gray-scale image) or a 3-dimensional (RGB image) matrix.

    If *limits* is a 2-element vector [*low, high*], the image is shown using a display range between *low* and *high*. If an empty matrix is passed for *limits*, the display range is computed as the range between the minimal and the maximal value in the image.

    If *map* is a valid color map, the image will be shown as an indexed image using the supplied color map.

    If a file name is given instead of an image, the file will be read and shown.

    If given, the parameter *string_param1* has value *value1*. *string_param1* can be any of the following:

"displayrange"
    *value1* is the display range as described above.

**See also:** image, imagesc, colormap, gray2ind, rgb2ind.

image (*img*)     Function File
image (*x, y, img*)     Function File
    Display a matrix as a color image. The elements of *x* are indices into the current colormap, and the colormap will be scaled so that the extremes of *x* are mapped to the extremes of the colormap.

    It first tries to use gnuplot, then display from ImageMagick, then xv, and then xloadimage. The actual program used can be changed using the image_viewer function.

    The axis values corresponding to the matrix elements are specified in *x* and *y*. If you're not using gnuplot 4.2 or later, these variables are ignored.

**See also:** imshow, imagesc, colormap, image_viewer.

imagesc (a)                                           Function File
imagesc (x, y, a)                                     Function File
imagesc (..., limits)                                 Function File
imagesc (h, ...)                                      Function File
h = imagesc (...)                                     Function File

Display a scaled version of the matrix *a* as a color image. The colormap is scaled so that the entries of the matrix occupy the entire colormap. If *limits* = [*lo, hi*] are given, then that range is set to the 'clim' of the current axes.

The axis values corresponding to the matrix elements are specified in *x* and *y*, either as pairs giving the minimum and maximum values for the respective axes, or as values for each row and column of the matrix *a*.

**See also:** image, imshow, clim, caxis.

[fcn, default_zoom] = image_viewer (fcn, default_zoom)     Function File
Change the program or function used for viewing images and return the previous values.

When the image or imshow function is called it will launch an external program to display the image. The default behaviour is to use gnuplot if the installed version supports image viewing, and otherwise try the programs display, xv, and xloadimage. Using this function it is possible to change that behaviour.

When called with one input argument images will be displayed by saving the image to a file and the system command *command* will be called to view the image. The *command* must be a string containing %s and possibly %f. The %s will be replaced by the filename of the image, and the %f will (if present) be replaced by the zoom factor given to the image function. For example,

    image_viewer ("eog %s");

changes the image viewer to the eog program.

With two input arguments, images will be displayed by calling the function *function_handle*. For example,

    image_viewer (data, @my_image_viewer);

sets the image viewer function to my_image_viewer. The image viewer function is called with

    my_image_viewer (x, y, im, zoom, data)

where *x* and *y* are the axis of the image, *im* is the image variable, and *data* is extra user-supplied data to be passed to the viewer function.

With three input arguments it is possible to change the zooming. Some programs (like xloadimage) require the zoom factor to be between 0 and 100, and not 0 and 1 like Octave assumes. This is solved by setting the third argument to 100.

**See also:** image, imshow.

## 30.3 Representing Images

In general Octave supports four different kinds of images, gray-scale images, RGB images, binary images, and indexed images. A gray-scale image is represented with an M-by-N matrix in which each element corresponds to the intensity of a pixel. An RGB image is represented with an M-by-N-by-3 array where each 3-vector corresponds to the red, green, and blue intensities of each pixel.

The actual meaning of the value of a pixel in a gray-scale or RGB image depends on the class of the matrix. If the matrix is of class double pixel intensities are between 0 and 1, if it is of class uint8 intensities are between 0 and 255, and if it is of class uint16 intensities are between 0 and 65535.

A binary image is an M-by-N matrix of class logical. A pixel in a binary image is black if it is false and white if it is true.

An indexed image consists of an M-by-N matrix of integers and a C-by-3 color map. Each integer corresponds to an index in the color map, and each row in the color map corresponds to an RGB color. The color map must be of class double with values between 0 and 1.

[img, map] = gray2ind ()　　　　　　　　　　　　　　　　　　Function File
    Convert a gray scale intensity image to an Octave indexed image.

ind2gray (x, map)　　　　　　　　　　　　　　　　　　　　　Function File
    Convert an Octave indexed image to a gray scale intensity image. If map is omitted, the current colormap is used to determine the intensities.

    See also: gray2ind, rgb2ntsc, image, colormap.

[x, map] = rgb2ind (rgb)　　　　　　　　　　　　　　　　　　Function File
[x, map] = rgb2ind (r, g, b)　　　　　　　　　　　　　　　　　Function File
    Convert an RGB image to an Octave indexed image.

    See also: ind2rgb, rgb2ntsc.

rgb = ind2rgb (x, map)　　　　　　　　　　　　　　　　　　Function File
[r, g, b] = ind2rgb (x, map)　　　　　　　　　　　　　　　　Function File
    Convert an indexed image to red, green, and blue color components. If the colormap doesn't contain enough colors, pad it with the last color in the map. If map is omitted, the current colormap is used for the conversion.

    See also: rgb2ind, image, imshow, ind2gray, gray2ind.

colormap (map)　　　　　　　　　　　　　　　　　　　　　Function File
colormap ("default")　　　　　　　　　　　　　　　　　　　Function File
    Set the current colormap.

    colormap (map) sets the current colormap to map. The color map should be an n row by 3 column matrix. The columns contain red, green, and blue intensities respectively. All entries should be between 0 and 1 inclusive. The new colormap is returned.

Chapter 30: Image Processing    431

colormap ("default") restores the default colormap (the jet map with 64 entries). The default colormap is returned.

With no arguments, colormap returns the current color map.

See also: jet.

autumn (n)    Function File
Create color colormap. This colormap is red through orange to yellow. The argument n should be a scalar. If it is omitted, the length of the current colormap or 64 is assumed.

See also: colormap.

bone (n)    Function File
Create color colormap. This colormap is a gray colormap with a light blue tone. The argument n should be a scalar. If it is omitted, the length of the current colormap or 64 is assumed.

See also: colormap.

cool (n)    Function File
Create color colormap. The colormap is cyan to magenta. The argument n should be a scalar. If it is omitted, the length of the current colormap or 64 is assumed.

See also: colormap.

copper (n)    Function File
Create color colormap. This colormap is black to a light copper tone. The argument n should be a scalar. If it is omitted, the length of the current colormap or 64 is assumed.

See also: colormap.

gray (n)    Function File
Return a gray colormap with n entries corresponding to values from 0 to n-1. The argument n should be a scalar. If it is omitted, the length of the current colormap or 64 is assumed.

hot (n)    Function File
Create color colormap. This colormap is black through dark red, red, orange, yellow to white. The argument n should be a scalar. If it is omitted, the length of the current colormap or 64 is assumed.

See also: colormap.

hsv (n)    Function File
Create color colormap. This colormap is red through yellow, green, cyan, blue, magenta to red. The argument n should be a scalar. If it is omitted, the length of the current colormap or 64 is assumed.

See also: colormap.

jet (n)                                                        Function File
    Create color colormap. This colormap is dark blue through blue, cyan, green, yellow, red to dark red. The argument n should be a scalar. If it is omitted, the length of the current colormap or 64 is assumed.

    See also: colormap.

ocean (n)                                                      Function File
    Create color colormap. The argument n should be a scalar. If it is omitted, the length of the current colormap or 64 is assumed.

pink (n)                                                       Function File
    Create color colormap. This colormap gives a sephia tone on black and white images. The argument n should be a scalar. If it is omitted, the length of the current colormap or 64 is assumed.

    See also: colormap.

prism (n)                                                      Function File
    Create color colormap. This colormap cycles trough red, orange, yellow, green, blue and violet. The argument n should be a scalar. If it is omitted, the length of the current colormap or 64 is assumed.

    See also: colormap.

rainbow (n)                                                    Function File
    Create color colormap. This colormap is red through orange, yellow, green, blue to violet. The argument n should be a scalar. If it is omitted, the length of the current colormap or 64 is assumed.

    See also: colormap.

spring (n)                                                     Function File
    Create color colormap. This colormap is magenta to yellow. The argument n should be a scalar. If it is omitted, the length of the current colormap or 64 is assumed.

    See also: colormap.

summer (n)                                                     Function File
    Create color colormap. This colormap is green to yellow. The argument n should be a scalar. If it is omitted, the length of the current colormap or 64 is assumed.

    See also: colormap.

white (n)                                                      Function File
    Create color colormap. This colormap is completely white. The argument n should be a scalar. If it is omitted, the length of the current colormap or 64 is assumed.

    See also: colormap.

winter (*n*)                                                              Function File
    Create color colormap. This colormap is blue to green. The argument *n* should be a scalar. If it is omitted, the length of the current colormap or 64 is assumed.

    See also: colormap.

    An additional colormap is gmap40. This code map contains only colors with integer values of the red, green and blue components. This is a workaround for a limitation of gnuplot 4.0, that does not allow the color of line or patch objects to be set, and so gmap40 is useful for gnuplot 4.0 users, and in particular in conjunction with the *bar*, *barh* or *contour* functions.

gmap40 (*n*)                                                              Function File
    Create a color colormap. The colormap is red, green, blue, yellow, magenta and cyan. These are the colors that are allowed with patch objects using gnuplot 4.0, and so this colormap function is specially designed for users of gnuplot 4.0. The argument *n* should be a scalar. If it is omitted, a length of 6 is assumed. Larger values of *n* result in a repetition of the above colors

    See also: colormap.

## 30.4 Plotting on top of Images

If gnuplot is being used to display images it is possible to plot on top of images. Since an image is a matrix it is indexed by row and column values. The plotting system is, however, based on the traditional $(x, y)$ system. To minimize the difference between the two systems Octave places the origin of the coordinate system in the point corresponding to the pixel at $(1, 1)$. So, to plot points given by row and column values on top of an image, one should simply call plot with the column values as the first argument and the row values as the second. As an example the following code generates an image with random intensities between 0 and 1, and shows the image with red circles over pixels with an intensity above 0.99.

```
I = rand (100, 100);
[row, col] = find (I > 0.99);
hold ("on");
imshow (I);
plot (col, row, "ro");
hold ("off");
```

## 30.5 Color Conversion

Octave supports conversion from the RGB color system to NTSC and HSV and vice versa.

*hsv_map* = rgb2hsv (*rgb_map*)       Function File
    Transform a colormap from the rgb space to the hsv space.

    A color n the RGB space consists of the red, green and blue intensities.

    In the HSV space each color is represented by their hue, saturation and value (brightness). Value gives the amount of light in the color. Hue describes the dominant wavelength. Saturation is the amount of Hue mixed into the color.

    **See also:** hsv2rgb.

*rgb_map* = hsv2rgb (*hsv_map*)       Function File
    Transform a colormap from the hsv space to the rgb space.

    **See also:** rgb2hsv.

rgb2ntsc (*rgb*)       Function File
    Image format conversion.

ntsc2rgb (*yiq*)       Function File
    Image format conversion.

# 31 Audio Processing

Octave provides a few functions for dealing with audio data. An audio 'sample' is a single output value from an A/D converter, i.e., a small integer number (usually 8 or 16 bits), and audio data is just a series of such samples. It can be characterized by three parameters: the sampling rate (measured in samples per second or Hz, e.g. 8000 or 44100), the number of bits per sample (e.g. 8 or 16), and the number of channels (1 for mono, 2 for stereo, etc.).

There are many different formats for representing such data. Currently, only the two most popular, *linear encoding* and *mu-law encoding*, are supported. Octave simply treats audio data as vectors of samples (non-mono data are not supported yet). It is assumed that audio files using linear encoding have one of the extensions 'lin' or 'raw', and that files holding data in mu-law encoding end in 'au', 'mu', or 'snd'.

## 31.1 Audio Conversion Functions

lin2mu (*x, n*)    Function File
Converts audio data from linear to mu-law. Mu-law values use 8-bit unsigned integers. Linear values use $n$-bit signed integers or floating point values in the range $-1 \leq x \leq 1$ if $n$ is 0. If $n$ is not specified it defaults to 0, 8 or 16 depending on the range values in $x$.

See also: mu2lin, loadaudio, saveaudio, playaudio, setaudio, record.

mu2lin (*x, bps*)    Function File
Converts audio data from linear to mu-law. Mu-law values are 8-bit unsigned integers. Linear values use $n$-bit signed integers or floating point values in the range $-1 \leq y \leq 1$ if $n$ is 0. If $n$ is not specified it defaults to 8.

See also: lin2mu, loadaudio, saveaudio, playaudio, setaudio, record.

## 31.2 Loading and Saving Audio Files

loadaudio (*name, ext, bps*)    Function File
Loads audio data from the file '*name.ext*' into the vector x.

The extension *ext* determines how the data in the audio file is interpreted; the extensions 'lin' (default) and 'raw' correspond to linear, the extensions 'au', 'mu', or 'snd' to mu-law encoding.

The argument *bps* can be either 8 (default) or 16, and specifies the number of bits per sample used in the audio file.

See also: lin2mu, mu2lin, saveaudio, playaudio, setaudio, record.

saveaudio (*name, x, ext, bps*)             Function File
    Saves a vector *x* of audio data to the file '*name.ext*'. The optional parameters *ext* and *bps* determine the encoding and the number of bits per sample used in the audio file (see loadaudio); defaults are 'lin' and 8, respectively.

    See also: lin2mu, mu2lin, loadaudio, playaudio, setaudio, record.

The following functions for audio I/O require special A/D hardware and operating system support. It is assumed that audio data in linear encoding can be played and recorded by reading from and writing to '/dev/dsp', and that similarly '/dev/audio' is used for mu-law encoding. These file names are system-dependent. Improvements so that these functions will work without modification on a wide variety of hardware are welcome.

playaudio (*name, ext*)             Function File
playaudio (*x*)             Function File
    Plays the audio file '*name.ext*' or the audio data stored in the vector *x*.

    See also: lin2mu, mu2lin, loadaudio, saveaudio, setaudio, record.

record (*sec, sampling_rate*)             Function File
    Records *sec* seconds of audio input into the vector *x*. The default value for *sampling_rate* is 8000 samples per second, or 8kHz. The program waits until the user types (RET) and then immediately starts to record.

    See also: lin2mu, mu2lin, loadaudio, saveaudio, playaudio, setaudio.

setaudio ([*w_type* [, *value*]])             Function File
    Execute the shell command mixer [w_type [, value]]

y = wavread (*filename*)             Function File
    Load the RIFF/WAVE sound file *filename*, and return the samples in vector *y*. If the file contains multichannel data, then *y* is a matrix with the channels represented as columns.

[y, Fs, bits] = wavread (*filename*)             Function File
    Additionally return the sample rate (*fs*) in Hz and the number of bits per sample (*bits*).

[...] = wavread (*filename, n*)             Function File
    Read only the first *n* samples from each channel.

[...] = wavread (*filename,[n1 n2]*)             Function File
    Read only samples *n1* through *n2* from each channel.

[samples, channels] = wavread (*filename*, "size")             Function File
    Return the number of samples (*n*) and channels (*ch*) instead of the audio data.

    See also: wavwrite.

Chapter 31: Audio Processing

wavwrite (*y*, *filename*) — Function File
wavwrite (*y*, *fs*, *filename*) — Function File
wavwrite (*y*, *fs*, *bits*, *filename*) — Function File

Write *y* to the canonical RIFF/WAVE sound file *filename* with sample rate *fs* and bits per sample *bits*. The default sample rate is 8000 Hz with 16-bits per sample. Each column of the data represents a separate channel.

See also: wavread.

# 32 System Utilities

This chapter describes the functions that are available to allow you to get information about what is happening outside of Octave, while it is still running, and use this information in your program. For example, you can get information about environment variables, the current time, and even start other programs from the Octave prompt.

## 32.1 Timing Utilities

Octave's core set of functions for manipulating time values are patterned after the corresponding functions from the standard C library. Several of these functions use a data structure for time that includes the following elements:

`usec`
: Microseconds after the second (0-999999).

`sec`
: Seconds after the minute (0-61). This number can be 61 to account for leap seconds.

`min`
: Minutes after the hour (0-59).

`hour`
: Hours since midnight (0-23).

`mday`
: Day of the month (1-31).

`mon`
: Months since January (0-11).

`year`
: Years since 1900.

`wday`
: Days since Sunday (0-6).

`yday`
: Days since January 1 (0-365).

`isdst`
: Daylight Savings Time flag.

`zone`
: Time zone.

In the descriptions of the following functions, this structure is referred to as a *tm_struct*.

time ()  Loadable Function
Return the current time as the number of seconds since the epoch. The epoch is referenced to 00:00:00 UTC (Coordinated Universal Time) 1 Jan 1970. For example, on Monday February 17, 1997 at 07:15:06 UTC, the value returned by time was 856163706.

See also: strftime, strptime, localtime, gmtime, mktime, now, date, clock, datenum, datestr, datevec, calendar, weekday.

t = now ()  Function File
Returns the current local time as the number of days since Jan 1, 0000. By this reckoning, Jan 1, 1970 is day number 719529.

The integral part, floor (now) corresponds to 00:00:00 today.

The fractional part, rem (now, 1) corresponds to the current time on Jan 1, 0000.

The returned value is also called a "serial date number" (see datenum).

See also: clock, date, datenum.

ctime (t)  Function File
Convert a value returned from time (or any other nonnegative integer), to the local time and return a string of the same form as asctime. The function ctime (time) is equivalent to asctime (localtime (time)). For example,

```
ctime (time ())
⇒ "Mon Feb 17 01:15:06 1997\n"
```

gmtime (t)  Loadable Function
Given a value returned from time (or any nonnegative integer), return a time structure corresponding to UTC. For example,

```
gmtime (time ())
⇒ {
    usec = 0
    year = 97
    mon = 1
    mday = 17
    sec = 6
    zone = CST
    min = 15
    wday = 1
    hour = 7
    isdst = 0
    yday = 47
  }
```

See also: strftime, strptime, localtime, mktime, time, now, date, clock, datenum, datestr, datevec, calendar, weekday.

Chapter 32: System Utilities                                                441

localtime (t)                                                Loadable Function
   Given a value returned from time (or any nonnegative integer), return a
   time structure corresponding to the local time zone.
      localtime (time ())
          ⇒ {
                 usec = 0
                 year = 97
                 mon = 1
                 mday = 17
                 sec = 6
                 zone = CST
                 min = 15
                 wday = 1
                 hour = 1
                 isdst = 0
                 yday = 47
             }
   See also: strftime, strptime, gmtime, mktime, time, now, date, clock,
   datenum, datestr, datevec, calendar, weekday.

mktime (tm_struct)                                           Loadable Function
   Convert a time structure corresponding to the local time to the number of
   seconds since the epoch. For example,
      mktime (localtime (time ()))
          ⇒ 856163706
   See also: strftime, strptime, localtime, gmtime, time, now, date, clock,
   datenum, datestr, datevec, calendar, weekday.

asctime (tm_struct)                                              Function File
   Convert a time structure to a string using the following five-field format:
   Thu Mar 28 08:40:14 1996. For example,
      asctime (localtime (time ()))
          ⇒ "Mon Feb 17 01:15:06 1997\n"
   This is equivalent to ctime (time ()).

strftime (fmt, tm_struct)                                    Loadable Function
   Format the time structure tm_struct in a flexible way using the format
   string fmt that contains % substitutions similar to those in printf. Except
   where noted, substituted fields have a fixed size; numeric fields are padded
   if necessary. Padding is with zeros by default; for fields that display a single
   number, padding can be changed or inhibited by following the % with one of
   the modifiers described below. Unknown field specifiers are copied as normal
   characters. All other characters are copied to the output without change.
   For example,

```
strftime ("%r (%Z) %A %e %B %Y", localtime (time ()))
     ⇒ "01:15:06 AM (CST) Monday 17 February 1997"
```
Octave's strftime function supports a superset of the ANSI C field specifiers.

Literal character fields:

% % character.

n Newline character.

t Tab character.

Numeric modifiers (a nonstandard extension):

- (dash)
 Do not pad the field.

_ (underscore)
 Pad the field with spaces.

Time fields:

%H Hour (00-23).

%I Hour (01-12).

%k Hour (0-23).

%l Hour (1-12).

%M Minute (00-59).

%p Locale's AM or PM.

%r Time, 12-hour (hh:mm:ss [AP]M).

%R Time, 24-hour (hh:mm).

%s Time in seconds since 00:00:00, Jan 1, 1970 (a nonstandard extension).

%S Second (00-61).

%T Time, 24-hour (hh:mm:ss).

%X Locale's time representation (%H:%M:%S).

%Z Time zone (EDT), or nothing if no time zone is determinable.

Date fields:

%a Locale's abbreviated weekday name (Sun-Sat).

%A Locale's full weekday name, variable length (Sunday-Saturday).

%b Locale's abbreviated month name (Jan-Dec).

%B Locale's full month name, variable length (January-December).

%c Locale's date and time (Sat Nov 04 12:02:33 EST 1989).

%C Century (00-99).

Chapter 32: System Utilities                                                    443

%d    Day of month (01-31).

%e    Day of month ( 1-31).

%D    Date (mm/dd/yy).

%h    Same as %b.

%j    Day of year (001-366).

%m    Month (01-12).

%U    Week number of year with Sunday as first day of week (00-53).

%w    Day of week (0-6).

%W    Week number of year with Monday as first day of week (00-53).

%x    Locale's date representation (mm/dd/yy).

%y    Last two digits of year (00-99).

%Y    Year (1970-).

**See also:** strptime, localtime, gmtime, mktime, time, now, date, clock, datenum, datestr, datevec, calendar, weekday.

[tm_struct, nchars] = strptime (str, fmt)                        Loadable Function
   Convert the string *str* to the time structure *tm_struct* under the control of the format string *fmt*.

   If *fmt* fails to match, *nchars* is 0; otherwise it is set to the position of last matched character plus 1. Always check for this unless you're absolutely sure the date string will be parsed correctly.

   **See also:** strftime, localtime, gmtime, mktime, time, now, date, clock, datenum, datestr, datevec, calendar, weekday.

   Most of the remaining functions described in this section are not patterned after the standard C library. Some are available for compatibility with MATLAB and others are provided because they are useful.

clock ()                                                              Function File
   Return a vector containing the current year, month (1-12), day (1-31), hour (0-23), minute (0-59) and second (0-61). For example,
      clock ()
      ⇒ [ 1993, 8, 20, 4, 56, 1 ]
   The function clock is more accurate on systems that have the gettimeofday function.

date ()                                                               Function File
   Return the date as a character string in the form DD-MMM-YY. For example,
      date ()
      ⇒ "20-Aug-93"

etime (*t1*, *t2*)                                                                                              Function File
    Return the difference (in seconds) between two time values returned from
    clock. For example:

        t0 = clock ();
        many computations later...
        elapsed_time = etime (clock (), t0);

    will set the variable elapsed_time to the number of seconds since the variable t0 was set.

    **See also:** tic, toc, clock, cputime.

[*total*, *user*, *system*] = cputime ();                                                              Built-in Function
    Return the CPU time used by your Octave session. The first output is the
    total time spent executing your process and is equal to the sum of second
    and third outputs, which are the number of CPU seconds spent executing in
    user mode and the number of CPU seconds spent executing in system mode,
    respectively. If your system does not have a way to report CPU time usage,
    cputime returns 0 for each of its output values. Note that because Octave
    used some CPU time to start, it is reasonable to check to see if cputime
    works by checking to see if the total CPU time used is nonzero.

is_leap_year (*year*)                                                                                    Function File
    Return 1 if the given year is a leap year and 0 otherwise. If no arguments
    are provided, is_leap_year will use the current year. For example,

        is_leap_year (2000)
        ⇒ 1

tic ()                                                                                                    Built-in Function
toc ()                                                                                                    Built-in Function
    Set or check a wall-clock timer. Calling tic without an output argument
    sets the timer. Subsequent calls to toc return the number of seconds since
    the timer was set. For example,

        tic ();
        # many computations later...
        elapsed_time = toc ();

    will set the variable elapsed_time to the number of seconds since the most
    recent call to the function tic.

    If called with one output argument then this function returns a scalar of
    type uint64 and the wall-clock timer is not started.

        t = tic; sleep (5); (double (tic ()) - double (t)) * 1e-6
        ⇒ 5

    Nested timing with tic and toc is not supported. Therefore toc will always
    return the elapsed time from the most recent call to tic.

    If you are more interested in the CPU time that your process used, you
    should use the cputime function instead. The tic and toc functions report
    the actual wall clock time that elapsed between the calls. This may include
    time spent processing other jobs or doing nothing at all. For example,

        tic (); sleep (5); toc ()
            ⇒ 5
        t = cputime (); sleep (5); cputime () - t
            ⇒ 0

(This example also illustrates that the CPU timer may have a fairly coarse resolution.)

pause (*seconds*)                                                         Built-in Function
   Suspend the execution of the program. If invoked without any arguments, Octave waits until you type a character. With a numeric argument, it pauses for the given number of seconds. For example, the following statement prints a message and then waits 5 seconds before clearing the screen.
        fprintf (stderr, "wait please...\n");
        pause (5);
        clc;

sleep (*seconds*)                                                        Built-in Function
   Suspend the execution of the program for the given number of seconds.

usleep (*microseconds*)                                      Built-in Function
   Suspend the execution of the program for the given number of microseconds. On systems where it is not possible to sleep for periods of time less than one second, usleep will pause the execution for round (*microseconds* / 1e6) seconds.

str = datestr (*date*, [*f*, [*p*]])                                     Function File
   Format the given date/time according to the format *f* and return the result in *str*. *date* is a serial date number (see datenum) or a date vector (see datevec). The value of *date* may also be a string or cell array of strings.

   *f* can be an integer which corresponds to one of the codes in the table below, or a date format string.

   *p* is the year at the start of the century in which two-digit years are to be interpreted in. If not specified, it defaults to the current year minus 50.

   For example, the date 730736.65149 (2000-09-07 15:38:09.0934) would be formatted as follows:

| Code | Format | Example |
| --- | --- | --- |
| 0 | dd-mmm-yyyy HH:MM:SS | 07-Sep-2000 15:38:09 |
| 1 | dd-mmm-yyyy | 07-Sep-2000 |
| 2 | mm/dd/yy | 09/07/00 |
| 3 | mmm | Sep |
| 4 | m | S |
| 5 | mm | 09 |
| 6 | mm/dd | 09/07 |
| 7 | dd | 07 |
| 8 | ddd | Thu |

| | | |
|---|---|---|
| 9  | d                      | T                   |
| 10 | yyyy                   | 2000                |
| 11 | yy                     | 00                  |
| 12 | mmmyy                  | Sep00               |
| 13 | HH:MM:SS               | 15:38:09            |
| 14 | HH:MM:SS PM            | 03:38:09 PM         |
| 15 | HH:MM                  | 15:38               |
| 16 | HH:MM PM               | 03:38 PM            |
| 17 | QQ-YY                  | Q3-00               |
| 18 | QQ                     | Q3                  |
| 19 | dd/mm                  | 13/03               |
| 20 | dd/mm/yy               | 13/03/95            |
| 21 | mmm.dd.yyyy HH:MM:SS   | Mar.03.1962 13:53:06|
| 22 | mmm.dd.yyyy            | Mar.03.1962         |
| 23 | mm/dd/yyyy             | 03/13/1962          |
| 24 | dd/mm/yyyy             | 12/03/1962          |
| 25 | yy/mm/dd               | 95/03/13            |
| 26 | yyyy/mm/dd             | 1995/03/13          |
| 27 | QQ-YYYY                | Q4-2132             |
| 28 | mmmyyyy                | Mar2047             |
| 29 | yyyymmdd               | 20470313            |
| 30 | yyyymmddTHHMMSS        | 20470313T132603     |
| 31 | yyyy-mm-dd HH:MM:SS    | 1047-03-13 13:26:03 |

If f is a format string, the following symbols are recognised:

| Symbol | Meaning | Example |
|---|---|---|
| yyyy  | Full year | 2005 |
| yy    | Two-digit year | 2005 |
| mmmm  | Full month name | December |
| mmm   | Abbreviated month name | Dec |
| mm    | Numeric month number (padded with zeros) | 01, 08, 12 |
| m     | First letter of month name (capitalized) | D |
| dddd  | Full weekday name | Sunday |
| ddd   | Abbreviated weekday name | Sun |
| dd    | Numeric day of month (padded with zeros) | 11 |
| d     | First letter of weekday name (capitalized) | S |
| HH    | Hour of day, padded with zeros if PM is set and not padded with zeros otherwise | 09:00 / 9:00 AM |
| MM    | Minute of hour (padded with zeros) | 10:05 |
| SS    | Second of minute (padded with zeros) | 10:05:03 |
| PM    | Use 12-hour time format | 11:30 PM |

If f is not specified or is -1, then use 0, 1 or 16, depending on whether the date portion or the time portion of *date* is empty.

If p is nor specified, it defaults to the current year minus 50.

If a matrix or cell array of dates is given, a vector of date strings is returned.

**See also:** datenum, datevec, date, clock, now, datetick.

Chapter 32: System Utilities    447

    v = datevec (*date*)    Function File
    v = datevec (*date, f*)    Function File
    v = datevec (*date, p*)    Function File
    v = datevec (*date, f, p*)    Function File
    [y, m, d, h, mi, s] = datevec (...)    Function File

        Convert a serial date number (see datenum) or date string (see datestr) into a date vector.

        A date vector is a row vector with six members, representing the year, month, day, hour, minute, and seconds respectively.

        *f* is the format string used to interpret date strings (see datestr).

        *p* is the year at the start of the century in which two-digit years are to be interpreted in. If not specified, it defaults to the current year minus 50.

        **See also:** datenum, datestr, date, clock, now.

    calendar (...)    Function File
    c = calendar ()    Function File
    c = calendar (*d*)    Function File
    c = calendar (*y, m*)    Function File

        If called with no arguments, return the current monthly calendar in a 6x7 matrix.

        If *d* is specified, return the calendar for the month containing the day *d*, which must be a serial date number or a date string.

        If *y* and *m* are specified, return the calendar for year *y* and month *m*.

        If no output arguments are specified, print the calendar on the screen instead of returning a matrix.

        **See also:** datenum.

    [n, s] = weekday (*d*, [*form*])    Function File

        Return the day of week as a number in *n* and a string in *s*, for example [1, "Sun"], [2, "Mon"], ..., or [7, "Sat"].

        *d* is a serial date number or a date string.

        If the string *form* is given and is "long", *s* will contain the full name of the weekday; otherwise (or if *form* is "short"), *s* will contain the abbreviated name of the weekday.

        **See also:** datenum, datevec, eomday.

    e = eomday (*y, m*)    Function File

        Return the last day of the month *m* for the year *y*.

        **See also:** datenum, datevec, weekday, eomdate.

## 32.2 Filesystem Utilities

Octave includes the following functions for renaming and deleting files, creating, deleting, and reading directories, and for getting information about the status of files.

[err, msg] = rename (*old*, *new*)  Built-in Function
    Change the name of file *old* to *new*.

    If successful, *err* is 0 and *msg* is an empty string. Otherwise, *err* is nonzero and *msg* contains a system-dependent error message.

    See also: ls, dir.

[err, msg] = link (*old*, *new*)  Built-in Function
    Create a new link (also known as a hard link) to an existing file.

    If successful, *err* is 0 and *msg* is an empty string. Otherwise, *err* is nonzero and *msg* contains a system-dependent error message.

    See also: symlink.

[err, msg] = symlink (*old*, *new*)  Built-in Function
    Create a symbolic link *new* which contains the string *old*.

    If successful, *err* is 0 and *msg* is an empty string. Otherwise, *err* is nonzero and *msg* contains a system-dependent error message.

    See also: link, readlink.

[*result*, err, msg] = readlink (*symlink*)  Built-in Function
    Read the value of the symbolic link *symlink*.

    If successful, *result* contains the contents of the symbolic link *symlink*, *err* is 0 and *msg* is an empty string. Otherwise, *err* is nonzero and *msg* contains a system-dependent error message.

    See also: link, symlink.

[err, msg] = unlink (*file*)  Built-in Function
    Delete the file named *file*.

    If successful, *err* is 0 and *msg* is an empty string. Otherwise, *err* is nonzero and *msg* contains a system-dependent error message.

[*files*, err, msg] = readdir (*dir*)  Built-in Function
    Return names of the files in the directory *dir* as a cell array of strings. If an error occurs, return an empty cell array in *files*.

    If successful, *err* is 0 and *msg* is an empty string. Otherwise, *err* is nonzero and *msg* contains a system-dependent error message.

    See also: dir, glob.

# Chapter 32: System Utilities

[*status*, *msg*, *msgid*] = mkdir (*dir*)     Built-in Function
[*status*, *msg*, *msgid*] = mkdir (*parent*, *dir*)     Built-in Function
    Create a directory named *dir*.

    If successful, *status* is 1, with *msg* and *msgid* empty character strings. Otherwise, *status* is 0, *msg* contains a system-dependent error message, and *msgid* contains a unique message identifier.

    **See also:** rmdir.

[*status*, *msg*, *msgid*] = rmdir (*dir*)     Built-in Function
[*status*, *msg*, *msgid*] = rmdir (*dir*, "s")     Built-in Function
    Remove the directory named *dir*.

    If successful, *status* is 1, with *msg* and *msgid* empty character strings. Otherwise, *status* is 0, *msg* contains a system-dependent error message, and *msgid* contains a unique message identifier.

    If the optional second parameter is supplied with value "s", recursively remove all subdirectories as well.

    **See also:** mkdir, confirm_recursive_rmdir.

*val* = confirm_recursive_rmdir ()     Built-in Function
*old_val* = confirm_recursive_rmdir (*new_val*)     Built-in Function
    Query or set the internal variable that controls whether Octave will ask for confirmation before recursively removing a directory tree.

[*err*, *msg*] = mkfifo (*name*, *mode*)     Built-in Function
    Create a fifo special file named *name* with file mode *mode*

    If successful, *err* is 0 and *msg* is an empty string. Otherwise, *err* is nonzero and *msg* contains a system-dependent error message.

umask (*mask*)     Built-in Function
    Set the permission mask for file creation. The parameter *mask* is an integer, interpreted as an octal number. If successful, returns the previous value of the mask (as an integer to be interpreted as an octal number); otherwise an error message is printed.

[*info*, *err*, *msg*] = stat (*file*)     Built-in Function
[*info*, *err*, *msg*] = lstat (*file*)     Built-in Function
    Return a structure *s* containing the following information about *file*.

    dev
        ID of device containing a directory entry for this file.

    ino
        File number of the file.

mode
: File mode, as an integer. Use the functions S_ISREG, S_ISDIR, S_ISCHR, S_ISBLK, S_ISFIFO, S_ISLNK, or S_ISSOCK to extract information from this value.

modestr
: File mode, as a string of ten letters or dashes as would be returned by `ls -l`.

nlink
: Number of links.

uid
: User ID of file's owner.

gid
: Group ID of file's group.

rdev
: ID of device for block or character special files.

size
: Size in bytes.

atime
: Time of last access in the same form as time values returned from time. See Section 32.1 [Timing Utilities], page 439.

mtime
: Time of last modification in the same form as time values returned from time. See Section 32.1 [Timing Utilities], page 439.

ctime
: Time of last file status change in the same form as time values returned from time. See Section 32.1 [Timing Utilities], page 439.

blksize
: Size of blocks in the file.

blocks
: Number of blocks allocated for file.

If the call is successful *err* is 0 and *msg* is an empty string. If the file does not exist, or some other error occurs, *s* is an empty matrix, *err* is $-1$, and *msg* contains the corresponding system error message.

If *file* is a symbolic link, stat will return information about the actual file that is referenced by the link. Use lstat if you want information about the symbolic link itself.

For example,

```
[s, err, msg] = stat ("/vmlinuz")
    ⇒ s =
      {
        atime = 855399756
        rdev = 0
        ctime = 847219094
        uid = 0
        size = 389218
        blksize = 4096
        mtime = 847219094
        gid = 6
        nlink = 1
        blocks = 768
        mode = -rw-r--r--
        modestr = -rw-r--r--
        ino = 9316
        dev = 2049
      }
    ⇒ err = 0
    ⇒ msg =
```

[*info*, err, *msg*] = lstat (*file*)      Built-in Function
     See stat.

[*status*, *msg*, *msgid*] = fileattrib (*file*)      Function File
     Return information about *file*.

     If successful, *status* is 1, with *result* containing a structure with the following fields:

Name
     Full name of *file*.

archive
     True if *file* is an archive (Windows).

system
     True if *file* is a system file (Windows).

hidden
     True if *file* is a hidden file (Windows).

directory
     True if *file* is a directory.

UserRead
GroupRead
OtherRead
     True if the user (group; other users) has read permission for *file*.

UserWrite
GroupWrite
OtherWrite
: True if the user (group; other users) has write permission for *file*.

UserExecute
GroupExecute
OtherExecute
: True if the user (group; other users) has execute permission for *file*.

If an attribute does not apply (i.e., archive on a Unix system) then the field is set to NaN.

With no input arguments, return information about the current directory.

If *file* contains globbing characters, return information about all the matching files.

**See also:** glob.

isdir (*f*)                                                                            Function File
: Return true if *f* is a directory.

glob (*pattern*)                                                  Built-in Function
: Given an array of strings (as a char array or a cell array) in *pattern*, return a cell array of file names that match any of them, or an empty cell array if no patterns match. Tilde expansion is performed on each of the patterns before looking for matching file names. For example,

```
glob ("/vm*")
    ⇒ "/vmlinuz"
```

**See also:** dir, ls, stat, readdir.

fnmatch (*pattern, string*)                                  Built-in Function
: Return 1 or zero for each element of *string* that matches any of the elements of the string array *pattern*, using the rules of filename pattern matching. For example,

```
fnmatch ("a*b", {"ab"; "axyzb"; "xyzab"})
    ⇒ [ 1; 1; 0 ]
```

file_in_path (*path, file*)                                 Built-in Function
file_in_path (*path, file*, "all")                     Built-in Function
: Return the absolute name of *file* if it can be found in *path*. The value of *path* should be a colon-separated list of directories in the format described for path. If no file is found, return an empty matrix. For example,

Chapter 32: System Utilities    453

        file_in_path (EXEC_PATH, "sh")
            ⇒ "/bin/sh"
If the second argument is a cell array of strings, search each directory of the path for element of the cell array and return the first that matches.

If the third optional argument "all" is supplied, return a cell array containing the list of all files that have the same name in the path. If no files are found, return an empty cell array.

See also: file_in_loadpath.

tilde_expand (*string*)                                    Built-in Function
Performs tilde expansion on *string*. If *string* begins with a tilde character, (~), all of the characters preceding the first slash (or all characters, if there is no slash) are treated as a possible user name, and the tilde and the following characters up to the slash are replaced by the home directory of the named user. If the tilde is followed immediately by a slash, the tilde is replaced by the home directory of the user running Octave. For example,

        tilde_expand ("~joeuser/bin")
            ⇒ "/home/joeuser/bin"
        tilde_expand ("~/bin")
            ⇒ "/home/jwe/bin"

[*cname, status, msg*] canonicalize_file_name (*name*)       Built-in Function
Return the canonical name of file *name*.

[*status, msg, msgid*] = movefile (*f1, f2*)                      Function File
Move the file *f1* to the new name *f2*. The name *f1* may contain globbing patterns. If *f1* expands to multiple file names, *f2* must be a directory.

If successful, *status* is 1, with *msg* and *msgid* empty\n\ character strings. Otherwise, *status* is 0, *msg* contains a\n\ system-dependent error message, and *msgid* contains a unique\n\ message identifier.\n\

See also: glob.

[*status, msg, msgid*] = copyfile (*f1, f2, force*)                Function File
Copy the file *f1* to the new name *f2*. The name *f1* may contain globbing patterns. If *f1* expands to multiple file names, *f2* must be a directory. If *force* is given and equals the string "f" the copy operation will be forced.

If successful, *status* is 1, with *msg* and *msgid* empty\n\ character strings. Otherwise, *status* is 0, *msg* contains a\n\ system-dependent error message, and *msgid* contains a unique\n\ message identifier.\n\

See also: glob, movefile.

[*dir, name, ext, ver*] = fileparts (*filename*)                   Function File
Return the directory, name, extension, and version components of *filename*.

See also: fullfile.

filesep ()  — Built-in Function
Return the system-dependent character used to separate directory names.

See also: pathsep, dir, ls.

*filename* = fullfile (*dir1*, *dir2*, ..., *file*)  — Function File
Return a complete filename constructed from the given components.

See also: fileparts.

*dir* = tempdir ()  — Function File
Return the name of the system's directory for temporary files.

*filename* = tempname ()  — Function File
This function is an alias for tmpnam.

P_tmpdir ()  — Built-in Function
Return the default name of the directory for temporary files on this system. The name of this directory is system dependent.

## 32.3 File Archiving Utilities

bunzip2 (*bzfile*, *dir*)  — Function File
Unpack the bzip2 archive *bzfile* to the directory *dir*. If *dir* is not specified, it defaults to the current directory.

See also: unpack, bzip2, tar, untar, gzip, gunzip, zip, unzip.

*entries* = gzip (*files*)  — Function File
*entries* = gzip (*files*, *outdir*)  — Function File
Compress the list of files and/or directories specified in *files*. Each file is compressed separately and a new file with a '.gz' extension is created. The original file is not touched. If *outdir* is defined the compressed versions of the files are placed in this directory.

See also: gunzip, zip, tar.

gunzip (*gzfile*, *dir*)  — Function File
Unpack the gzip archive *gzfile* to the directory *dir*. If *dir* is not specified, it defaults to the current directory. If the *gzfile* is a directory, all files in the directory will be recursively gunzipped.

See also: unpack, bzip2, bunzip2, tar, untar, gzip, gunzip, zip, unzip.

## Chapter 32: System Utilities

*entries* = tar (*tarfile*, *files*, *root*)       Function File
Pack *files* *files* into the TAR archive *tarfile*. The list of files must be a string or a cell array of strings.

The optional argument *root* changes the relative path of *files* from the current directory.

If an output argument is requested the entries in the archive are returned in a cell array.

See also: untar, gzip, gunzip, zip, unzip.

untar (*tarfile*, *dir*)       Function File
Unpack the TAR archive *tarfile* to the directory *dir*. If *dir* is not specified, it defaults to the current directory.

See also: unpack, bzip2, bunzip2, tar, gzip, gunzip, zip, unzip.

*entries* = zip (*zipfile*, *files*)       Function File
*entries* = zip (*zipfile*, *files*, *rootdir*)       Function File
Compress the list of files and/or directories specified in *files* into the archive *zipfiles* in the same directory. If *rootdir* is defined the *files* is located relative to *rootdir* rather than the current directory

See also: unzip,tar.

unzip (*zipfile*, *dir*)       Function File
Unpack the ZIP archive *zipfile* to the directory *dir*. If *dir* is not specified, it defaults to the current directory.

See also: unpack, bzip2, bunzip2, tar, untar, gzip, gunzip, zip.

pack ()       Function File
This function is provided for compatibility with MATLAB, but it doesn't actually do anything.

*files* = unpack (*file*, *dir*)       Function File
*files* = unpack (*file*, *dir*, *filetype*)       Function File
Unpack the archive *file* based on its extension to the directory *dir*. If *file* is a cellstr, then all files will be handled individually. If *dir* is not specified, it defaults to the current directory. It returns a list of *files* unpacked. If a directory is in the file list, then the *filetype* to unpack must also be specified. The *files* includes the entire path to the output files.

See also: bzip2,bunzip2,tar,untar,gzip,gunzip,zip,unzip.

## 32.4 Networking Utilities

*s* = urlread (*url*)     Loadable Function
[*s, success*] = urlread (*url*)     Loadable Function
[*s, success, message*] = urlread (*url*)     Loadable Function
[...] = urlread (*url, method, param*)     Loadable Function

Download a remote file specified by its *URL* and return its content in string *s*. For example,

     s = urlread ("ftp://ftp.octave.org/pub/octave/README");

The variable *success* is 1 if the download was successful, otherwise it is 0 in which case *message* contains an error message. If no output argument is specified and if an error occurs, then the error is signaled through Octave's error handling mechanism.

This function uses libcurl. Curl supports, among others, the HTTP, FTP and FILE protocols. Username and password may be specified in the URL. For example,

     s = urlread ("http://user:password@example.com/file.txt");

GET and POST requests can be specified by *method* and *param*. The parameter *method* is either get or post and *param* is a cell array of parameter and value pairs. For example,

     s = urlread ("http://www.google.com/search", "get",
               {"query", "octave"});

See also: urlwrite.

urlwrite (*URL, localfile*)     Loadable Function
*f* = urlwrite (*url, localfile*)     Loadable Function
[*f, success*] = urlwrite (*url, localfile*)     Loadable Function
[*f, success, message*] = urlwrite (*url, localfile*)     Loadable Function

Download a remote file specified by its *URL* and save it as *localfile*. For example,

     urlwrite ("ftp://ftp.octave.org/pub/octave/README",
              "README.txt");

The full path of the downloaded file is returned in *f*. The variable *success* is 1 if the download was successful, otherwise it is 0 in which case *message* contains an error message. If no output argument is specified and if an error occurs, then the error is signaled through Octave's error handling mechanism.

This function uses libcurl. Curl supports, among others, the HTTP, FTP and FILE protocols. Username and password may be specified in the URL, for example:

     urlwrite ("http://username:password@example.com/file.txt",
              "file.txt");

GET and POST requests can be specified by *method* and *param*. The parameter *method* is either get or post and *param* is a cell array of parameter and value pairs. For example:

```
urlwrite ("http://www.google.com/search", "search.html",
          "get", {"query", "octave"});
```
See also: urlread.

## 32.5 Controlling Subprocesses

Octave includes some high-level commands like system and popen for starting subprocesses. If you want to run another program to perform some task and then look at its output, you will probably want to use these functions.

Octave also provides several very low-level Unix-like functions which can also be used for starting subprocesses, but you should probably only use them if you can't find any way to do what you need with the higher-level functions.

system (*string, return_output, type*)  Built-in Function

Execute a shell command specified by *string*. The second argument is optional. If *type* is "async", the process is started in the background and the process id of the child process is returned immediately. Otherwise, the process is started, and Octave waits until it exits. If the *type* argument is omitted, a value of "sync" is assumed.

If two input arguments are given (the actual value of *return_output* is irrelevant) and the subprocess is started synchronously, or if *system* is called with one input argument and one or more output arguments, the output from the command is returned. Otherwise, if the subprocess is executed synchronously, its output is sent to the standard output. To send the output of a command executed with *system* through the pager, use a command like

```
disp (system (cmd, 1));
```
or
```
printf ("%s\n", system (cmd, 1));
```

The system function can return two values. The first is the exit status of the command and the second is any output from the command that was written to the standard output stream. For example,

```
[status, output] = system ("echo foo; exit 2");
```

will set the variable output to the string foo, and the variable status to the integer 2.

[status, text] = unix (*command*)  Function File
[status, text] = unix (*command*, "-echo")  Function File

Execute a system command if running under a Unix-like operating system, otherwise do nothing. Return the exit status of the program in *status* and any output sent to the standard output in *text*. If the optional second argument "-echo" is given, then also send the output from the command to the standard output.

See also: isunix, ispc, system.

[*status*, *text*] = dos (*command*)                             Function File
[*status*, *text*] = dos (*command*, "-echo")            Function File

    Execute a system command if running under a Windows-like operating system, otherwise do nothing. Return the exit status of the program in *status* and any output sent to the standard output in *text*. If the optional second argument "-echo" is given, then also send the output from the command to the standard output.

    See also: unix, isunix, ispc, system.

*fid* = popen (*command*, *mode*)                          Built-in Function

    Start a process and create a pipe. The name of the command to run is given by *command*. The file identifier corresponding to the input or output stream of the process is returned in *fid*. The argument *mode* may be

"r"
    The pipe will be connected to the standard output of the process, and open for reading.

"w"
    The pipe will be connected to the standard input of the process, and open for writing.

For example,
```
    fid = popen ("ls -ltr / | tail -3", "r");
    while (isstr (s = fgets (fid)))
      fputs (stdout, s);
    endwhile
        ⊣ drwxr-xr-x   33 root   root   3072 Feb 15 13:28 etc
        ⊣ drwxr-xr-x    3 root   root   1024 Feb 15 13:28 lib
        ⊣ drwxrwxrwt   15 root   root   2048 Feb 17 14:53 tmp
```

pclose (*fid*)                                                 Built-in Function

    Close a file identifier that was opened by popen. You may also use fclose for the same purpose.

[*in*, *out*, *pid*] = popen2 (*command*, *args*)                   Built-in Function

    Start a subprocess with two-way communication. The name of the process is given by *command*, and *args* is an array of strings containing options for the command. The file identifiers for the input and output streams of the subprocess are returned in *in* and *out*. If execution of the command is successful, *pid* contains the process ID of the subprocess. Otherwise, *pid* is −1.

    For example,

Chapter 32: System Utilities                                                    459

```
[in, out, pid] = popen2 ("sort", "-r");
fputs (in, "these\nare\nsome\nstrings\n");
fclose (in);
EAGAIN = errno ("EAGAIN");
done = false;
do
  s = fgets (out);
  if (ischar (s))
    fputs (stdout, s);
  elseif (errno () == EAGAIN)
    sleep (0.1);
    fclear (out);
  else
    done = true;
  endif
until (done)
fclose (out);
  ⊣ are
  ⊣ some
  ⊣ strings
  ⊣ these
```

*val* = EXEC_PATH ()                                              Built-in Function
*old_val* = EXEC_PATH (*new_val*)                                 Built-in Function
Query or set the internal variable that specifies a colon separated list of directories to search when executing external programs. Its initial value is taken from the environment variable OCTAVE_EXEC_PATH (if it exists) or PATH, but that value can be overridden by the command line argument --exec-path PATH. At startup, an additional set of directories (including the shell PATH) is appended to the path specified in the environment or on the command line. If you use the EXEC_PATH function to modify the path, you should take care to preserve these additional directories.

In most cases, the following functions simply decode their arguments and make the corresponding Unix system calls. For a complete example of how they can be used, look at the definition of the function popen2.

[*pid*, *msg*] = fork ()                                          Built-in Function
Create a copy of the current process.
Fork can return one of the following values:

> 0   You are in the parent process. The value returned from fork is the process id of the child process. You should probably arrange to wait for any child processes to exit.

0     You are in the child process. You can call exec to start another process. If that fails, you should probably call exit.

< 0   The call to fork failed for some reason. You must take evasive action. A system dependent error message will be waiting in *msg*.

[err, msg] = exec (*file*, *args*)  Built-in Function
Replace current process with a new process. Calling exec without first calling fork will terminate your current Octave process and replace it with the program named by *file*. For example,

    exec ("ls" "-l")

will run ls and return you to your shell prompt.

If successful, exec does not return. If exec does return, *err* will be nonzero, and *msg* will contain a system-dependent error message.

[read_fd, write_fd, err, msg] = pipe ()  Built-in Function
Create a pipe and return the reading and writing ends of the pipe into *read_fd* and *write_fd* respectively.

If successful, *err* is 0 and *msg* is an empty string. Otherwise, *err* is nonzero and *msg* contains a system-dependent error message.

[fid, msg] = dup2 (*old*, *new*)  Built-in Function
Duplicate a file descriptor.

If successful, *fid* is greater than zero and contains the new file ID. Otherwise, *fid* is negative and *msg* contains a system-dependent error message.

[pid, status, msg] = waitpid (*pid*, *options*)  Built-in Function
Wait for process *pid* to terminate. The *pid* argument can be:

−1  Wait for any child process.

0   Wait for any child process whose process group ID is equal to that of the Octave interpreter process.

\> 0  Wait for termination of the child process with ID *pid*.

The *options* argument can be a bitwise OR of zero or more of the following constants:

0   Wait until signal is received or a child process exits (this is the default if the *options* argument is missing).

WNOHANG
    Do not hang if status is not immediately available.

WUNTRACED
    Report the status of any child processes that are stopped, and whose status has not yet been reported since they stopped.

WCONTINUED
    Return if a stopped child has been resumed by delivery of SIGCONT. This value may not be meaningful on all systems.

If the returned value of *pid* is greater than 0, it is the process ID of the child process that exited. If an error occurs, *pid* will be less than zero and *msg* will contain a system-dependent error message. The value of *status* contains additional system-dependent information about the subprocess that exited.

**See also:** WCONTINUED, WCOREDUMP, WEXITSTATUS, WIFCONTINUED, WIFSIGNALED, WIFSTOPPED, WNOHANG, WSTOPSIG, WTERMSIG, WUNTRACED.

[err, msg] = fcntl (*fid*, *request*, *arg*)  Built-in Function
Change the properties of the open file *fid*. The following values may be passed as *request*:

F_DUPFD
: Return a duplicate file descriptor.

F_GETFD
: Return the file descriptor flags for *fid*.

F_SETFD
: Set the file descriptor flags for *fid*.

F_GETFL
: Return the file status flags for *fid*. The following codes may be returned (some of the flags may be undefined on some systems).

  O_RDONLY
  : Open for reading only.

  O_WRONLY
  : Open for writing only.

  O_RDWR
  : Open for reading and writing.

  O_APPEND
  : Append on each write.

  O_CREAT
  : Create the file if it does not exist.

  O_NONBLOCK
  : Nonblocking mode.

  O_SYNC
  : Wait for writes to complete.

  O_ASYNC
  : Asynchronous I/O.

F_SETFL
: Set the file status flags for *fid* to the value specified by *arg*. The only flags that can be changed are O_APPEND and O_NONBLOCK.

If successful, *err* is 0 and *msg* is an empty string. Otherwise, *err* is nonzero and *msg* contains a system-dependent error message.

[err, msg] = kill (*pid*, *sig*)  Built-in Function
    Send signal *sig* to process *pid*.

    If *pid* is positive, then signal *sig* is sent to *pid*.

    If *pid* is 0, then signal *sig* is sent to every process in the process group of the current process.

    If *pid* is -1, then signal *sig* is sent to every process except process 1.

    If *pid* is less than -1, then signal *sig* is sent to every process in the process group -*pid*.

    If *sig* is 0, then no signal is sent, but error checking is still performed.

    Return 0 if successful, otherwise return -1.

SIG ()  Built-in Function
    Return a structure containing Unix signal names and their defined values.

## 32.6 Process, Group, and User IDs

pgid = getpgrp ()  Built-in Function
    Return the process group id of the current process.

pid = getpid ()  Built-in Function
    Return the process id of the current process.

pid = getppid ()  Built-in Function
    Return the process id of the parent process.

euid = geteuid ()  Built-in Function
    Return the effective user id of the current process.

uid = getuid ()  Built-in Function
    Return the real user id of the current process.

egid = getegid ()  Built-in Function
    Return the effective group id of the current process.

gid = getgid ()  Built-in Function
    Return the real group id of the current process.

## 32.7 Environment Variables

getenv (*var*)  Built-in Function
    Return the value of the environment variable *var*. For example,

        getenv ("PATH")

    returns a string containing the value of your path.

putenv (*var*, *value*)  Built-in Function
setenv (*var*, *value*)  Built-in Function
    Set the value of the environment variable *var* to *value*.

## 32.8 Current Working Directory

**cd** *dir*     Command
**chdir** *dir*     Command

    Change the current working directory to *dir*. If *dir* is omitted, the current directory is changed to the user's home directory. For example,

        cd ~/octave

    Changes the current working directory to '~/octave'. If the directory does not exist, an error message is printed and the working directory is not changed.

    **See also:** mkdir, rmdir, dir.

**ls** *options*     Command

    List directory contents. For example,

```
    ls -l
    ⊣ total 12
    ⊣ -rw-r--r--   1 jwe   users   4488 Aug 19 04:02 foo.m
    ⊣ -rw-r--r--   1 jwe   users   1315 Aug 17 23:14 bar.m
```

    The dir and ls commands are implemented by calling your system's directory listing command, so the available options may vary from system to system.

    **See also:** dir, stat, readdir, glob, filesep, ls_command.

*old_cmd* = **ls_command** (*cmd*)     Function File

    Set or return the shell command used by Octave's ls command. The value of *cmd* must be a character string. With no arguments, simply return the previous value.

    **See also:** ls.

**dir** (*directory*)     Function File
[*list*] = **dir** (*directory*)     Function File

    Display file listing for directory *directory*. If a return value is requested, return a structure array with the fields

        name
        bytes
        date
        isdir
        statinfo

    in which statinfo is the structure returned from stat.

    If *directory* is not a directory, return information about the named *filename*. *directory* may be a list of directories specified either by name or with wildcard characters (like * and ?) which will be expanded with glob.

    Note that for symbolic links, dir returns information about the file that a symbolic link points to instead of the link itself. However, if the link points to a nonexistent file, dir returns information about the link.

    **See also:** ls, stat, lstat, readdir, glob, filesep.

pwd ()                                                                Built-in Function
    Return the current working directory.
    See also: dir, ls.

## 32.9 Password Database Functions

Octave's password database functions return information in a structure with the following fields.

name
    The user name.

passwd
    The encrypted password, if available.

uid
    The numeric user id.

gid
    The numeric group id.

gecos
    The GECOS field.

dir
    The home directory.

shell
    The initial shell.

In the descriptions of the following functions, this data structure is referred to as a *pw_struct*.

pw_struct = getpwent ()                                              Loadable Function
    Return a structure containing an entry from the password database, opening it if necessary. Once the end of the data has been reached, getpwent returns 0.

pw_struct = getpwuid (*uid*).                                         Loadable Function
    Return a structure containing the first entry from the password database with the user ID *uid*. If the user ID does not exist in the database, getpwuid returns 0.

pw_struct = getpwnam (*name*)                                         Loadable Function
    Return a structure containing the first entry from the password database with the user name *name*. If the user name does not exist in the database, getpwname returns 0.

setpwent ()                                                          Loadable Function
    Return the internal pointer to the beginning of the password database.

endpwent ()                                                          Loadable Function
    Close the password database.

## 32.10 Group Database Functions

Octave's group database functions return information in a structure with the following fields.

name
: The user name.

passwd
: The encrypted password, if available.

gid
: The numeric group id.

mem
: The members of the group.

In the descriptions of the following functions, this data structure is referred to as a *grp_struct*.

*grp_struct* = getgrent ()  Loadable Function
: Return an entry from the group database, opening it if necessary. Once the end of the data has been reached, getgrent returns 0.

*grp_struct* = getgrgid (*gid*).  Loadable Function
: Return the first entry from the group database with the group ID *gid*. If the group ID does not exist in the database, getgrgid returns 0.

*grp_struct* = getgrnam (*name*)  Loadable Function
: Return the first entry from the group database with the group name *name*. If the group name does not exist in the database, getgrname returns 0.

setgrent ()  Loadable Function
: Return the internal pointer to the beginning of the group database.

endgrent ()  Loadable Function
: Close the group database.

## 32.11 System Information

[c, maxsize, endian] = computer ()      *Function File*
    Print or return a string of the form *cpu-vendor-os* that identifies the kind of computer Octave is running on. If invoked with an output argument, the value is returned instead of printed. For example,

        computer ()
        ⊣ i586-pc-linux-gnu

        x = computer ()
        ⇒ x = "i586-pc-linux-gnu"

    If two output arguments are requested, also return the maximum number of elements for an array.

    If three output arguments are requested, also return the byte order of the current system as a character ("B" for big-endian or "L" for little-endian).

[uts, err, msg] = uname ()      *Built-in Function*
    Return system information in the structure. For example,

        uname ()
        ⇒ {
            sysname = x86_64
            nodename = segfault
            release = 2.6.15-1-amd64-k8-smp
            version = Linux
            machine = #2 SMP Thu Feb 23 04:57:49 UTC 2006
        }

    If successful, *err* is 0 and *msg* is an empty string. Otherwise, *err* is nonzero and *msg* contains a system-dependent error message.

ispc ()      *Function File*
    Return 1 if Octave is running on a Windows system and 0 otherwise.

    **See also:** ismac, isunix.

isunix ()      *Function File*
    Return 1 if Octave is running on a Unix-like system and 0 otherwise.

    **See also:** ismac, ispc.

isieee ()      *Built-in Function*
    Return 1 if your computer claims to conform to the IEEE standard for floating point calculations.

OCTAVE_HOME ()      *Built-in Function*
    Return the name of the top-level Octave installation directory.

OCTAVE_VERSION ()  Built-in Function
　　Return the version number of Octave, as a string.

version ()  Function File
　　Return Octave's version number as a string. This is also the value of the built-in variable OCTAVE_VERSION.

ver ()  Function File
　　Display a header containing the current Octave version number, license string and operating system.

　　**See also:** license, version.

octave_config_info (*option*)  Built-in Function
　　Return a structure containing configuration and installation information for Octave.

　　if *option* is a string, return the configuration information for the specified option.

getrusage ()  Loadable Function
　　Return a structure containing a number of statistics about the current Octave process. Not all fields are available on all systems. If it is not possible to get CPU time statistics, the CPU time slots are set to zero. Other missing data are replaced by NaN. Here is a list of all the possible fields that can be present in the structure returned by getrusage:

idrss
　　Unshared data size.

inblock
　　Number of block input operations.

isrss
　　Unshared stack size.

ixrss
　　Shared memory size.

majflt
　　Number of major page faults.

maxrss
　　Maximum data size.

minflt
　　Number of minor page faults.

msgrcv
　　Number of messages received.

msgsnd
　　Number of messages sent.

**nivcsw**
    Number of involuntary context switches.

**nsignals**
    Number of signals received.

**nswap**
    Number of swaps.

**nvcsw**
    Number of voluntary context switches.

**oublock**
    Number of block output operations.

**stime**
    A structure containing the system CPU time used. The structure has the elements sec (seconds) usec (microseconds).

**utime**
    A structure containing the user CPU time used. The structure has the elements sec (seconds) usec (microseconds).

## 32.12 Hashing Functions

It is often necessary to find if two strings or files are identical. This might be done by comparing them character by character and looking for differences. However, this can be slow, and so comparing a hash of the string or file can be a rapid way of finding if the files differ.

Another use of the hashing function is to check for file integrity. The user can check the hash of the file against a known value and find if the file they have is the same as the one that the original hash was produced with.

Octave supplies the md5sum function to perform MD5 hashes on strings and files. An example of the use of md5sum function might be

```
if exist (file, "file")
  hash = md5sum (file);
else
  # Treat the variable "file" as a string
  hash = md5sum (file, true);
endif
```

md5sum (*file*)                                              Loadable Function
md5sum (*str, opt*)                                          Loadable Function
    Calculates the MD5 sum of the file *file*. If the second parameter *opt* exists and is true, then calculate the MD5 sum of the string *str*.

# 33 Packages

Since Octave is Free Software users are encouraged to share their programs amongst each other. To aid this sharing Octave supports the installation of extra packages.

The 'Octave-Forge' project is a community-maintained set of packages that can be downloaded and installed in Octave. At the time of writing the 'Octave-Forge' project can be found on-line at http://octave.sourceforge.net, but since the Internet is an ever-changing place this may not be true at the time of reading—check the Octave website for the current location.

## 33.1 Installing and Removing Packages

Assuming a package is available in the file image-1.0.0.tar.gz it can be installed from the Octave prompt with the command

    pkg install image-1.0.0.tar.gz

If the package is installed successfully nothing will be printed on the prompt, but if an error occurred during installation it will be reported. It is possible to install several packages at once by writing several package files after the pkg install command. If a different version of the package is already installed it will be removed prior to installing the new package. This makes it easy to upgrade and downgrade the version of a package, but makes it impossible to have several versions of the same package installed at once.

To see which packages are installed type

    pkg list
    ⊣ Package Name  | Version | Installation directory
    ⊣ --------------+---------+-----------------------
    ⊣        image *|  1.0.0  | /home/jwe/octave/image-1.0.0

In this case only version 1.0.0 of the image package is installed. The '*' character next to the package name shows that the image package is loaded and ready for use.

It is possible to remove a package from the system using the pkg uninstall command like this

    pkg uninstall image

If the package is removed successfully nothing will be printed in the prompt, but if an error occurred it will be reported. It should be noted that the package file used for installation is not needed for removal, and that only the package name as reported by pkg list should be used when removing a package. It is possible to remove several packages at once by writing several package names after the pkg uninstall command.

To minimize the amount of code duplication between packages it is possible that one package depends on another one. If a package depends on another, it will check if that package is installed during installation. If it is not, an error will be reported and the package will not be installed. This behaviour can be disabled by passing the -nodeps flag to the pkg install command

```
pkg install -nodeps my_package_with_dependencies.tar.gz
```
Since the installed package expects its dependencies to be installed it may not function correctly. Because of this it is not recommended to disable dependency checking.

## 33.2 Using Packages

By default installed packages are available from the Octave prompt, but it is possible to control this using the pkg load and pkg unload commands. The functions from a package can be removed from the Octave path by typing
```
pkg unload package_name
```
where package_name is the name of the package to be removed from the path.

In much the same way a package can be added to the Octave path by typing
```
pkg load package_name
```

## 33.3 Administrating Packages

On UNIX-like systems it is possible to make both per-user and system-wide installations of a package. If the user performing the installation is root the packages will be installed in a system-wide directory that defaults to OCTAVE_HOME/share/octave/packages/. If the user is not root the default installation directory is ~/octave/. Packages will be installed in a subdirectory of the installation directory that will be named after the package. It is possible to change the installation directory by using the pkg prefix command
```
pkg prefix new_installation_directory
```
The current installation directory can be retrieved by typing
```
current_installation_directory = pkg prefix
```

To function properly the package manager needs to keep some information about the installed packages. For per-user packages this information is by default stored in the file ~/.octave_packages and for system-wide installations it is stored in OCTAVE_HOME/share/octave/octave_packages. The path to the per-user file can be changed with the pkg local_list command
```
pkg local_list /path/to/new_file
```
For system-wide installations this can be changed in the same way using the pkg global_list command. If these commands are called without a new path, the current path will be returned.

## 33.4 Creating Packages

Internally a package is simply a gzipped tar file that contains a top level directory of any given name. This directory will in the following be referred to as package and may contain the following files

package/DESCRIPTION
> This is a required file containing information about the package. See Section 33.4.1 [The DESCRIPTION File], page 472, for details on this file.

package/COPYING
> This is a required file containing the license of the package. No restrictions is made on the license in general. If however the package contains dynamically linked functions the license must be compatible with the GNU General Public License.

package/INDEX
> This is an optional file describing the functions provided by the package. If this file is not given then one with be created automatically from the functions in the package and the Categories keyword in the DESCRIPTION file. See Section 33.4.2 [The INDEX file], page 474, for details on this file.

package/PKG_ADD
> An optional file that includes commands that are run when the package is added to the users path. Note that PKG_ADD directives in the source code of the package will also be added to this file by the Octave package manager. Note that symbolic links are to be avoided in packages, as symbolic links do not exist on some file systems, and so a typical use for this file is the replacement of the symbolic link
>
>     ln -s foo.oct bar.oct
>
> with an autoload directive like
>
>     autoload ('bar', which ('foo'));
>
> See Section 33.4.3 [PKG_ADD and PKG_DEL directives], page 475, for details on PKG_ADD directives.

package/PKG_DEL
> An optional file that includes commands that are run when the package is removed from the users path. Note that PKG_DEL directives in the source code of the package will also be added to this file by the Octave package manager. See Section 33.4.3 [PKG_ADD and PKG_DEL directives], page 475, for details on PKG_DEL directives.

package/pre_install.m
> This is an optional script that is run prior to the installation of a package.

package/post_install.m
> This is an optional script that is run after the installation of a package.

package/on_uninstall.m
> This is an optional script that is run prior to the removal of a package.

Besides the above mentioned files, a package can also contain on or more of the following directories

package/inst
: An optional directory containing any files that are directly installed by the package. Typically this will include any m-files.

package/src
: An optional directory containing code that must be built prior to the packages installation. The Octave package manager will execute ./configure in this directory if this script exists, and will then call make if a file Makefile exists in this directory. make install will however not be called. If a file called FILES exist all files listed there will be copied to the inst directory, so they also will be installed. If the FILES file doesn't exist, src/*.m and src/*.oct will be copied to the inst directory.

package/doc
: An optional directory containing documentation for the package. The files in this directory will be directly installed in a sub-directory of the installed package for future reference.

package/bin
: An optional directory containing files that will be added to the Octave EXEC_PATH when the package is loaded. This might contain external scripts, etc, called by functions within the package.

### 33.4.1 The DESCRIPTION File

The DESCRIPTION file contains various information about the package, such as its name, author, and version. This file has a very simple format

- Lines starting with # are comments.
- Lines starting with a blank character are continuations from the previous line.
- Everything else is of the form NameOfOption: ValueOfOption.

The following is a simple example of a DESCRIPTION file

```
Name: The name of my package
Version: 1.0.0
Date: 2007-18-04
Author: The name (and possibly email) of the package author.
Maintainer: The name (and possibly email) of the current
 package maintainer.
Title: The title of the package
Description: A short description of the package. If this
 description gets too long for one line it can continue
 on the next by adding a space to the beginning of the
 following lines.
License: GPL version 3 or later
```

The package manager currently recognizes the following keywords

Name
: Name of the package.

Version
: Version of the package.

Date
: Date of last update.

Author
: Original author of the package.

Maintainer
: Maintainer of the package.

Title
: A one line description of the package.

Description
: A one paragraph description of the package.

Categories
: Optional keyword describing the package (if no INDEX file is given this is mandatory).

Problems
: Optional list of known problems.

Url
: Optional list of homepages related to the package.

Autoload
: Optional field that sets the default loading behavior for the package. If set to yes, true or on, then Octave will automatically load the package when starting. Otherwise the package must be manually loaded with the pkg load command. This default behavior can be overridden when the package is installed.

Depends
: A list of other Octave packages that this package depends on. This can include dependencies on particular versions, with a format

    Depends: package (>= 1.0.0)

    Possible operators are <, <=, ==, >= or >. If the part of the dependency in () is missing, any version of the package is acceptable. Multiple dependencies can be defined either as a comma separated list or on separate Depends lines.

License
: An optional short description of the used license (e.g. GPL version 3 or newer). This is optional since the file COPYING is mandatory.

SystemRequirements
: These are the external install dependencies of the package and are not checked by the package manager. This is here as a hint to the distribution packager. They follow the same conventions as the Depends keyword.

BuildRequires
: These are the external build dependencies of the package and are not checked by the package manager. This is here as a hint to the distribution packager. They follow the same conventions as the Depends keyword. Note that in general, packaging systems such as rpm or deb and autoprobe the install dependencies from the build dependencies, and therefore the often a BuildRequires dependency removes the need for a SystemRequirements dependency.

The developer is free to add additional arguments to the DESCRIPTION file for their own purposes. One further detail to aid the packager is that the SystemRequirements and BuildRequires keywords can have a distribution dependent section, and the automatic build process will use these. An example of the format of this is

```
BuildRequires: libtermcap-devel [Mandriva] libtermcap2-devel
```

where the first package name will be used as a default and if the RPMs are built on a Mandriva distribution, then the second package name will be used instead.

### 33.4.2 The INDEX file

The optional INDEX file provides a categorical view of the functions in the package. This file has a very simple format

- Lines beginning with # are comments.
- The first non-comment line should look like this

    ```
    toolbox >> Toolbox name
    ```

- Lines beginning with an alphabetical character indicates a new category of functions.
- Lines starting with a white space character indicate that the function names on the line belong to the last mentioned category.

The format can be summarized with the following example

```
# A comment
toolbox >> Toolbox name
Category Name 1
 function1 function2 function3
 function4
Category Name 2
 function2 function5
```

## 33.4.3 PKG_ADD and PKG_DEL directives

If the package contains files called PKG_ADD or PKG_DEL the commands in these files will be executed when the package is added or removed from the users path. In some situations such files are a bit cumbersome to maintain, so the package manager supports automatic creation of such files. If a source file in the package contains a PKG_ADD or PKG_DEL directive they will be added to either the PKG_ADD or PKG_DEL files.

In m-files a PKG_ADD directive looks like this

```
## PKG_ADD: some_octave_command
```

Such lines should be added before the function keyword. In C++ files a PKG_ADD directive looks like this

```
// PKG_ADD: some_octave_command
```

In both cases some_octave_command should be replaced by the command that should be placed in the PKG_ADD file. PKG_DEL directives work in the same way, except the PKG_ADD keyword is replaced with PKG_DEL and the commands get added to the PKG_DEL file.

# Appendix A  Command Line Editing

Octave uses the GNU readline library to provide an extensive set of command-line editing and history features. Only the most common features are described in this manual. Please see The GNU Readline Library manual for more information.

To insert printing characters (letters, digits, symbols, etc.), simply type the character. Octave will insert the character at the cursor and advance the cursor forward.

Many of the command-line editing functions operate using control characters. For example, the character *Control-a* moves the cursor to the beginning of the line. To type *C-a*, hold down (CTRL) and then press (a). In the following sections, control characters such as *Control-a* are written as *C-a*.

Another set of command-line editing functions use Meta characters. On some terminals, you type *M-u* by holding down (META) and pressing (u). If your terminal does not have a (META) key, you can still type Meta characters using two-character sequences starting with *ESC*. Thus, to enter *M-u*, you could type (ESC)(u). The *ESC* character sequences are also allowed on terminals with real Meta keys. In the following sections, Meta characters such as *Meta-u* are written as *M-u*.

## A.0.1 Cursor Motion

The following commands allow you to position the cursor.

*C-b*
> Move back one character.

*C-f*
> Move forward one character.

(DEL)
> Delete the character to the left of the cursor.

*C-d*
> Delete the character underneath the cursor.

*M-f*
> Move forward a word.

*M-b*
> Move backward a word.

*C-a*
> Move to the start of the line.

*C-e*
> Move to the end of the line.

*C-l*
> Clear the screen, reprinting the current line at the top.

C-_
C-/
: Undo the last thing that you did. You can undo all the way back to an empty line.

M-r
: Undo all changes made to this line. This is like typing the 'undo' command enough times to get back to the beginning.

The above table describes the most basic possible keystrokes that you need in order to do editing of the input line. On most terminals, you can also use the arrow keys in place of C-f and C-b to move forward and backward.

Notice how C-f moves forward a character, while M-f moves forward a word. It is a loose convention that control keystrokes operate on characters while meta keystrokes operate on words.

The function clc will allow you to clear the screen from within Octave programs.

## A.0.2 Killing and Yanking

*Killing* text means to delete the text from the line, but to save it away for later use, usually by *yanking* it back into the line. If the description for a command says that it 'kills' text, then you can be sure that you can get the text back in a different (or the same) place later.

Here is the list of commands for killing text.

C-k
: Kill the text from the current cursor position to the end of the line.

M-d
: Kill from the cursor to the end of the current word, or if between words, to the end of the next word.

M-[DEL]
: Kill from the cursor to the start of the previous word, or if between words, to the start of the previous word.

C-w
: Kill from the cursor to the previous whitespace. This is different than M-[DEL] because the word boundaries differ.

And, here is how to *yank* the text back into the line. Yanking means to copy the most-recently-killed text from the kill buffer.

C-y
: Yank the most recently killed text back into the buffer at the cursor.

M-y
: Rotate the kill-ring, and yank the new top. You can only do this if the prior command is C-y or M-y.

Appendix A: Command Line Editing

When you use a kill command, the text is saved in a *kill-ring*. Any number of consecutive kills save all of the killed text together, so that when you yank it back, you get it in one clean sweep. The kill ring is not line specific; the text that you killed on a previously typed line is available to be yanked back later, when you are typing another line.

### A.0.3 Commands For Changing Text

The following commands can be used for entering characters that would otherwise have a special meaning (e.g. *TAB*, *C-q*, etc.), or for quickly correcting typing mistakes.

*C-q*
*C-v*
> Add the next character that you type to the line verbatim. This is how to insert things like *C-q* for example.

*M-*TAB
> Insert a tab character.

*C-t*
> Drag the character before the cursor forward over the character at the cursor, also moving the cursor forward. If the cursor is at the end of the line, then transpose the two characters before it.

*M-t*
> Drag the word behind the cursor past the word in front of the cursor moving the cursor over that word as well.

*M-u*
> Uppercase the characters following the cursor to the end of the current (or following) word, moving the cursor to the end of the word.

*M-l*
> Lowercase the characters following the cursor to the end of the current (or following) word, moving the cursor to the end of the word.

*M-c*
> Uppercase the character following the cursor (or the beginning of the next word if the cursor is between words), moving the cursor to the end of the word.

### A.0.4 Letting Readline Type For You

The following commands allow Octave to complete command and variable names for you.

TAB
> Attempt to do completion on the text before the cursor. Octave can complete the names of commands and variables.

*M-?*
> List the possible completions of the text before the cursor.

`val = completion_append_char ()`  Built-in Function
`old_val = completion_append_char (new_val)`  Built-in Function
>    Query or set the internal character variable that is appended to successful command-line completion attempts. The default value is " " (a single space).

`completion_matches (hint)`  Built-in Function
>    Generate possible completions given *hint*.
>
>    This function is provided for the benefit of programs like Emacs which might be controlling Octave and handling user input. The current command number is not incremented when this function is called. This is a feature, not a bug.

## A.0.5 Commands For Manipulating The History

Octave normally keeps track of the commands you type so that you can recall previous commands to edit or execute them again. When you exit Octave, the most recent commands you have typed, up to the number specified by the variable `history_size`, are saved in a file. When Octave starts, it loads an initial list of commands from the file named by the variable `history_file`.

Here are the commands for simple browsing and searching the history list.

LFD
RET
>    Accept the line regardless of where the cursor is. If this line is non-empty, add it to the history list. If this line was a history line, then restore the history line to its original state.

C-p
>    Move 'up' through the history list.

C-n
>    Move 'down' through the history list.

M-<
>    Move to the first line in the history.

M->
>    Move to the end of the input history, i.e., the line you are entering!

C-r
>    Search backward starting at the current line and moving 'up' through the history as necessary. This is an incremental search.

C-s
>    Search forward starting at the current line and moving 'down' through the history as necessary.

On most terminals, you can also use the arrow keys in place of C-p and C-n to move through the history list.

In addition to the keyboard commands for moving through the history list, Octave provides three functions for viewing, editing, and re-running chunks of commands from the history list.

Appendix A: Command Line Editing                                             481

history options                                                        Command
    If invoked with no arguments, history displays a list of commands that you
    have executed. Valid options are:

    -w *file*

        Write the current history to the file *file*. If the name is omitted, use
        the default history file (normally '~/.octave_hist').

    -r *file*

        Read the file *file*, replacing the current history list with its con-
        tents. If the name is omitted, use the default history file (normally
        '~/.octave_hist').

    *n*    Only display the most recent *n* lines of history.

    -q   Don't number the displayed lines of history. This is useful for cutting
        and pasting commands if you are using the X Window System.

    For example, to display the five most recent commands that you have typed
    without displaying line numbers, use the command *history -q 5*.

edit_history options                                                   Command
    If invoked with no arguments, edit_history allows you to edit the history
    list using the editor named by the variable EDITOR. The commands to be
    edited are first copied to a temporary file. When you exit the editor, Octave
    executes the commands that remain in the file. It is often more convenient
    to use edit_history to define functions rather than attempting to enter
    them directly on the command line. By default, the block of commands is
    executed as soon as you exit the editor. To avoid executing any commands,
    simply delete all the lines from the buffer before exiting the editor.

    The edit_history command takes two optional arguments specifying the
    history numbers of first and last commands to edit. For example, the com-
    mand

        edit_history 13

    extracts all the commands from the 13th through the last in the history list.
    The command

        edit_history 13 169

    only extracts commands 13 through 169. Specifying a larger number for the
    first command than the last command reverses the list of commands before
    placing them in the buffer to be edited. If both arguments are omitted, the
    previous command in the history list is used.

run_history [first] [last]                                             Command
    Similar to edit_history, except that the editor is not invoked, and the
    commands are simply executed as they appear in the history list.

    Octave also allows you customize the details of how and where the history is
saved.

`val = history_file ()`   Built-in Function
`old_val = history_file (new_val)`   Built-in Function
    Query or set the internal variable that specifies the name of the file used to store command history. The default value is `"~/.octave_hist"`, but may be overridden by the environment variable OCTAVE_HISTFILE.

    See also: history_size, saving_history, history_timestamp_format_string.

`val = history_size ()`   Built-in Function
`old_val = history_size (new_val)`   Built-in Function
    Query or set the internal variable that specifies how many entries to store in the history file. The default value is 1024, but may be overridden by the environment variable OCTAVE_HISTSIZE.

    See also: history_file, history_timestamp_format, saving_history.

`val = saving_history ()`   Built-in Function
`old_val = saving_history (new_val)`   Built-in Function
    Query or set the internal variable that controls whether commands entered on the command line are saved in the history file.

    See also: history_file, history_size, history_timestamp_format.

`val = history_timestamp_format_string ()`   Built-in Function
`old_val = history_timestamp_format_string (new_val)`   Built-in Function
    Query or set the internal variable that specifies the format string for the comment line that is written to the history file when Octave exits. The format string is passed to strftime. The default value is

        `"# Octave VERSION, %a %b %d %H:%M:%S %Y %Z <USER@HOST>"`

    See also: strftime, history_file, history_size, saving_history.

`val = EDITOR ()`   Built-in Function
`old_val = EDITOR (new_val)`   Built-in Function
    Query or set the internal variable that specifies the editor to use with the edit_history command. If the environment variable EDITOR is set when Octave starts, its value is used as the default. Otherwise, EDITOR is set to `"emacs"`.

    See also: edit_history.

### A.0.6 Customizing readline

As mentioned earlier Octave uses the GNU readline library for command-line editing and history features. It is possible to customize how readline works through a configuration file.

`read_readline_init_file (file)`   Built-in Function
    Read the readline library initialization file *file*. If *file* is omitted, read the default initialization file (normally '`~/.inputrc`').
    See section "Readline Init File" in *GNU Readline Library*, for details.

Appendix A: Command Line Editing                                    483

## A.0.7 Customizing the Prompt

The following variables are available for customizing the appearance of the command-line prompts. Octave allows the prompt to be customized by inserting a number of backslash-escaped special characters that are decoded as follows:

\t    The time.

\d    The date.

\n    Begins a new line by printing the equivalent of a carriage return followed by a line feed.

\s    The name of the program (usually just octave).

\w    The current working directory.

\W    The basename of the current working directory.

\u    The username of the current user.

\h    The hostname, up to the first '.'.

\H    The hostname.

\#    The command number of this command, counting from when Octave starts.

\!    The history number of this command. This differs from \# by the number of commands in the history list when Octave starts.

\$    If the effective UID is 0, a #, otherwise a $.

\nnn
      The character whose character code in octal is *nnn*.

\\    A backslash.

*val* = PS1 ()                                                Built-in Function
*old_val* = PS1 (*new_val*)                                   Built-in Function
   Query or set the primary prompt string. When executing interactively, Octave displays the primary prompt when it is ready to read a command. The default value of the primary prompt string is "\s:\#> ". To change it, use a command like

            octave:13> PS1 ("\\u@\\H> ")

   which will result in the prompt boris@kremvax> for the user boris logged in on the host kremvax.kgb.su. Note that two backslashes are required to enter a backslash into a double-quoted character string. See Chapter 5 [Strings], page 47.

   See also: PS2, PS4.

*val* = PS2 ()  Built-in Function
*old_val* = PS2 (*new_val*)  Built-in Function
    Query or set the secondary prompt string. The secondary prompt is printed when Octave is expecting additional input to complete a command. For example, if you are typing a for loop that spans several lines, Octave will print the secondary prompt at the beginning of each line after the first. The default value of the secondary prompt string is "> ".

    See also: PS1, PS4.

*val* = PS4 ()  Built-in Function
*old_val* = PS4 (*new_val*)  Built-in Function
    Query or set the character string used to prefix output produced when echoing commands when echo_executing_commands is enabled. The default value is "+ ". See Section 2.1 [Invoking Octave from the Command Line], page 17, for a description of --echo-commands.

    See also: echo_executing_commands, PS1, PS2.

## A.0.8 Diary and Echo Commands

Octave's diary feature allows you to keep a log of all or part of an interactive session by recording the input you type and the output that Octave produces in a separate file.

diary options  Command
    Create a list of all commands *and* the output they produce, mixed together just as you see them on your terminal. Valid options are:

    on    Start recording your session in a file called 'diary' in your current working directory.

    off
        Stop recording your session in the diary file.

    *file*
        Record your session in the file named *file*.

    Without any arguments, diary toggles the current diary state.

Sometimes it is useful to see the commands in a function or script as they are being evaluated. This can be especially helpful for debugging some kinds of problems.

echo options  Command
    Control whether commands are displayed as they are executed. Valid options are:

    on    Enable echoing of commands as they are executed in script files.

    off
        Disable echoing of commands as they are executed in script files.

Appendix A: Command Line Editing

**on all**
> Enable echoing of commands as they are executed in script files and functions.

**off all**
> Disable echoing of commands as they are executed in script files and functions.

If invoked without any arguments, echo toggles the current echo state.

*val* = echo_executing_commands ()  Built-in Function
*old_val* = echo_executing_commands (*new_val*)  Built-in Function
> Query or set the internal variable that controls the echo state. It may be the sum of the following values:
>
> 1    Echo commands read from script files.
>
> 2    Echo commands from functions.
>
> 4    Echo commands read from command line.
>
> More than one state can be active at once. For example, a value of 3 is equivalent to the command *echo on all*.
>
> The value of *echo_executing_commands* is set by the *echo* command and the command line option --echo-input.

# Appendix B  Test and Demo Functions

Octave includes a number of functions to allow the integration of testing and demonstration code in the source code of the functions themselves.

## B.1  Test Functions

| | |
|---|---|
| test *name* | Function File |
| test *name* quiet\|normal\|verbose | Function File |
| test ('*name*', 'quiet\|normal\|verbose', *fid*) | Function File |
| test ([], 'explain', *fid*) | Function File |
| *success* = test (...) | Function File |
| [*n*, *max*] = test (...) | Function File |
| [*code*, *idx*] = test ('*name*','grabdemo') | Function File |

Perform tests from the first file in the loadpath matching *name*. test can be called as a command or as a function. Called with a single argument *name*, the tests are run interactively and stop after the first error is encountered.

With a second argument the tests which are performed and the amount of output is selected.

'quiet'
    Don't report all the tests as they happen, just the errors.

'normal'
    Report all tests as they happen, but don't do tests which require user interaction.

'verbose'
    Do tests which require user interaction.

The argument *fid* can be used to allow batch processing. Errors can be written to the already open file defined by *fid*, and hopefully when Octave crashes this file will tell you what was happening when it did. You can use stdout if you want to see the results as they happen. You can also give a file name rather than an *fid*, in which case the contents of the file will be replaced with the log from the current test.

Called with a single output argument *success*, test returns true if all of the tests were successful. Called with two output arguments *n* and *max*, the number of successful tests and the total number of tests in the file *name* are returned.

If the second argument is the string 'grabdemo', the contents of the demo blocks are extracted but not executed. Code for all code blocks is concatenated and returned as *code* with *idx* being a vector of positions of the ends of the demo blocks.

If the second argument is 'explain', then *name* is ignored and an explanation of the line markers used is written to the file *fid*.

See also: error, assert, fail, demo, example.

test scans the named script file looking for lines which start with %!. The prefix is stripped off and the rest of the line is processed through the Octave interpreter. If the code generates an error, then the test is said to fail.

Since eval() will stop at the first error it encounters, you must divide your tests up into blocks, with anything in a separate block evaluated separately. Blocks are introduced by the keyword test immediately following the %!. For example,

```
%!test error ("this test fails!");
%!test "test doesn't fail. it doesn't generate an error";
```

When a test fails, you will see something like:

```
***** test error ('this test fails!')
!!!!! test failed
this test fails!
```

Generally, to test if something works, you want to assert that it produces a correct value. A real test might look something like

```
%!test
%! a = [1, 2, 3; 4, 5, 6]; B = [1; 2];
%! expect = [ a ; 2*a ];
%! get = kron (b, a);
%! if (any(size(expect) != size(get)))
%!    error ("wrong size: expected %d,%d but got %d,%d",
%!           size(expect), size(get));
%! elseif (any(any(expect!=get)))
%!    error ("didn't get what was expected.");
%! endif
```

To make the process easier, use the assert function. For example, with assert the previous test is reduced to:

```
%!test
%! a = [1, 2, 3; 4, 5, 6]; b = [1; 2];
%! assert (kron (b, a), [ a; 2*a ]);
```

assert can accept a tolerance so that you can compare results absolutely or relatively. For example, the following all succeed:

```
%!test assert (1+eps, 1, 2*eps)         # absolute error
%!test assert (100+100*eps, 100, -2*eps) # relative error
```

You can also do the comparison yourself, but still have assert generate the error:

```
%!test assert (isempty([]))
%!test assert ([ 1,2; 3,4 ] > 0)
```

Because assert is so frequently used alone in a test block, there is a shorthand form:

```
%!assert (...)
```

which is equivalent to:

Appendix B: Test and Demo Functions

```
%!test assert (...)
```

Sometimes during development there is a test that should work but is known to fail. You still want to leave the test in because when the final code is ready the test should pass, but you may not be able to fix it immediately. To avoid unnecessary bug reports for these known failures, mark the block with xtest rather than test:

```
%!xtest assert (1==0)
%!xtest fail ('success=1','error'))
```

Another use of xtest is for statistical tests which should pass most of the time but are known to fail occasionally.

Each block is evaluated in its own function environment, which means that variables defined in one block are not automatically shared with other blocks. If you do want to share variables, then you must declare them as shared before you use them. For example, the following declares the variable a, gives it an initial value (default is empty), then uses it in several subsequent tests.

```
%!shared a
%! a = [1, 2, 3; 4, 5, 6];
%!assert (kron ([1; 2], a), [ a; 2*a ]);
%!assert (kron ([1, 2], a), [ a, 2*a ]);
%!assert (kron ([1,2; 3,4], a), [ a,2*a; 3*a,4*a ]);
```

You can share several variables at the same time:

```
%!shared a, b
```

You can also share test functions:

```
%!function a = fn(b)
%!   a = 2*b;
%!assert (a(2),4);
```

Note that all previous variables and values are lost when a new shared block is declared.

Error and warning blocks are like test blocks, but they only succeed if the code generates an error. You can check the text of the error is correct using an optional regular expression <pattern>. For example:

```
%!error <passes!> error('this test passes!');
```

If the code doesn't generate an error, the test fails. For example,

```
%!error "this is an error because it succeeds.";
```

produces

```
***** error "this is an error because it succeeds.";
!!!!! test failed: no error
```

It is important to automate the tests as much as possible, however some tests require user interaction. These can be isolated into demo blocks, which if you are in batch mode, are only run when called with demo or verbose. The code is displayed before it is executed. For example,

```
%!demo
%! t=[0:0.01:2*pi]; x=sin(t);
%! plot(t,x);
%! you should now see a sine wave in your figure window
```
produces
```
> t=[0:0.01:2*pi]; x=sin(t);
> plot(t,x);
> you should now see a sine wave in your figure window
Press <enter> to continue:
```

Note that demo blocks cannot use any shared variables. This is so that they can be executed by themselves, ignoring all other tests.

If you want to temporarily disable a test block, put # in place of the block type. This creates a comment block which is echoed in the log file, but is not executed. For example:

```
%!#demo
%! t=[0:0.01:2*pi]; x=sin(t);
%! plot(t,x);
%! you should now see a sine wave in your figure window
```

Block type summary:

%!test
> check that entire block is correct

%!error
> check for correct error message

%!warning
> check for correct warning message

%!demo
> demo only executes in interactive mode

%!#
> comment: ignore everything within the block

%!shared x,y,z
> declares variables for use in multiple tests

%!function
> defines a function value for a shared variable

%!assert (x, y, tol)
> shorthand for %!test assert (x, y, tol)

You can also create test scripts for builtins and your own C++ functions. Just put a file of the function name on your path without any extension and it will be picked up by the test procedure. You can even embed tests directly in your C++ code:

Appendix B: Test and Demo Functions                                          491

```
                #if 0
                %!test disp('this is a test')
                #endif
```
or
```
                /*
                %!test disp('this is a test')
                */
```
but then the code will have to be on the load path and the user will have to remember to type test('name.cc'). Conversely, you can separate the tests from normal Octave script files by putting them in plain files with no extension rather than in script files.

assert (*cond*)                                                  Function File
assert (*observed,expected*)                                     Function File
assert (*observed,expected,tol*)                                 Function File
    Produces an error if the condition is not met. assert can be called in three different ways.

    assert (*cond*)
        Called with a single argument *cond*, assert produces an error if *cond* is zero.

    assert (*observed, expected*)
        Produce an error if observed is not the same as expected. Note that observed and expected can be strings, scalars, vectors, matrices, lists or structures.

    assert(*observed, expected, tol*)
        Accept a tolerance when comparing numbers. If *tol* is positive use it as an absolute tolerance, will produce an error if abs(*observed - expected*) > abs(*tol*). If *tol* is negative use it as a relative tolerance, will produce an error if abs(*observed - expected*) > abs(*tol * expected*). If *expected* is zero *tol* will always be used as an absolute tolerance.

    See also: test.

fail (*code,pattern*)                                            Function File
fail (*code,'warning',pattern*)                                  Function File
    Return true if *code* fails with an error message matching *pattern*, otherwise produce an error. Note that *code* is a string and if *code* runs successfully, the error produced is:

        expected error but got none

    If the code fails with a different error, the message produced is:

expected <pattern>
but got <text of actual error>

The angle brackets are not part of the output.

Called with three arguments, the behavior is similar to fail(code, pattern), but produces an error if no warning is given during code execution or if the code fails.

## B.2 Demonstration Functions

demo ('name', n)                                               Function File
Runs any examples associated with the function 'name'. Examples are stored in the script file, or in a file with the same name but no extension somewhere on your path. To keep them separate from the usual script code, all lines are prefixed by %!. Each example is introduced by the keyword 'demo' flush left to the prefix, with no intervening spaces. The remainder of the example can contain arbitrary Octave code. For example:

```
%!demo
%! t=0:0.01:2*pi; x = sin(t);
%! plot(t,x);
%! %-------------------------------------------------
%! % the figure window shows one cycle of a sine wave
```

Note that the code is displayed before it is executed, so a simple comment at the end suffices. It is generally not necessary to use disp or printf within the demo.

Demos are run in a function environment with no access to external variables. This means that all demos in your function must use separate initialization code. Alternatively, you can combine your demos into one huge demo, with the code:

```
%! input("Press <enter> to continue: ","s");
```

between the sections, but this is discouraged. Other techniques include using multiple plots by saying figure between each, or using subplot to put multiple plots in the same window.

Also, since demo evaluates inside a function context, you cannot define new functions inside a demo. Instead you will have to use eval(example('function',n)) to see them. Because eval only evaluates one line, or one statement if the statement crosses multiple lines, you must wrap your demo in "if 1 <demo stuff> endif" with the 'if' on the same line as 'demo'. For example,

```
%!demo if 1
%!   function y=f(x)
%!     y=x;
%!   endfunction
%!   f(3)
%! endif
```

See also: test, example.

Appendix B: Test and Demo Functions 493

example ('name',n)     Function File
[x, idx] = example ('name',n)     Function File
    Display the code for example n associated with the function 'name', but do not run it. If n is not given, all examples are displayed.

    Called with output arguments, the examples are returned in the form of a string x, with idx indicating the ending position of the various examples.

    See demo for a complete explanation.

    **See also:** demo, test.

speed (f, init, max_n, f2, tol)     Function File
[order, n, T_f, T_f2] = speed (...)     Function File
    Determine the execution time of an expression for various n. The n are log-spaced from 1 to max_n. For each n, an initialization expression is computed to create whatever data are needed for the test. If a second expression is given, the execution times of the two expressions will be compared. Called without output arguments the results are presented graphically.

    f     The expression to evaluate.

    max_n
        The maximum test length to run. Default value is 100. Alternatively, use [min_n,max_n] or for complete control, [n1,n2,...,nk].

    init
        Initialization expression for function argument values. Use k for the test number and n for the size of the test. This should compute values for all variables listed in args. Note that init will be evaluated first for k=0, so things which are constant throughout the test can be computed then. The default value is x = randn (n, 1);.

    f2     An alternative expression to evaluate, so the speed of the two can be compared. Default is [].

    tol
        If tol is Inf, then no comparison will be made between the results of expression f and expression f2. Otherwise, expression f should produce a value v and expression f2 should produce a value v2, and these shall be compared using assert(v,v2,tol). If tol is positive, the tolerance is assumed to be absolute. If tol is negative, the tolerance is assumed to be relative. The default is eps.

    order
        The time complexity of the expression O(a n^p). This is a structure with fields a and p.

    n     The values n for which the expression was calculated and the execution time was greater than zero.

    T_f
        The nonzero execution times recorded for the expression f in seconds.

`T_f2`
: The nonzero execution times recorded for the expression f2 in seconds.

If it is needed, the mean time ratio is just mean(T_f./T_f2).

The slope of the execution time graph shows the approximate power of the asymptotic running time O(n^p). This power is plotted for the region over which it is approximated (the latter half of the graph). The estimated power is not very accurate, but should be sufficient to determine the general order of your algorithm. It should indicate if for example your implementation is unexpectedly O(n^2) rather than O(n) because it extends a vector each time through the loop rather than preallocating one which is big enough. For example, in the current version of Octave, the following is not the expected O(n):

```
speed("for i=1:n,y{i}=x(i); end", "", [1000,10000])
```

but it is if you preallocate the cell array y:

```
speed("for i=1:n,y{i}=x(i);end", ...
      "x=rand(n,1);y=cell(size(x));", [1000,10000])
```

An attempt is made to approximate the cost of the individual operations, but it is wildly inaccurate. You can improve the stability somewhat by doing more work for each n. For example:

```
speed("airy(x)", "x=rand(n,10)", [10000,100000])
```

When comparing a new and original expression, the line on the speedup ratio graph should be larger than 1 if the new expression is faster. Better algorithms have a shallow slope. Generally, vectorizing an algorithm will not change the slope of the execution time graph, but it will shift it relative to the original. For example:

```
speed("v=sum(x)", "", [10000,100000], ...
      "v=0;for i=1:length(x),v+=x(i);end")
```

A more complex example, if you had an original version of xcorr using for loops and another version using an FFT, you could compare the run speed for various lags as follows, or for a fixed lag with varying vector lengths as follows:

```
speed("v=xcorr(x,n)", "x=rand(128,1);", 100, ...
      "v2=xcorr_orig(x,n)", -100*eps)
speed("v=xcorr(x,15)", "x=rand(20+n,1);", 100, ...
      "v2=xcorr_orig(x,n)", -100*eps)
```

Assuming one of the two versions is in xcorr_orig, this would compare their speed and their output values. Note that the FFT version is not exact, so we specify an acceptable tolerance on the comparison 100*eps, and the errors should be computed relatively, as abs((x - y)./y) rather than absolutely as abs(x - y).

Type example('speed') to see some real examples. Note for obscure reasons, you can't run examples 1 and 2 directly using demo('speed'). Instead use, eval(example('speed',1)) and eval(example('speed',2)).

# Appendix C  Tips and Standards

This chapter describes no additional features of Octave. Instead it gives advice on making effective use of the features described in the previous chapters.

## C.1  Writing Clean Octave Programs

Here are some tips for avoiding common errors in writing Octave code intended for widespread use:

- Since all global variables share the same name space, and all functions share another name space, you should choose a short word to distinguish your program from other Octave programs. Then take care to begin the names of all global variables, constants, and functions with the chosen prefix. This helps avoid name conflicts.

  If you write a function that you think ought to be added to Octave under a certain name, such as fiddle_matrix, don't call it by that name in your program. Call it mylib_fiddle_matrix in your program, and send mail to maintainers@octave.org suggesting that it be added to Octave. If and when it is, the name can be changed easily enough.

  If one prefix is insufficient, your package may use two or three alternative common prefixes, so long as they make sense.

  Separate the prefix from the rest of the symbol name with an underscore _. This will be consistent with Octave itself and with most Octave programs.

- When you encounter an error condition, call the function error (or usage). The error and usage functions do not return. See Section 2.5 [Errors], page 24.

- Please put a copyright notice on the file if you give copies to anyone. Use the same lines that appear at the top of the function files distributed with Octave. If you have not signed papers to assign the copyright to anyone else, then place your name in the copyright notice.

## C.2  Tips for Making Code Run Faster.

Here are some ways of improving the execution speed of Octave programs.

- Avoid looping wherever possible.
- Use iteration rather than recursion whenever possible. Function calls are slow in Octave.
- Avoid resizing matrices unnecessarily. When building a single result matrix from a series of calculations, set the size of the result matrix first, then insert values into it. Write

```
        result = zeros (big_n, big_m)
        for i = over:and_over
          r1 = ...
          r2 = ...
          result (r1, r2) = new_value ();
        endfor
```
instead of
```
        result = [];
        for i = ever:and_ever
          result = [ result, new_value() ];
        endfor
```

- Avoid calling `eval` or `feval` whenever possible, because they require Octave to parse input or look up the name of a function in the symbol table.

  If you are using `eval` as an exception handling mechanism and not because you need to execute some arbitrary text, use the `try` statement instead. See Section 10.9 [The try Statement], page 123.

- If you are calling lots of functions but none of them will need to change during your run, set the variable `ignore_function_time_stamp` to "all" so that Octave doesn't waste a lot of time checking to see if you have updated your function files.

## C.3 Tips on Writing Comments

Here are the conventions to follow when writing comments.

`#`     Comments that start with a single sharp-sign, `#`, should all be aligned to the same column on the right of the source code. Such comments usually explain how the code on the same line does its job. In the Emacs mode for Octave, the `M-;` (indent-for-comment) command automatically inserts such a `#` in the right place, or aligns such a comment if it is already present.

`##`    Comments that start with a double sharp-sign, `##`, should be aligned to the same level of indentation as the code. Such comments usually describe the purpose of the following lines or the state of the program at that point.

The indentation commands of the Octave mode in Emacs, such as `M-;` (indent-for-comment) and `TAB` (`octave-indent-line`) automatically indent comments according to these conventions, depending on the number of semicolons. See section "Manipulating Comments" in *The GNU Emacs Manual*.

# C.4 Conventional Headers for Octave Functions

Octave has conventions for using special comments in function files to give information such as who wrote them. This section explains these conventions.

The top of the file should contain a copyright notice, followed by a block of comments that can be used as the help text for the function. Here is an example:

```
## Copyright (C) 1996, 1997, 2007 John W. Eaton
##
## This file is part of Octave.
##
## Octave is free software; you can redistribute it and/or
## modify it under the terms of the GNU General Public
## License as published by the Free Software Foundation;
## either version 3 of the License, or (at your option) any
## later version.
##
## Octave is distributed in the hope that it will be useful,
## but WITHOUT ANY WARRANTY; without even the implied
## warranty of MERCHANTABILITY or FITNESS FOR A PARTICULAR
## PURPOSE.  See the GNU General Public License for more
## details.
##
## You should have received a copy of the GNU General Public
## License along with Octave; see the file COPYING.  If not,
## see <http://www.gnu.org/licenses/>.

## usage: [IN, OUT, PID] = popen2 (COMMAND, ARGS)
##
## Start a subprocess with two-way communication.  COMMAND
## specifies the name of the command to start.  ARGS is an
## array of strings containing options for COMMAND.  IN and
## OUT are the file ids of the input and streams for the
## subprocess, and PID is the process id of the subprocess,
## or -1 if COMMAND could not be executed.
##
## Example:
##
##   [in, out, pid] = popen2 ("sort", "-nr");
##   fputs (in, "these\nare\nsome\nstrings\n");
##   fclose (in);
##   while (isstr (s = fgets (out)))
##     fputs (stdout, s);
##   endwhile
##   fclose (out);
```

Octave uses the first block of comments in a function file that do not appear to be a copyright notice as the help text for the file. For Octave to recognize the first comment block as a copyright notice, it must start with the word 'Copyright' after stripping the leading comment characters.

After the copyright notice and help text come several *header comment* lines, each beginning with ## *header-name* :. For example,

    ## Author: jwe
    ## Keywords: subprocesses input-output
    ## Maintainer: jwe

Here is a table of the conventional possibilities for *header-name*:

Author
: This line states the name and net address of at least the principal author of the library.

    ## Author: John W. Eaton <jwe@bevo.che.wisc.edu>

Maintainer
: This line should contain a single name/address as in the Author line, or an address only, or the string jwe. If there is no maintainer line, the person(s) in the Author field are presumed to be the maintainers. The example above is mildly bogus because the maintainer line is redundant.

    The idea behind the Author and Maintainer lines is to make possible a function to "send mail to the maintainer" without having to mine the name out by hand.

    Be sure to surround the network address with <...> if you include the person's full name as well as the network address.

Created
: This optional line gives the original creation date of the file. For historical interest only.

Version
: If you wish to record version numbers for the individual Octave program, put them in this line.

Adapted-By
: In this header line, place the name of the person who adapted the library for installation (to make it fit the style conventions, for example).

Keywords
: This line lists keywords. Eventually, it will be used by an apropos command to allow people will find your package when they're looking for things by topic area. To separate the keywords, you can use spaces, commas, or both.

Just about every Octave function ought to have the Author and Keywords header comment lines. Use the others if they are appropriate. You can also put in header lines with other header names—they have no standard meanings, so they can't do any harm.

## C.5 Tips for Documentation Strings

As noted above, documentation is typically in a commented header block on an Octave function following the copyright statement. The help string shown above is an unformatted string and will be displayed as is by Octave. Here are some tips for the writing of documentation strings.

- Every command, function, or variable intended for users to know about should have a documentation string.
- An internal variable or subroutine of an Octave program might as well have a documentation string.
- The first line of the documentation string should consist of one or two complete sentences that stand on their own as a summary.

  The documentation string can have additional lines that expand on the details of how to use the function or variable. The additional lines should also be made up of complete sentences.

- For consistency, phrase the verb in the first sentence of a documentation string as an infinitive with "to" omitted. For instance, use "Return the frob of A and B." in preference to "Returns the frob of A and B." Usually it looks good to do likewise for the rest of the first paragraph. Subsequent paragraphs usually look better if they have proper subjects.
- Write documentation strings in the active voice, not the passive, and in the present tense, not the future. For instance, use "Return a list containing A and B." instead of "A list containing A and B will be returned."
- Avoid using the word "cause" (or its equivalents) unnecessarily. Instead of, "Cause Octave to display text in boldface," write just "Display text in boldface."
- Do not start or end a documentation string with whitespace.
- Format the documentation string so that it fits in an Emacs window on an 80-column screen. It is a good idea for most lines to be no wider than 60 characters.

  However, rather than simply filling the entire documentation string, you can make it much more readable by choosing line breaks with care. Use blank lines between topics if the documentation string is long.

- **Do not** indent subsequent lines of a documentation string so that the text is lined up in the source code with the text of the first line. This looks nice in the source code, but looks bizarre when users view the documentation. Remember that the indentation before the starting double-quote is not part of the string!
- The documentation string for a variable that is a yes-or-no flag should start with words such as "Nonzero means...", to make it clear that all nonzero values are equivalent and indicate explicitly what zero and nonzero mean.

- When a function's documentation string mentions the value of an argument of the function, use the argument name in capital letters as if it were a name for that value. Thus, the documentation string of the operator / refers to its second argument as DIVISOR, because the actual argument name is divisor.

  Also use all caps for meta-syntactic variables, such as when you show the decomposition of a list or vector into subunits, some of which may vary.

Octave also allows extensive formatting of the help string of functions using Texinfo. The effect on the online documentation is relatively small, but makes the help string of functions conform to the help of Octave's own functions. However, the effect on the appearance of printed or online documentation will be greatly improved.

The fundamental building block of Texinfo documentation strings is the Texinfo-macro @deftypefn, which takes three arguments: The class the function is in, its output arguments, and the function's signature. Typical classes for functions include Function File for standard Octave functions, and Loadable Function for dynamically linked functions. A skeletal Texinfo documentation string therefore looks like this

```
-*- texinfo -*-
@deftypefn{Function File} {@var{ret} = } fn (...)
@cindex index term
Help text in Texinfo format.  Code samples should be marked
like @code{sample of code} and variables should be marked
as @var{variable}.
@seealso{fn2}
@end deftypefn
```

This help string must be commented in user functions, or in the help string of the DEFUN_DLD macro for dynamically loadable functions. The important aspects of the documentation string are

-*- texinfo -*-

> This string signals Octave that the following text is in Texinfo format, and should be the first part of any help string in Texinfo format.

@deftypefn{class} ... @end deftypefn

> The entire help string should be enclosed within the block defined by deftypefn.

@cindex index term

> This generates an index entry, and can be useful when the function is included as part of a larger piece of documentation. It is ignored within Octave's help viewer.

@var{variable}

> All variables should be marked with this macro. The markup of variables is then changed appropriately for display.

@code{sample of code}

> All samples of code should be marked with this macro for the same reasons as the @var macro.

# Appendix C: Tips and Standards

@seealso{function2}
: This is a comma separated list of function names that allows cross referencing from one function documentation string to another.

Texinfo format has been designed to generate output for online viewing with text-terminals as well as generating high-quality printed output. To these ends, Texinfo has commands which control the diversion of parts of the document into a particular output processor. Three formats are of importance: info, html and TeX. These are selected with

```
@ifinfo
Text area for info only
@end ifinfo
@ifhtml
Text area for html only
@end ifhtml
@iftex
@tex
text for TeX only
@end tex
@end iftex
```

Note that often TeX output can be used in html documents and so often the @ifhtml blocks are unnecessary. If no specific output processor is chosen, by default, the text goes into all output processors. It is usual to have the above blocks in pairs to allow the same information to be conveyed in all output formats, but with a different markup.

Another important feature of Texinfo that is often used in Octave help strings is the @example environment. An example of its use is

```
@example
@group
@code{2 * 2}
@result{} 4
@end group
@end example
```

which produces

```
2 * 2
⇒ 4
```

The @group block prevents the example from being split across a page boundary, while the @result{} macro produces a right arrow signifying the result of a command.

In many cases a function has multiple ways in which it can be called, and the @deftypefnx macro can be used to give alternatives. For example

```
-*- texinfo -*-
@deftypefn{Function File} {@var{a} = } fn (@var{x}, ...)
@deftypefnx{Function File} {@var{a} = } fn (@var{y}, ...)
Help text in Texinfo format.
@end deftypefn
```

Many complete examples of Texinfo documentation can be taken from the help strings for the Octave functions themselves. A relatively complete example of which is the nchoosek function. The Texinfo documentation string of nchoosek is

```
-*- texinfo -*-
@deftypefn {Function File} {} nchoosek (@var{n}, @var{k})

Compute the binomial coefficient or all combinations of
@var{n}. If @var{n} is a scalar then, calculate the
binomial coefficient of @var{n} and @var{k}, defined as

@iftex
@tex
$$
 {n \choose k} = {n (n-1) (n-2) \cdots (n-k+1) \over k!}
$$
@end tex
@end iftex
@ifinfo

@example
@group
 /   \
 | n |    n (n-1) (n-2) ... (n-k+1)
 |   |  = -------------------------
 | k |               k!
 \   /
@end group
@end example
@end ifinfo

If @var{n} is a vector, this generates all combinations
of the elements of @var{n}, taken @var{k} at a time,
one row per combination. The resulting @var{c} has size
@code{[nchoosek (length (@var{n}),@var{k}), @var{k}]}.

@seealso{bincoeff}
@end deftypefn
```

which demonstrates most of the concepts discussed above. This documentation string renders as

```
-- Function File: C = nchoosek (N, K)
    Compute the binomial coefficient or all combinations
    of N. If N is a scalar then, calculate the binomial
    coefficient of N and K, defined as

        /   \
        | n |     n (n-1) (n-2) ... (n-k+1)           n!
        |   |  = ---------------------------  =   ---------
        | k |                k!                   k! (n-k)!
        \   /

    If N is a vector generate all combinations of the
    elements of N, taken K at a time, one row per
    combination. The resulting C has size '[nchoosek
    (length (N), K), K]'.

    See also: bincoeff.
```
using info, whereas in a printed documentation using TeX it will appear as

c = nchoosek (n, k)                                          Function File
Compute the binomial coefficient or all combinations of n. If n is a scalar then, calculate the binomial coefficient of n and k, defined as

$$\binom{n}{k} = \frac{n(n-1)(n-2)\cdots(n-k+1)}{k!}$$

If n is a vector generate all combinations of the elements of n, taken k at a time, one row per combination. The resulting c has size [nchoosek (length (n), k), k].

**See also:** bincoeff.

# Appendix D  Known Causes of Trouble

This section describes known problems that affect users of Octave. Most of these are not Octave bugs per se—if they were, we would fix them. But the result for a user may be like the result of a bug.

Some of these problems are due to bugs in other software, some are missing features that are too much work to add, and some are places where people's opinions differ as to what is best.

## D.1  Actual Bugs We Haven't Fixed Yet

- Output that comes directly from Fortran functions is not sent through the pager and may appear out of sequence with other output that is sent through the pager. One way to avoid this is to force pending output to be flushed before calling a function that will produce output from within Fortran functions. To do this, use the command

    fflush (stdout)

  Another possible workaround is to use the command

    page_screen_output = "false"

  to turn the pager off.

A list of ideas for future enhancements is distributed with Octave. See the file 'PROJECTS' in the top level directory in the source distribution.

## D.2  Reporting Bugs

Your bug reports play an essential role in making Octave reliable.

When you encounter a problem, the first thing to do is to see if it is already known. See Appendix D [Trouble], page 505. If it isn't known, then you should report the problem.

Reporting a bug may help you by bringing a solution to your problem, or it may not. In any case, the principal function of a bug report is to help the entire community by making the next version of Octave work better. Bug reports are your contribution to the maintenance of Octave.

In order for a bug report to serve its purpose, you must include the information that makes it possible to fix the bug.

If you have Octave working at all, the easiest way to prepare a complete bug report is to use the Octave function bug_report. When you execute this function, Octave will prompt you for a subject and then invoke the editor on a file that already contains all the configuration information. When you exit the editor, Octave will mail the bug report for you.

bug_report ()                                                    Function File

  Have Octave create a bug report template file, invoke your favorite editor, and submit the report to the bug-octave mailing list when you are finished editing.

## D.3 Have You Found a Bug?

If you are not sure whether you have found a bug, here are some guidelines:

- If Octave gets a fatal signal, for any input whatever, that is a bug. Reliable interpreters never crash.
- If Octave produces incorrect results, for any input whatever, that is a bug.
- Some output may appear to be incorrect when it is in fact due to a program whose behavior is undefined, which happened by chance to give the desired results on another system. For example, the range operator may produce different results because of differences in the way floating point arithmetic is handled on various systems.
- If Octave produces an error message for valid input, that is a bug.
- If Octave does not produce an error message for invalid input, that is a bug. However, you should note that your idea of "invalid input" might be my idea of "an extension" or "support for traditional practice".
- If you are an experienced user of programs like Octave, your suggestions for improvement are welcome in any case.

## D.4 Where to Report Bugs

If you have Octave working at all, the easiest way to prepare a complete bug report is to use the Octave function bug_report. When you execute this function, Octave will prompt you for a subject and then invoke the editor on a file that already contains all the configuration information. When you exit the editor, Octave will mail the bug report for you.

If for some reason you cannot use Octave's bug_report function, send bug reports for Octave to bug@octave.org.

**Do not send bug reports** to help-octave. Most users of Octave do not want to receive bug reports. Those that do have asked to be on the mailing list.

As a last resort, send bug reports on paper to:

```
Octave Bugs c/o John W. Eaton
University of Wisconsin-Madison
Department of Chemical Engineering
1415 Engineering Drive
Madison, Wisconsin 53706  USA
```

Appendix D: Known Causes of Trouble    507

## D.5 How to Report Bugs

Send bug reports for Octave to one of the addresses listed in Section D.4 [Bug Lists], page 506.

The fundamental principle of reporting bugs usefully is this: **report all the facts**. If you are not sure whether to state a fact or leave it out, state it!

Often people omit facts because they think they know what causes the problem and they conclude that some details don't matter. Thus, you might assume that the name of the variable you use in an example does not matter. Well, probably it doesn't, but one cannot be sure. Perhaps the bug is a stray memory reference which happens to fetch from the location where that name is stored in memory; perhaps, if the name were different, the contents of that location would fool the interpreter into doing the right thing despite the bug. Play it safe and give a specific, complete example.

Keep in mind that the purpose of a bug report is to enable someone to fix the bug if it is not known. Always write your bug reports on the assumption that the bug is not known.

Sometimes people give a few sketchy facts and ask, "Does this ring a bell?" This cannot help us fix a bug. It is better to send a complete bug report to begin with.

Try to make your bug report self-contained. If we have to ask you for more information, it is best if you include all the previous information in your response, as well as the information that was missing.

To enable someone to investigate the bug, you should include all these things:

- The version of Octave. You can get this by noting the version number that is printed when Octave starts, or running it with the -v option.

- A complete input file that will reproduce the bug.

  A single statement may not be enough of an example—the bug might depend on other details that are missing from the single statement where the error finally occurs.

- The command arguments you gave Octave to execute that example and observe the bug. To guarantee you won't omit something important, list all the options.

  If we were to try to guess the arguments, we would probably guess wrong and then we would not encounter the bug.

- The type of machine you are using, and the operating system name and version number.

- The command-line arguments you gave to the `configure` command when you installed the interpreter.

- A complete list of any modifications you have made to the interpreter source.

  Be precise about these changes—show a context diff for them.

- Details of any other deviations from the standard procedure for installing Octave.

- A description of what behavior you observe that you believe is incorrect. For example, "The interpreter gets a fatal signal," or, "The output produced at line 208 is incorrect."

  Of course, if the bug is that the interpreter gets a fatal signal, then one can't miss it. But if the bug is incorrect output, we might not notice unless it is glaringly wrong.

  Even if the problem you experience is a fatal signal, you should still say so explicitly. Suppose something strange is going on, such as, your copy of the interpreter is out of synch, or you have encountered a bug in the C library on your system. Your copy might crash and the copy here would not. If you said to expect a crash, then when the interpreter here fails to crash, we would know that the bug was not happening. If you don't say to expect a crash, then we would not know whether the bug was happening. We would not be able to draw any conclusion from our observations.

  Often the observed symptom is incorrect output when your program is run. Unfortunately, this is not enough information unless the program is short and simple. It is very helpful if you can include an explanation of the expected output, and why the actual output is incorrect.

- If you wish to suggest changes to the Octave source, send them as context diffs. If you even discuss something in the Octave source, refer to it by context, not by line number, because the line numbers in the development sources probably won't match those in your sources.

Here are some things that are not necessary:

- A description of the envelope of the bug.

  Often people who encounter a bug spend a lot of time investigating which changes to the input file will make the bug go away and which changes will not affect it. Such information is usually not necessary to enable us to fix bugs in Octave, but if you can find a simpler example to report *instead* of the original one, that is a convenience. Errors in the output will be easier to spot, running under the debugger will take less time, etc. Most Octave bugs involve just one function, so the most straightforward way to simplify an example is to delete all the function definitions except the one in which the bug occurs.

  However, simplification is not vital; if you don't want to do this, report the bug anyway and send the entire test case you used.

- A patch for the bug. Patches can be helpful, but if you find a bug, you should report it, even if you cannot send a fix for the problem.

## D.6 Sending Patches for Octave

If you would like to write bug fixes or improvements for Octave, that is very helpful. When you send your changes, please follow these guidelines to avoid causing extra work for us in studying the patches.

If you don't follow these guidelines, your information might still be useful, but using it will take extra work. Maintaining Octave is a lot of work in the best of circumstances, and we can't keep up unless you do your best to help.

- Send an explanation with your changes of what problem they fix or what improvement they bring about. For a bug fix, just include a copy of the bug report, and explain why the change fixes the bug.

- Always include a proper bug report for the problem you think you have fixed. We need to convince ourselves that the change is right before installing it. Even if it is right, we might have trouble judging it if we don't have a way to reproduce the problem.

- Include all the comments that are appropriate to help people reading the source in the future understand why this change was needed.

- Don't mix together changes made for different reasons. Send them *individually*.

  If you make two changes for separate reasons, then we might not want to install them both. We might want to install just one.

- Use diff -c to make your diffs. Diffs without context are hard for us to install reliably. More than that, they make it hard for us to study the diffs to decide whether we want to install them. Unidiff format is better than contextless diffs, but not as easy to read as -c format.

  If you have GNU diff, use diff -cp, which shows the name of the function that each change occurs in.

- Write the change log entries for your changes.

  Read the 'ChangeLog' file to see what sorts of information to put in, and to learn the style that we use. The purpose of the change log is to show people where to find what was changed. So you need to be specific about what functions you changed; in large functions, it's often helpful to indicate where within the function the change was made.

  On the other hand, once you have shown people where to find the change, you need not explain its purpose. Thus, if you add a new function, all you need to say about it is that it is new. If you feel that the purpose needs explaining, it probably does—but the explanation will be much more useful if you put it in comments in the code.

  If you would like your name to appear in the header line for who made the change, send us the header line.

## D.7 How To Get Help with Octave

The mailing list help@octave.org exists for the discussion of matters related to using and installing Octave. If would like to join the discussion, please send a short note to help-request@octave.org.

**Please do not** send requests to be added or removed from the mailing list, or other administrative trivia to the list itself.

If you think you have found a bug in the installation procedure, however, you should send a complete bug report for the problem to bug@octave.org. See Section D.5 [Bug Reporting], page 507, for information that will help you to submit a useful report.

# Appendix E  Installing Octave

Here is the procedure for installing Octave from scratch on a Unix system.
- Run the shell script 'configure'. This will determine the features your system has (or doesn't have) and create a file named 'Makefile' from each of the files named 'Makefile.in'.

  Here is a summary of the configure options that are most frequently used when building Octave:

  --prefix=*prefix*
  : Install Octave in subdirectories below *prefix*. The default value of *prefix* is '/usr/local'.

  --srcdir=*dir*
  : Look for Octave sources in the directory *dir*.

  --enable-bounds-check
  : Enable bounds checking for indexing operators in the internal array classes. This option is primarily used for debugging Octave. Building Octave with this option has a negative impact on performance and is not recommended for general use.

  --enable-64
  : This is an **experimental** option to enable Octave to use 64-bit integers for array dimensions and indexing on 64-bit platforms. You probably don't want to use this option unless you know what you are doing.

    If you use --enable-64, you must ensure that your Fortran compiler generates code with 8 byte signed INTEGER values, and that your BLAS and LAPACK libraries are compiled to use 8 byte signed integers for array dimensions and indexing.

  --enable-shared
  : Create shared libraries (this is the default). If you are planning to use the dynamic loading features, you will probably want to use this option. It will make your '.oct' files much smaller and on some systems it may be necessary to build shared libraries in order to use dynamically linked functions.

    You may also want to build a shared version of libstdc++, if your system doesn't already have one.

  --enable-dl
  : Use dlopen and friends to make Octave capable of dynamically linking externally compiled functions (this is the default if --enable-shared is specified). This option only works on systems that actually have these functions. If you plan on using this feature, you should probably also use --enable-shared to reduce the size of your '.oct' files.

--without-blas
: Compile and use the generic BLAS and LAPACK versions included with Octave. By default, configure first looks for BLAS and LAPACK matrix libraries on your system, including optimized BLAS implementations such as the free ATLAS 3.0, as well as vendor-tuned libraries. (The use of an optimized BLAS will generally result in several-times faster matrix operations.) Only use this option if your system has BLAS/LAPACK libraries that cause problems for some reason. You can also use --with-blas=lib to specify a particular BLAS library -llib that configure doesn't check for automatically.

--without-ccolamd
: Don't use CCOLAMD, disable some sparse matrix functionality.

--without-colamd
: Don't use COLAMD, disable some sparse matrix functionality.

--without-curl
: Don't use the cURL, disable the urlread and urlwrite functions.

--without-cxsparse
: Don't use CXSPARSE, disable some sparse matrix functionality.

--without-umfpack
: Don't use UMFPACK, disable some sparse matrix functionality.

--without-fftw
: Use the included fftpack library instead of the FFTW library.

--without-glpk
: Don't use the GLPK library for linear programming.

--without-hdf5
: Don't use the HDF5 library for reading and writing HDF5 files.

--without-zlib
: Don't use the zlib library, disable data file compression and support for recent MAT file formats.

--without-lapack
: Compile and use the generic BLAS and LAPACK versions included with Octave. By default, configure first looks for BLAS and LAPACK matrix libraries on your system, including optimized BLAS implementations such as the free ATLAS 3.0, as well as vendor-tuned libraries. (The use of an optimized BLAS will generally result in several-times faster matrix operations.) Only use this option if your system has BLAS/LAPACK libraries that cause problems for some reason. You can also use --with-blas=lib to specify a particular BLAS library -llib that configure doesn't check for automatically.

--help
: Print a summary of the options recognized by the configure script.

# Appendix E: Installing Octave

See the file 'INSTALL' for more information about the command line options used by configure. That file also contains instructions for compiling in a directory other than where the source is located.

- Run make.

  You will need a recent version of GNU Make. Modifying Octave's makefiles to work with other make programs is probably not worth your time. We recommend you get and compile GNU Make instead.

  For plotting, you will need to have gnuplot installed on your system. Gnuplot is a command-driven interactive function plotting program. Gnuplot is copyrighted, but freely distributable. The 'gnu' in gnuplot is a coincidence—it is not related to the GNU project or the FSF in any but the most peripheral sense.

  To compile Octave, you will need a recent version of GNU Make. You will also need a recent version of g++ or other ANSI C++ compiler. You will also need a Fortran 77 compiler or f2c. If you use f2c, you will need a script like fort77 that works like a normal Fortran compiler by combining f2c with your C compiler in a single script.

  If you plan to modify the parser you will also need GNU bison and flex. If you modify the documentation, you will need GNU Texinfo, along with the patch for the makeinfo program that is distributed with Octave.

  GNU Make, gcc, and libstdc++, gnuplot, bison, flex, and Texinfo are all available from many anonymous ftp archives. The primary site is ftp.gnu.org, but it is often very busy. A list of sites that mirror the software on ftp.gnu.org is available by anonymous ftp from ftp://ftp.gnu.org/pub/gnu/GNUinfo/FTP.

  You will need about 925 megabytes of disk storage to work with when building Octave from source (considerably less if you don't compile with debugging symbols). To do that, use the command

      make CFLAGS=-O CXXFLAGS=-O LDFLAGS=

  instead of just make.

- If you encounter errors while compiling Octave, first check the list of known problems below to see if there is a workaround or solution for your problem. If not, see Appendix D [Trouble], page 505, for information about how to report bugs.

- Once you have successfully compiled Octave, run make install.

  This will install a copy of Octave, its libraries, and its documentation in the destination directory. As distributed, Octave is installed in the following directories. In the table below, *prefix* defaults to '/usr/local', *version* stands for the current version number of the interpreter, and *arch* is the type of computer on which Octave is installed (for example, i586-unknown-gnu).

  '*prefix*/bin'
  Octave and other binaries that people will want to run directly.

'prefix/lib'
: Libraries like libcruft.a and liboctave.a.

'prefix/share'
: Architecture-independent data files.

'prefix/include/octave'
: Include files distributed with Octave.

'prefix/man/man1'
: Unix-style man pages describing Octave.

'prefix/info'
: Info files describing Octave.

'prefix/share/octave/version/m'
: Function files distributed with Octave. This includes the Octave version, so that multiple versions of Octave may be installed at the same time.

'prefix/lib/octave/version/exec/arch'
: Executables to be run by Octave rather than the user.

'prefix/lib/octave/version/oct/arch'
: Object files that will be dynamically loaded.

'prefix/share/octave/version/imagelib'
: Image files that are distributed with Octave.

## E.1 Installation Problems

This section contains a list of problems (and some apparent problems that don't really mean anything is wrong) that may show up during installation of Octave.

- On some SCO systems, info fails to compile if HAVE_TERMIOS_H is defined in 'config.h'. Simply removing the definition from 'info/config.h' should allow it to compile.

- If configure finds dlopen, dlsym, dlclose, and dlerror, but not the header file 'dlfcn.h', you need to find the source for the header file and install it in the directory 'usr/include'. This is reportedly a problem with Slackware 3.1. For Linux/GNU systems, the source for 'dlfcn.h' is in the ldso package.

- Building '.oct' files doesn't work.

    You should probably have a shared version of libstdc++. A patch is needed to build shared versions of version 2.7.2 of libstdc++ on the HP-PA architecture. You can find the patch at ftp://ftp.cygnus.com/pub/g++/libg++-2.7.2-hppa-gcc-fix.

- On some alpha systems there may be a problem with the libdxml library, resulting in floating point errors and/or segmentation faults in the linear algebra routines called by Octave. If you encounter such problems, then you should modify the configure script so that SPECIAL_MATH_LIB is not set to -ldxml.

Appendix E: Installing Octave                                              515

- On FreeBSD systems Octave may hang while initializing some internal constants. The fix appears to be to use

    options      GPL_MATH_EMULATE

  rather than

    options      MATH_EMULATE

  in the kernel configuration files (typically found in the directory '/sys/i386/conf'. After making this change, you'll need to rebuild the kernel, install it, and reboot.

- If you encounter errors like

    passing 'void (*)()' as argument 2 of
      'octave_set_signal_handler(int, void (*)(int))'

  or

    warning: ANSI C++ prohibits conversion from '(int)'
        to '(...)'

  while compiling 'sighandlers.cc', you may need to edit some files in the gcc include subdirectory to add proper prototypes for functions there. For example, Ultrix 4.2 needs proper declarations for the signal function and the SIG_IGN macro in the file 'signal.h'.

  On some systems the SIG_IGN macro is defined to be something like this:

    #define  SIG_IGN   (void (*)())1

  when it should really be something like:

    #define  SIG_IGN   (void (*)(int))1

  to match the prototype declaration for the signal function. This change should also be made for the SIG_DFL and SIG_ERR symbols. It may be necessary to change the definitions in 'sys/signal.h' as well.

  The gcc fixincludes and fixproto scripts should probably fix these problems when gcc installs its modified set of header files, but I don't think that's been done yet.

  **You should not change the files in '/usr/include'.** You can find the gcc include directory tree by running the command

    gcc -print-libgcc-file-name

  The directory of gcc include files normally begins in the same directory that contains the file 'libgcc.a'.

- Some of the Fortran subroutines may fail to compile with older versions of the Sun Fortran compiler. If you get errors like

    zgemm.f:
       zgemm:
    warning: unexpected parent of complex expression subtree
    zgemm.f, line 245: warning: unexpected parent of complex
      expression subtree
    warning: unexpected parent of complex expression subtree
    zgemm.f, line 304: warning: unexpected parent of complex
      expression subtree

```
warning: unexpected parent of complex expression subtree
zgemm.f, line 327: warning: unexpected parent of complex
  expression subtree
pcc_binval: missing IR_CONV in complex op
make[2]: *** [zgemm.o] Error 1
```
when compiling the Fortran subroutines in the 'libcruft' subdirectory, you should either upgrade your compiler or try compiling with optimization turned off.

- On NeXT systems, if you get errors like this:
  ```
  /usr/tmp/cc007458.s:unknown:Undefined local
      symbol LBB7656
  /usr/tmp/cc007458.s:unknown:Undefined local
      symbol LBE7656
  ```
  when compiling 'Array.cc' and 'Matrix.cc', try recompiling these files without -g.

- Some people have reported that calls to shell_cmd and the pager do not work on SunOS systems. This is apparently due to having G_HAVE_SYS_WAIT defined to be 0 instead of 1 when compiling libg++.

- On NeXT systems, linking to 'libsys_s.a' may fail to resolve the following functions
  ```
  _tcgetattr
  _tcsetattr
  _tcflow
  ```
  which are part of 'libposix.a'. Unfortunately, linking Octave with -posix results in the following undefined symbols.
  ```
  .destructors_used
  .constructors_used
  _objc_msgSend
  _NXGetDefaultValue
  _NXRegisterDefaults
  .objc_class_name_NXStringTable
  .objc_class_name_NXBundle
  ```
  One kluge around this problem is to extract 'termios.o' from 'libposix.a', put it in Octave's 'src' directory, and add it to the list of files to link together in the makefile. Suggestions for better ways to solve this problem are welcome!

- If Octave crashes immediately with a floating point exception, it is likely that it is failing to initialize the IEEE floating point values for infinity and NaN.

  If your system actually does support IEEE arithmetic, you should be able to fix this problem by modifying the function octave_ieee_init in the file 'lo-ieee.cc' to correctly initialize Octave's internal infinity and NaN variables.

Appendix E: Installing Octave    517

If your system does not support IEEE arithmetic but Octave's configure script incorrectly determined that it does, you can work around the problem by editing the file 'config.h' to not define HAVE_ISINF, HAVE_FINITE, and HAVE_ISNAN.

In any case, please report this as a bug since it might be possible to modify Octave's configuration script to automatically determine the proper thing to do.

- If Octave is unable to find a header file because it is installed in a location that is not normally searched by the compiler, you can add the directory to the include search path by specifying (for example) CPPFLAGS=-I/some/nonstandard/directory as an argument to configure. Other variables that can be specified this way are CFLAGS, CXXFLAGS, FFLAGS, and LDFLAGS. Passing them as options to the configure script also records them in the 'config.status' file. By default, CPPFLAGS and LDFLAGS are empty, CFLAGS and CXXFLAGS are set to "-g -O" and FFLAGS is set to "-O".

# Appendix F  Emacs Octave Support

The development of Octave code can greatly be facilitated using Emacs with Octave mode, a major mode for editing Octave files which can e.g. automatically indent the code, do some of the typing (with Abbrev mode) and show keywords, comments, strings, etc. in different faces (with Font-lock mode on devices that support it).

It is also possible to run Octave from within Emacs, either by directly entering commands at the prompt in a buffer in Inferior Octave mode, or by interacting with Octave from within a file with Octave code. This is useful in particular for debugging Octave code.

Finally, you can convince Octave to use the Emacs info reader for `help -i`.

All functionality is provided by the Emacs Lisp package EOS (for "Emacs Octave Support"). This chapter describes how to set up and use this package.

Please contact <Kurt.Hornik@wu-wien.ac.at> if you have any questions or suggestions on using EOS.

## F.1 Installing EOS

The Emacs package EOS consists of the three files 'octave-mod.el', 'octave-inf.el', and 'octave-hlp.el'. These files, or better yet their byte-compiled versions, should be somewhere in your Emacs load-path.

If you have GNU Emacs with a version number at least as high as 19.35, you are all set up, because EOS is respectively will be part of GNU Emacs as of version 19.35.

Otherwise, copy the three files from the 'emacs' subdirectory of the Octave distribution to a place where Emacs can find them (this depends on how your Emacs was installed). Byte-compile them for speed if you want.

## F.2 Using Octave Mode

If you are lucky, your sysadmins have already arranged everything so that Emacs automatically goes into Octave mode whenever you visit an Octave code file as characterized by its extension '.m'. If not, proceed as follows.

1. To begin using Octave mode for all '.m' files you visit, add the following lines to a file loaded by Emacs at startup time, typically your '~/.emacs' file:

    ```
    (autoload 'octave-mode "octave-mod" nil t)
    (setq auto-mode-alist
          (cons '("\\.m$" . octave-mode) auto-mode-alist))
    ```

2. Finally, to turn on the abbrevs, auto-fill and font-lock features automatically, also add the following lines to one of the Emacs startup files:

```
(add-hook 'octave-mode-hook
          (lambda ()
            (abbrev-mode 1)
            (auto-fill-mode 1)
            (if (eq window-system 'x)
                (font-lock-mode 1))))
```

See the Emacs manual for more information about how to customize Font-lock mode.

In Octave mode, the following special Emacs commands can be used in addition to the standard Emacs commands.

C-h m
: Describe the features of Octave mode.

LFD
: Reindent the current Octave line, insert a newline and indent the new line (octave-reindent-then-newline-and-indent). An abbrev before point is expanded if abbrev-mode is non-nil.

TAB
: Indents current Octave line based on its contents and on previous lines (indent-according-to-mode).

;
: Insert an "electric" semicolon (octave-electric-semi). If octave-auto-indent is non-nil, reindent the current line. If octave-auto-newline is non-nil, automagically insert a newline and indent the new line.

'
: Start entering an abbreviation (octave-abbrev-start). If Abbrev mode is turned on, typing 'C-h or '? lists all abbrevs. Any other key combination is executed normally. Note that all Octave abbrevs start with a grave accent.

M-LFD
: Break line at point and insert continuation marker and alignment (octave-split-line).

M-TAB
: Perform completion on Octave symbol preceding point, comparing that symbol against Octave's reserved words and built-in variables (octave-complete-symbol).

M-C-a
: Move backward to the beginning of a function (octave-beginning-of-defun). With prefix argument $N$, do it that many times if $N$ is positive; otherwise, move forward to the $N$-th following beginning of a function.

M-C-e
: Move forward to the end of a function (octave-end-of-defun). With prefix argument $N$, do it that many times if $N$ is positive; otherwise, move back to the $N$-th preceding end of a function.

M-C-h
: Puts point at beginning and mark at the end of the current Octave function, i.e., the one containing point or following point (octave-mark-defun).

M-C-q
: Properly indents the Octave function which contains point (octave-indent-defun).

M-;
: If there is no comment already on this line, create a code-level comment (started by two comment characters) if the line is empty, or an in-line comment (started by one comment character) otherwise (octave-indent-for-comment). Point is left after the start of the comment which is properly aligned.

C-c ;
: Puts the comment character # (more precisely, the string value of octave-comment-start) at the beginning of every line in the region (octave-comment-region). With just C-u prefix argument, uncomment each line in the region. A numeric prefix argument $N$ means use $N$ comment characters.

C-c :
: Uncomments every line in the region (octave-uncomment-region).

C-c C-p
: Move one line of Octave code backward, skipping empty and comment lines (octave-previous-code-line). With numeric prefix argument $N$, move that many code lines backward (forward if $N$ is negative).

C-c C-n
: Move one line of Octave code forward, skipping empty and comment lines (octave-next-code-line). With numeric prefix argument $N$, move that many code lines forward (backward if $N$ is negative).

C-c C-a
: Move to the 'real' beginning of the current line (octave-beginning-of-line). If point is in an empty or comment line, simply go to its beginning; otherwise, move backwards to the beginning of the first code line which is not inside a continuation statement, i.e., which does not follow a code line ending in ... or \, or is inside an open parenthesis list.

C-c C-e
: Move to the 'real' end of the current line (octave-end-of-line). If point is in a code line, move forward to the end of the first Octave code line which does not end in ... or \ or is inside an open parenthesis list. Otherwise, simply go to the end of the current line.

C-c M-C-n
: Move forward across one balanced begin-end block of Octave code (octave-forward-block). With numeric prefix argument $N$, move forward across $n$ such blocks (backward if $N$ is negative).

*C-c M-C-p*
: Move back across one balanced begin-end block of Octave code (octave-backward-block). With numeric prefix argument N, move backward across N such blocks (forward if N is negative).

*C-c M-C-d*
: Move forward down one begin-end block level of Octave code (octave-down-block). With numeric prefix argument, do it that many times; a negative argument means move backward, but still go down one level.

*C-c M-C-u*
: Move backward out of one begin-end block level of Octave code (octave-backward-up-block). With numeric prefix argument, do it that many times; a negative argument means move forward, but still to a less deep spot.

*C-c M-C-h*
: Put point at the beginning of this block, mark at the end (octave-mark-block). The block marked is the one that contains point or follows point.

*C-c ]*
: Close the current block on a separate line (octave-close-block). An error is signaled if no block to close is found.

*C-c f*
: Insert a function skeleton, prompting for the function's name, arguments and return values which have to be entered without parens (octave-insert-defun).

*C-c C-h*
: Search the function, operator and variable indices of all info files with documentation for Octave for entries (octave-help). If used interactively, the entry is prompted for with completion. If multiple matches are found, one can cycle through them using the standard , (Info-index-next) command of the Info reader.

  The variable octave-help-files is a list of files to search through and defaults to '("octave"). If there is also an Octave Local Guide with corresponding info file, say, 'octave-LG', you can have octave-help search both files by

        (setq octave-help-files '("octave" "octave-LG"))

  in one of your Emacs startup files.

A common problem is that the (RET) key does *not* indent the line to where the new text should go after inserting the newline. This is because the standard Emacs convention is that (RET) (aka *C-m*) just adds a newline, whereas (LFD) (aka *C-j*) adds a newline and indents it. This is particularly inconvenient for users with keyboards which do not have a special (LFD) key at all; in such cases, it is typically more convenient to use (RET) as the (LFD) key (rather than typing *C-j*).

You can make (RET) do this by adding

# Appendix F: Emacs Octave Support

```
(define-key octave-mode-map "\C-m"
  'octave-reindent-then-newline-and-indent)
```
to one of your Emacs startup files. Another, more generally applicable solution is
```
(defun RET-behaves-as-LFD ()
  (let ((x (key-binding "\C-j")))
    (local-set-key "\C-m" x)))
(add-hook 'octave-mode-hook 'RET-behaves-as-LFD)
```
(this works for all modes by adding to the startup hooks, without having to know the particular binding of RET in that mode!). Similar considerations apply for using M-RET as M-LFD. As Barry A. Warsaw <bwarsaw@cnri.reston.va.us> says in the documentation for his cc-mode, "This is a very common question. :-) If you want this to be the default behavior, don't lobby me, lobby RMS!"

The following variables can be used to customize Octave mode.

octave-auto-indent
: Non-nil means auto-indent the current line after a semicolon or space. Default is nil.

octave-auto-newline
: Non-nil means auto-insert a newline and indent after semicolons are typed. The default value is nil.

octave-blink-matching-block
: Non-nil means show matching begin of block when inserting a space, newline or ; after an else or end keyword. Default is t. This is an extremely useful feature for automatically verifying that the keywords match—if they don't, an error message is displayed.

octave-block-offset
: Extra indentation applied to statements in block structures. Default is 2.

octave-continuation-offset
: Extra indentation applied to Octave continuation lines. Default is 4.

octave-continuation-string
: String used for Octave continuation lines. Normally \.

octave-mode-startup-message
: If t (default), a startup message is displayed when Octave mode is called.

If Font Lock mode is enabled, Octave mode will display

- strings in font-lock-string-face
- comments in font-lock-comment-face
- the Octave reserved words (such as all block keywords) and the text functions (such as cd or who) which are also reserved using font-lock-keyword-face
- the built-in operators (&&, ==, ...) using font-lock-reference-face
- and the function names in function declarations in font-lock-function-name-face

There is also rudimentary support for Imenu (currently, function names can be indexed).

You can generate TAGS files for Emacs from Octave '.m' files using the shell script `octave-tags` that is installed alongside your copy of Octave.

Customization of Octave mode can be performed by modification of the variable `octave-mode-hook`. If the value of this variable is non-nil, turning on Octave mode calls its value.

If you discover a problem with Octave mode, you can conveniently send a bug report using `C-c C-b` (`octave-submit-bug-report`). This automatically sets up a mail buffer with version information already added. You just need to add a description of the problem, including a reproducible test case and send the message.

## F.3 Running Octave From Within Emacs

The package 'octave' provides commands for running an inferior Octave process in a special Emacs buffer. Use

    `M-x run-octave`

to directly start an inferior Octave process. If Emacs does not know about this command, add the line

    `(autoload 'run-octave "octave-inf" nil t)`

to your '.emacs' file.

This will start Octave in a special buffer the name of which is specified by the variable `inferior-octave-buffer` and defaults to `"*Inferior Octave*"`. From within this buffer, you can interact with the inferior Octave process 'as usual', i.e., by entering Octave commands at the prompt. The buffer is in Inferior Octave mode, which is derived from the standard Comint mode, a major mode for interacting with an inferior interpreter. See the documentation for `comint-mode` for more details, and use `C-h b` to find out about available special keybindings.

You can also communicate with an inferior Octave process from within files with Octave code (i.e., buffers in Octave mode), using the following commands.

`C-c i l`
> Send the current line to the inferior Octave process (`octave-send-line`). With positive prefix argument N, send that many lines. If `octave-send-line-auto-forward` is non-nil, go to the next unsent code line.

`C-c i b`
> Send the current block to the inferior Octave process (`octave-send-block`).

`C-c i f`
> Send the current function to the inferior Octave process (`octave-send-defun`).

`C-c i r`
> Send the region to the inferior Octave process (`octave-send-region`).

# Appendix F: Emacs Octave Support

`C-c i s`
Make sure that 'inferior-octave-buffer' is displayed (`octave-show-process-buffer`).

`C-c i h`
Delete all windows that display the inferior Octave buffer (`octave-hide-process-buffer`).

`C-c i k`
Kill the inferior Octave process and its buffer (`octave-kill-process`).

The effect of the commands which send code to the Octave process can be customized by the following variables.

`octave-send-echo-input`
Non-nil means echo input sent to the inferior Octave process. Default is t.

`octave-send-show-buffer`
Non-nil means display the buffer running the Octave process after sending a command (but without selecting it). Default is t.

If you send code and there is no inferior Octave process yet, it will be started automatically.

The startup of the inferior Octave process is highly customizable. The variable `inferior-octave-startup-args` can be used for specifying command lines arguments to be passed to Octave on startup as a list of strings. For example, to suppress the startup message and use 'traditional' mode, set this to '("-q" "--traditional"). You can also specify a startup file of Octave commands to be loaded on startup; note that these commands will not produce any visible output in the process buffer. Which file to use is controlled by the variable `inferior-octave-startup-file`. If this is nil, the file '~/.emacs-octave' is used if it exists.

And finally, `inferior-octave-mode-hook` is run after starting the process and putting its buffer into Inferior Octave mode. Hence, if you like the up and down arrow keys to behave in the interaction buffer as in the shell, and you want this buffer to use nice colors, add

```
(add-hook 'inferior-octave-mode-hook
          (lambda ()
            (turn-on-font-lock)
            (define-key inferior-octave-mode-map [up]
              'comint-previous-input)
            (define-key inferior-octave-mode-map [down]
              'comint-next-input)))
```

to your '.emacs' file. You could also swap the roles of `C-a` (`beginning-of-line`) and C-c C-a (`comint-bol`) using this hook.

Note that if you set your Octave prompts to something different from the defaults, make sure that `inferior-octave-prompt` matches them. Otherwise, *nothing* will work, because Emacs will not know when Octave is waiting for input, or done sending output.

## F.4 Using the Emacs Info Reader for Octave

You may also use the Emacs Info reader with Octave's doc function. For this, the package 'gnuserv' needs to be installed.

If 'gnuserv' is installed, add the lines

```
(autoload 'octave-help "octave-hlp" nil t)
(require 'gnuserv)
(gnuserv-start)
```

to your '.emacs' file.

You can use either 'plain' Emacs Info or the function octave-help as your Octave info reader (for help -i). In the former case, use info_program ("info-emacs-info"). The latter is perhaps more attractive because it allows to look up keys in the indices of *several* info files related to Octave (provided that the Emacs variable octave-help-files is set correctly). In this case, use info_program ("info-emacs-octave-help").

If you use Octave from within Emacs, it is best to add these settings to your '~/.emacs-octave' startup file (or the file pointed to by the Emacs variable inferior-octave-startup-file).

# Appendix G  GNU GENERAL PUBLIC LICENSE

Version 3, 29 June 2007

Copyright © 2007 Free Software Foundation, Inc. http://fsf.org/

Everyone is permitted to copy and distribute verbatim copies of this license document, but changing it is not allowed.

## Preamble

The GNU General Public License is a free, copyleft license for software and other kinds of works.

The licenses for most software and other practical works are designed to take away your freedom to share and change the works. By contrast, the GNU General Public License is intended to guarantee your freedom to share and change all versions of a program—to make sure it remains free software for all its users. We, the Free Software Foundation, use the GNU General Public License for most of our software; it applies also to any other work released this way by its authors. You can apply it to your programs, too.

When we speak of free software, we are referring to freedom, not price. Our General Public Licenses are designed to make sure that you have the freedom to distribute copies of free software (and charge for them if you wish), that you receive source code or can get it if you want it, that you can change the software or use pieces of it in new free programs, and that you know you can do these things.

To protect your rights, we need to prevent others from denying you these rights or asking you to surrender the rights. Therefore, you have certain responsibilities if you distribute copies of the software, or if you modify it: responsibilities to respect the freedom of others.

For example, if you distribute copies of such a program, whether gratis or for a fee, you must pass on to the recipients the same freedoms that you received. You must make sure that they, too, receive or can get the source code. And you must show them these terms so they know their rights.

Developers that use the GNU GPL protect your rights with two steps: (1) assert copyright on the software, and (2) offer you this License giving you legal permission to copy, distribute and/or modify it.

For the developers' and authors' protection, the GPL clearly explains that there is no warranty for this free software. For both users' and authors' sake, the GPL requires that modified versions be marked as changed, so that their problems will not be attributed erroneously to authors of previous versions.

Some devices are designed to deny users access to install or run modified versions of the software inside them, although the manufacturer can do so. This is fundamentally incompatible with the aim of protecting users' freedom to change the software. The systematic pattern of such abuse occurs in the area of products for individuals to use, which is precisely where it is most unacceptable. Therefore, we have designed this version of the GPL to prohibit the practice for those products. If such problems arise substantially in other domains, we stand ready to extend this provision to those domains in future versions of the GPL, as needed to protect the freedom of users.

Finally, every program is threatened constantly by software patents. States should not allow patents to restrict development and use of software on general-purpose computers, but in those that do, we wish to avoid the special danger that patents applied to a free program could make it effectively proprietary. To prevent this, the GPL assures that patents cannot be used to render the program non-free.

The precise terms and conditions for copying, distribution and modification follow.

# TERMS AND CONDITIONS

0. Definitions.

    "This License" refers to version 3 of the GNU General Public License.

    "Copyright" also means copyright-like laws that apply to other kinds of works, such as semiconductor masks.

    "The Program" refers to any copyrightable work licensed under this License. Each licensee is addressed as "you". "Licensees" and "recipients" may be individuals or organizations.

    To "modify" a work means to copy from or adapt all or part of the work in a fashion requiring copyright permission, other than the making of an exact copy. The resulting work is called a "modified version" of the earlier work or a work "based on" the earlier work.

    A "covered work" means either the unmodified Program or a work based on the Program.

    To "propagate" a work means to do anything with it that, without permission, would make you directly or secondarily liable for infringement under applicable copyright law, except executing it on a computer or modifying a private copy. Propagation includes copying, distribution (with or without modification), making available to the public, and in some countries other activities as well.

    To "convey" a work means any kind of propagation that enables other parties to make or receive copies. Mere interaction with a user through a computer network, with no transfer of a copy, is not conveying.

    An interactive user interface displays "Appropriate Legal Notices" to the extent that it includes a convenient and prominently visible feature that (1) displays an appropriate copyright notice, and (2) tells the user that there is no warranty for the work (except to the extent that warranties are provided), that licensees may convey the work under this License, and how to view a copy of this License. If the interface presents a list of user commands or options, such as a menu, a prominent item in the list meets this criterion.

1. Source Code.

    The "source code" for a work means the preferred form of the work for making modifications to it. "Object code" means any non-source form of a work.

    A "Standard Interface" means an interface that either is an official standard defined by a recognized standards body, or, in the case of interfaces specified for a particular programming language, one that is widely used among developers working in that language.

    The "System Libraries" of an executable work include anything, other than the work as a whole, that (a) is included in the normal form of packaging a Major Component, but which is not part of that Major Component, and (b) serves only to enable use of the work with that Major Component, or to implement a Standard Interface for which an implementation is available to the public in source code form. A "Major Component", in this context, means a major essential component (kernel, window system, and so on) of the specific operating system (if any) on which the executable work runs, or a compiler used to produce the work, or an object code interpreter used to run it.

    The "Corresponding Source" for a work in object code form means all the source code needed to generate, install, and (for an executable work) run the object code and to modify the work, including scripts to control those activities. However, it does not include the work's System Libraries, or general-purpose tools or generally available free programs which are used unmodified in performing those activities but which are not part of the work. For example, Corresponding Source includes interface definition files associated with source files for the work, and the source code for shared libraries and dynamically linked subprograms that the work is specifically designed to require, such as by intimate data communication or control flow between those subprograms and other parts of the work.

# Appendix G: GNU GENERAL PUBLIC LICENSE

The Corresponding Source need not include anything that users can regenerate automatically from other parts of the Corresponding Source.

The Corresponding Source for a work in source code form is that same work.

2. Basic Permissions.

All rights granted under this License are granted for the term of copyright on the Program, and are irrevocable provided the stated conditions are met. This License explicitly affirms your unlimited permission to run the unmodified Program. The output from running a covered work is covered by this License only if the output, given its content, constitutes a covered work. This License acknowledges your rights of fair use or other equivalent, as provided by copyright law.

You may make, run and propagate covered works that you do not convey, without conditions so long as your license otherwise remains in force. You may convey covered works to others for the sole purpose of having them make modifications exclusively for you, or provide you with facilities for running those works, provided that you comply with the terms of this License in conveying all material for which you do not control copyright. Those thus making or running the covered works for you must do so exclusively on your behalf, under your direction and control, on terms that prohibit them from making any copies of your copyrighted material outside their relationship with you.

Conveying under any other circumstances is permitted solely under the conditions stated below. Sublicensing is not allowed; section 10 makes it unnecessary.

3. Protecting Users' Legal Rights From Anti-Circumvention Law.

No covered work shall be deemed part of an effective technological measure under any applicable law fulfilling obligations under article 11 of the WIPO copyright treaty adopted on 20 December 1996, or similar laws prohibiting or restricting circumvention of such measures.

When you convey a covered work, you waive any legal power to forbid circumvention of technological measures to the extent such circumvention is effected by exercising rights under this License with respect to the covered work, and you disclaim any intention to limit operation or modification of the work as a means of enforcing, against the work's users, your or third parties' legal rights to forbid circumvention of technological measures.

4. Conveying Verbatim Copies.

You may convey verbatim copies of the Program's source code as you receive it, in any medium, provided that you conspicuously and appropriately publish on each copy an appropriate copyright notice; keep intact all notices stating that this License and any non-permissive terms added in accord with section 7 apply to the code; keep intact all notices of the absence of any warranty; and give all recipients a copy of this License along with the Program.

You may charge any price or no price for each copy that you convey, and you may offer support or warranty protection for a fee.

5. Conveying Modified Source Versions.

You may convey a work based on the Program, or the modifications to produce it from the Program, in the form of source code under the terms of section 4, provided that you also meet all of these conditions:

   a. The work must carry prominent notices stating that you modified it, and giving a relevant date.

   b. The work must carry prominent notices stating that it is released under this License and any conditions added under section 7. This requirement modifies the requirement in section 4 to "keep intact all notices".

   c. You must license the entire work, as a whole, under this License to anyone who comes into possession of a copy. This License will therefore apply, along with any applicable section 7 additional terms, to the whole of the work, and all its parts, regardless of how they are packaged. This License gives no permission to license the work in any other way, but it does not invalidate such permission if you have separately received it.

d. If the work has interactive user interfaces, each must display Appropriate Legal Notices; however, if the Program has interactive interfaces that do not display Appropriate Legal Notices, your work need not make them do so.

A compilation of a covered work with other separate and independent works, which are not by their nature extensions of the covered work, and which are not combined with it such as to form a larger program, in or on a volume of a storage or distribution medium, is called an "aggregate" if the compilation and its resulting copyright are not used to limit the access or legal rights of the compilation's users beyond what the individual works permit. Inclusion of a covered work in an aggregate does not cause this License to apply to the other parts of the aggregate.

6. Conveying Non-Source Forms.

You may convey a covered work in object code form under the terms of sections 4 and 5, provided that you also convey the machine-readable Corresponding Source under the terms of this License, in one of these ways:

   a. Convey the object code in, or embodied in, a physical product (including a physical distribution medium), accompanied by the Corresponding Source fixed on a durable physical medium customarily used for software interchange.

   b. Convey the object code in, or embodied in, a physical product (including a physical distribution medium), accompanied by a written offer, valid for at least three years and valid for as long as you offer spare parts or customer support for that product model, to give anyone who possesses the object code either (1) a copy of the Corresponding Source for all the software in the product that is covered by this License, on a durable physical medium customarily used for software interchange, for a price no more than your reasonable cost of physically performing this conveying of source, or (2) access to copy the Corresponding Source from a network server at no charge.

   c. Convey individual copies of the object code with a copy of the written offer to provide the Corresponding Source. This alternative is allowed only occasionally and noncommercially, and only if you received the object code with such an offer, in accord with subsection 6b.

   d. Convey the object code by offering access from a designated place (gratis or for a charge), and offer equivalent access to the Corresponding Source in the same way through the same place at no further charge. You need not require recipients to copy the Corresponding Source along with the object code. If the place to copy the object code is a network server, the Corresponding Source may be on a different server (operated by you or a third party) that supports equivalent copying facilities, provided you maintain clear directions next to the object code saying where to find the Corresponding Source. Regardless of what server hosts the Corresponding Source, you remain obligated to ensure that it is available for as long as needed to satisfy these requirements.

   e. Convey the object code using peer-to-peer transmission, provided you inform other peers where the object code and Corresponding Source of the work are being offered to the general public at no charge under subsection 6d.

A separable portion of the object code, whose source code is excluded from the Corresponding Source as a System Library, need not be included in conveying the object code work.

A "User Product" is either (1) a "consumer product", which means any tangible personal property which is normally used for personal, family, or household purposes, or (2) anything designed or sold for incorporation into a dwelling. In determining whether a product is a consumer product, doubtful cases shall be resolved in favor of coverage. For a particular product received by a particular user, "normally used" refers to a typical or common use of that class of product, regardless of the status of the particular user or of the way in which the particular user actually uses, or expects or is expected to use, the product. A product is a consumer product regardless of whether the product has substantial commercial, industrial or non-consumer uses, unless such uses represent the only significant mode of use of the product.

# Appendix G: GNU GENERAL PUBLIC LICENSE      531

"Installation Information" for a User Product means any methods, procedures, authorization keys, or other information required to install and execute modified versions of a covered work in that User Product from a modified version of its Corresponding Source. The information must suffice to ensure that the continued functioning of the modified object code is in no case prevented or interfered with solely because modification has been made.

If you convey an object code work under this section in, or with, or specifically for use in, a User Product, and the conveying occurs as part of a transaction in which the right of possession and use of the User Product is transferred to the recipient in perpetuity or for a fixed term (regardless of how the transaction is characterized), the Corresponding Source conveyed under this section must be accompanied by the Installation Information. But this requirement does not apply if neither you nor any third party retains the ability to install modified object code on the User Product (for example, the work has been installed in ROM).

The requirement to provide Installation Information does not include a requirement to continue to provide support service, warranty, or updates for a work that has been modified or installed by the recipient, or for the User Product in which it has been modified or installed. Access to a network may be denied when the modification itself materially and adversely affects the operation of the network or violates the rules and protocols for communication across the network.

Corresponding Source conveyed, and Installation Information provided, in accord with this section must be in a format that is publicly documented (and with an implementation available to the public in source code form), and must require no special password or key for unpacking, reading or copying.

7. Additional Terms.

"Additional permissions" are terms that supplement the terms of this License by making exceptions from one or more of its conditions. Additional permissions that are applicable to the entire Program shall be treated as though they were included in this License, to the extent that they are valid under applicable law. If additional permissions apply only to part of the Program, that part may be used separately under those permissions, but the entire Program remains governed by this License without regard to the additional permissions.

When you convey a copy of a covered work, you may at your option remove any additional permissions from that copy, or from any part of it. (Additional permissions may be written to require their own removal in certain cases when you modify the work.) You may place additional permissions on material, added by you to a covered work, for which you have or can give appropriate copyright permission.

Notwithstanding any other provision of this License, for material you add to a covered work, you may (if authorized by the copyright holders of that material) supplement the terms of this License with terms:

   a. Disclaiming warranty or limiting liability differently from the terms of sections 15 and 16 of this License; or

   b. Requiring preservation of specified reasonable legal notices or author attributions in that material or in the Appropriate Legal Notices displayed by works containing it; or

   c. Prohibiting misrepresentation of the origin of that material, or requiring that modified versions of such material be marked in reasonable ways as different from the original version; or

   d. Limiting the use for publicity purposes of names of licensors or authors of the material; or

   e. Declining to grant rights under trademark law for use of some trade names, trademarks, or service marks; or

   f. Requiring indemnification of licensors and authors of that material by anyone who conveys the material (or modified versions of it) with contractual assumptions of liability to the recipient, for any liability that these contractual assumptions directly impose on those licensors and authors.

All other non-permissive additional terms are considered "further restrictions" within the meaning of section 10. If the Program as you received it, or any part of it, contains a notice stating that it is governed by this License along with a term that is a further restriction, you may remove that term. If a license document contains a further restriction but permits relicensing or conveying under this License, you may add to a covered work material governed by the terms of that license document, provided that the further restriction does not survive such relicensing or conveying.

If you add terms to a covered work in accord with this section, you must place, in the relevant source files, a statement of the additional terms that apply to those files, or a notice indicating where to find the applicable terms.

Additional terms, permissive or non-permissive, may be stated in the form of a separately written license, or stated as exceptions; the above requirements apply either way.

8. Termination.

You may not propagate or modify a covered work except as expressly provided under this License. Any attempt otherwise to propagate or modify it is void, and will automatically terminate your rights under this License (including any patent licenses granted under the third paragraph of section 11).

However, if you cease all violation of this License, then your license from a particular copyright holder is reinstated (a) provisionally, unless and until the copyright holder explicitly and finally terminates your license, and (b) permanently, if the copyright holder fails to notify you of the violation by some reasonable means prior to 60 days after the cessation.

Moreover, your license from a particular copyright holder is reinstated permanently if the copyright holder notifies you of the violation by some reasonable means, this is the first time you have received notice of violation of this License (for any work) from that copyright holder, and you cure the violation prior to 30 days after your receipt of the notice.

Termination of your rights under this section does not terminate the licenses of parties who have received copies or rights from you under this License. If your rights have been terminated and not permanently reinstated, you do not qualify to receive new licenses for the same material under section 10.

9. Acceptance Not Required for Having Copies.

You are not required to accept this License in order to receive or run a copy of the Program. Ancillary propagation of a covered work occurring solely as a consequence of using peer-to-peer transmission to receive a copy likewise does not require acceptance. However, nothing other than this License grants you permission to propagate or modify any covered work. These actions infringe copyright if you do not accept this License. Therefore, by modifying or propagating a covered work, you indicate your acceptance of this License to do so.

10. Automatic Licensing of Downstream Recipients.

Each time you convey a covered work, the recipient automatically receives a license from the original licensors, to run, modify and propagate that work, subject to this License. You are not responsible for enforcing compliance by third parties with this License.

An "entity transaction" is a transaction transferring control of an organization, or substantially all assets of one, or subdividing an organization, or merging organizations. If propagation of a covered work results from an entity transaction, each party to that transaction who receives a copy of the work also receives whatever licenses to the work the party's predecessor in interest had or could give under the previous paragraph, plus a right to possession of the Corresponding Source of the work from the predecessor in interest, if the predecessor has it or can get it with reasonable efforts.

You may not impose any further restrictions on the exercise of the rights granted or affirmed under this License. For example, you may not impose a license fee, royalty, or other charge for exercise of rights granted under this License, and you may not initiate litigation (including a cross-claim or counterclaim in a lawsuit) alleging that any patent claim is infringed by making, using, selling, offering for sale, or importing the Program or any portion of it.

# Appendix G: GNU GENERAL PUBLIC LICENSE 533

11. Patents.

A "contributor" is a copyright holder who authorizes use under this License of the Program or a work on which the Program is based. The work thus licensed is called the contributor's "contributor version".

A contributor's "essential patent claims" are all patent claims owned or controlled by the contributor, whether already acquired or hereafter acquired, that would be infringed by some manner, permitted by this License, of making, using, or selling its contributor version, but do not include claims that would be infringed only as a consequence of further modification of the contributor version. For purposes of this definition, "control" includes the right to grant patent sublicenses in a manner consistent with the requirements of this License.

Each contributor grants you a non-exclusive, worldwide, royalty-free patent license under the contributor's essential patent claims, to make, use, sell, offer for sale, import and otherwise run, modify and propagate the contents of its contributor version.

In the following three paragraphs, a "patent license" is any express agreement or commitment, however denominated, not to enforce a patent (such as an express permission to practice a patent or covenant not to sue for patent infringement). To "grant" such a patent license to a party means to make such an agreement or commitment not to enforce a patent against the party.

If you convey a covered work, knowingly relying on a patent license, and the Corresponding Source of the work is not available for anyone to copy, free of charge and under the terms of this License, through a publicly available network server or other readily accessible means, then you must either (1) cause the Corresponding Source to be so available, or (2) arrange to deprive yourself of the benefit of the patent license for this particular work, or (3) arrange, in a manner consistent with the requirements of this License, to extend the patent license to downstream recipients. "Knowingly relying" means you have actual knowledge that, but for the patent license, your conveying the covered work in a country, or your recipient's use of the covered work in a country, would infringe one or more identifiable patents in that country that you have reason to believe are valid.

If, pursuant to or in connection with a single transaction or arrangement, you convey, or propagate by procuring conveyance of, a covered work, and grant a patent license to some of the parties receiving the covered work authorizing them to use, propagate, modify or convey a specific copy of the covered work, then the patent license you grant is automatically extended to all recipients of the covered work and works based on it.

A patent license is "discriminatory" if it does not include within the scope of its coverage, prohibits the exercise of, or is conditioned on the non-exercise of one or more of the rights that are specifically granted under this License. You may not convey a covered work if you are a party to an arrangement with a third party that is in the business of distributing software, under which you make payment to the third party based on the extent of your activity of conveying the work, and under which the third party grants, to any of the parties who would receive the covered work from you, a discriminatory patent license (a) in connection with copies of the covered work conveyed by you (or copies made from those copies), or (b) primarily for and in connection with specific products or compilations that contain the covered work, unless you entered into that arrangement, or that patent license was granted, prior to 28 March 2007.

Nothing in this License shall be construed as excluding or limiting any implied license or other defenses to infringement that may otherwise be available to you under applicable patent law.

12. No Surrender of Others' Freedom.

If conditions are imposed on you (whether by court order, agreement or otherwise) that contradict the conditions of this License, they do not excuse you from the conditions of this License. If you cannot convey a covered work so as to satisfy simultaneously your obligations under this License and any other pertinent obligations, then as a consequence you may not convey it at all. For example, if you agree to terms that obligate you to collect a royalty for further conveying from those to whom you convey the Program, the only way you could satisfy both those terms and this License would be to refrain entirely from conveying the Program.

13. Use with the GNU Affero General Public License.

    Notwithstanding any other provision of this License, you have permission to link or combine any covered work with a work licensed under version 3 of the GNU Affero General Public License into a single combined work, and to convey the resulting work. The terms of this License will continue to apply to the part which is the covered work, but the special requirements of the GNU Affero General Public License, section 13, concerning interaction through a network will apply to the combination as such.

14. Revised Versions of this License.

    The Free Software Foundation may publish revised and/or new versions of the GNU General Public License from time to time. Such new versions will be similar in spirit to the present version, but may differ in detail to address new problems or concerns.

    Each version is given a distinguishing version number. If the Program specifies that a certain numbered version of the GNU General Public License "or any later version" applies to it, you have the option of following the terms and conditions either of that numbered version or of any later version published by the Free Software Foundation. If the Program does not specify a version number of the GNU General Public License, you may choose any version ever published by the Free Software Foundation.

    If the Program specifies that a proxy can decide which future versions of the GNU General Public License can be used, that proxy's public statement of acceptance of a version permanently authorizes you to choose that version for the Program.

    Later license versions may give you additional or different permissions. However, no additional obligations are imposed on any author or copyright holder as a result of your choosing to follow a later version.

15. Disclaimer of Warranty.

    THERE IS NO WARRANTY FOR THE PROGRAM, TO THE EXTENT PERMITTED BY APPLICABLE LAW. EXCEPT WHEN OTHERWISE STATED IN WRITING THE COPYRIGHT HOLDERS AND/OR OTHER PARTIES PROVIDE THE PROGRAM "AS IS" WITHOUT WARRANTY OF ANY KIND, EITHER EXPRESSED OR IMPLIED, INCLUDING, BUT NOT LIMITED TO, THE IMPLIED WARRANTIES OF MERCHANTABILITY AND FITNESS FOR A PARTICULAR PURPOSE. THE ENTIRE RISK AS TO THE QUALITY AND PERFORMANCE OF THE PROGRAM IS WITH YOU. SHOULD THE PROGRAM PROVE DEFECTIVE, YOU ASSUME THE COST OF ALL NECESSARY SERVICING, REPAIR OR CORRECTION.

16. Limitation of Liability.

    IN NO EVENT UNLESS REQUIRED BY APPLICABLE LAW OR AGREED TO IN WRITING WILL ANY COPYRIGHT HOLDER, OR ANY OTHER PARTY WHO MODIFIES AND/OR CONVEYS THE PROGRAM AS PERMITTED ABOVE, BE LIABLE TO YOU FOR DAMAGES, INCLUDING ANY GENERAL, SPECIAL, INCIDENTAL OR CONSEQUENTIAL DAMAGES ARISING OUT OF THE USE OR INABILITY TO USE THE PROGRAM (INCLUDING BUT NOT LIMITED TO LOSS OF DATA OR DATA BEING RENDERED INACCURATE OR LOSSES SUSTAINED BY YOU OR THIRD PARTIES OR A FAILURE OF THE PROGRAM TO OPERATE WITH ANY OTHER PROGRAMS), EVEN IF SUCH HOLDER OR OTHER PARTY HAS BEEN ADVISED OF THE POSSIBILITY OF SUCH DAMAGES.

17. Interpretation of Sections 15 and 16.

    If the disclaimer of warranty and limitation of liability provided above cannot be given local legal effect according to their terms, reviewing courts shall apply local law that most closely approximates an absolute waiver of all civil liability in connection with the Program, unless a warranty or assumption of liability accompanies a copy of the Program in return for a fee.

# END OF TERMS AND CONDITIONS

Appendix G: GNU GENERAL PUBLIC LICENSE

## How to Apply These Terms to Your New Programs

If you develop a new program, and you want it to be of the greatest possible use to the public, the best way to achieve this is to make it free software which everyone can redistribute and change under these terms.

To do so, attach the following notices to the program. It is safest to attach them to the start of each source file to most effectively state the exclusion of warranty; and each file should have at least the "copyright" line and a pointer to where the full notice is found.

> one line to give the program's name and a brief idea of what it does.
> Copyright (C) year name of author
>
> This program is free software: you can redistribute it and/or modify
> it under the terms of the GNU General Public License as published by
> the Free Software Foundation, either version 3 of the License, or (at
> your option) any later version.
>
> This program is distributed in the hope that it will be useful, but
> WITHOUT ANY WARRANTY; without even the implied warranty of
> MERCHANTABILITY or FITNESS FOR A PARTICULAR PURPOSE. See the GNU
> General Public License for more details.
>
> You should have received a copy of the GNU General Public License
> along with this program. If not, see http://www.gnu.org/licenses/.

Also add information on how to contact you by electronic and paper mail.

If the program does terminal interaction, make it output a short notice like this when it starts in an interactive mode:

> program Copyright (C) year name of author
> This program comes with ABSOLUTELY NO WARRANTY; for details type show w.
> This is free software, and you are welcome to redistribute it
> under certain conditions; type show c for details.

The hypothetical commands show w and show c should show the appropriate parts of the General Public License. Of course, your program's commands might be different; for a GUI interface, you would use an "about box".

You should also get your employer (if you work as a programmer) or school, if any, to sign a "copyright disclaimer" for the program, if necessary. For more information on this, and how to apply and follow the GNU GPL, see http://www.gnu.org/licenses/.

The GNU General Public License does not permit incorporating your program into proprietary programs. If your program is a subroutine library, you may consider it more useful to permit linking proprietary applications with the library. If this is what you want to do, use the GNU Lesser General Public License instead of this License. But first, please read http://www.gnu.org/philosophy/why-not-lgpl.html.

# Books from the publisher

Network Theory publishes books about free software under free documentation licenses. Our current catalogue includes the following titles:

- **GNU Bash Reference Manual** by Chet Ramey and Brian Fox (ISBN 0-9541617-7-7) $29.95 (£19.95)

  This manual is the definitive reference for GNU Bash, the standard GNU command-line interpreter. GNU Bash is a complete implementation of the POSIX.2 Bourne shell specification, with additional features from the C-shell and Korn shell. For each copy of this manual sold, $1 is donated to the Free Software Foundation.

- **Version Management with CVS** by Per Cederqvist et al. (ISBN 0-9541617-1-8) $29.95 (£19.95)

  This manual describes how to use CVS, the concurrent versioning system—one of the most widely-used source-code management systems available today. The manual provides tutorial examples for new users of CVS, as well as the definitive reference documentation for every CVS command and configuration option.

- **Comparing and Merging Files with GNU diff and patch** by David MacKenzie, Paul Eggert, and Richard Stallman (ISBN 0-9541617-5-0) $19.95 (£12.95)

  This manual describes how to compare and merge files using GNU diff and patch. It includes an extensive tutorial that guides the reader through all the options of the diff and patch commands. For each copy of this manual sold, $1 is donated to the Free Software Foundation.

- **An Introduction to GCC** by Brian J. Gough, foreword by Richard M. Stallman. (ISBN 0-9541617-9-3) $19.95 (£12.95)

  This manual provides a tutorial introduction to the GNU C and C++ compilers, gcc and g++. Many books teach the C and C++ languages, but this book explains how to use the compiler itself. Based on years of observation of questions posted on mailing lists, it guides the reader straight to the important options of GCC.

- **An Introduction to Python** by Guido van Rossum and Fred L. Drake, Jr. (ISBN 0-9541617-6-9) $19.95 (£12.95)

  This tutorial provides an introduction to Python, an easy to learn object oriented programming language. For each copy of this manual sold, $1 is donated to the Python Software Foundation.

- **Python Language Reference Manual** by Guido van Rossum and Fred L. Drake, Jr. (ISBN 0-9541617-8-5) $19.95 (£12.95)

  This manual is the official reference for the Python language itself. It describes the syntax of Python and its built-in datatypes in depth, This manual is suitable for readers who need to be familiar with the details and rules of the Python language and its object system. For each copy of this manual sold, $1 is donated to the Python Software Foundation.

- **GNU Scientific Library Reference Manual—Revised Second Edition** by M. Galassi, et al (ISBN 0-9541617-3-4) $39.99 (£24.99)

  This reference manual is the definitive guide to the GNU Scientific Library (GSL), a numerical library for C and C++ programmers. The manual documents over 1,000 mathematical routines needed for solving problems in science and engineering. All the money raised from the sale of this book supports the development of the GNU Scientific Library.

- **An Introduction to R** by W.N. Venables, D.M. Smith and the R Development Core Team (ISBN 0-9541617-4-2) $19.95 (£12.95)

  This tutorial manual provides a comprehensive introduction to GNU R, a free software package for statistical computing and graphics.

- **PostgreSQL Reference Manual: Volumes 1–3** (ISBN 0-9546120-2-7) $49.95 (£32.00), (ISBN 0-9546120-3-5) $34.95 (£19.95), (ISBN 0-9546120-4-3) $24.95 (£13.95)

  These manuals documents the SQL language and commands of PostgreSQL, its client and server programming interfaces, and the configuration and maintenance of PostgreSQL servers. For each copy of these manuals sold, $1 is donated to the PostgreSQL project.

All titles are available for order from bookstores worldwide.

Sales of the manuals fund the development of more free software and documentation.

For details, visit the website http://www.network-theory.co.uk/

# Index

## !
! .................................. 101
!= ................................. 100

## "
" ............................... 30, 47

## #
# ................................... 26
#! .................................. 25

## %
% ................................... 26

## &
& .................................. 101
&& ................................. 102

## '
' ........................... 30, 47, 99

## (
( ................................... 93
() ................................... 20

## )
) ................................... 93

## *
* ................................... 98
** .................................. 99
*= ................................. 104

## +
+ .............................. 98, 99

++ ................................. 106
+= ................................. 104

## ,
, ................................... 34

## -
- .............................. 98, 99
-- ................................. 106
--braindead ....................... 18
--debug ........................... 17
--echo-commands ................... 17
--exec-path path .................. 17
--help ............................ 17
--image-path path ................. 17
--info-file filename .............. 18
--info-program program ............ 18
--interactive ..................... 18
--no-init-file .................... 18
--no-line-editing ................. 18
--no-site-file .................... 18
--norc ............................ 18
--path path ....................... 18
--quiet ........................... 18
--silent .......................... 18
--traditional ..................... 18
--verbose ......................... 19
--version ......................... 19
-= ................................ 104
-? ................................. 17
-all ............................... 22
-d ................................. 17
-f ................................. 18
-h ................................. 17
-i ................................. 18
-p path ........................... 18
-q ................................. 18
-v ................................. 19
-V ................................. 19
-x ................................. 17

## .

.' . . . . . . . . . . . . . . . . . . . . . . . . . . . . . . . . . . . . 99
.* . . . . . . . . . . . . . . . . . . . . . . . . . . . . . . . . . . . . . 98
.** . . . . . . . . . . . . . . . . . . . . . . . . . . . . . . . . . . . . 99
.+ . . . . . . . . . . . . . . . . . . . . . . . . . . . . . . . . . . . . . 98
... . . . . . . . . . . . . . . . . . . . . . . . . . . . . . . . . . . . 129
... continuation marker . . . . . . . . . . . 123
./ . . . . . . . . . . . . . . . . . . . . . . . . . . . . . . . . . . . . . 98
.^ . . . . . . . . . . . . . . . . . . . . . . . . . . . . . . . . . . . . . 99
.\ . . . . . . . . . . . . . . . . . . . . . . . . . . . . . . . . . . . . . 98
.octaverc . . . . . . . . . . . . . . . . . . . . . . . . . . 21

## /

/ . . . . . . . . . . . . . . . . . . . . . . . . . . . . . . . . . . . . . 98
/= . . . . . . . . . . . . . . . . . . . . . . . . . . . . . . . . . . . 104

## ;

; . . . . . . . . . . . . . . . . . . . . . . . . . . . . . . . . . . . . . 34

## <

< . . . . . . . . . . . . . . . . . . . . . . . . . . . . . . . . . . . . 100
<= . . . . . . . . . . . . . . . . . . . . . . . . . . . . . . . . . . . 100

## =

= . . . . . . . . . . . . . . . . . . . . . . . . . . . . . . . 103, 457
== . . . . . . . . . . . . . . . . . . . . . . . . . . . . . . . . . . . 100

## >

> . . . . . . . . . . . . . . . . . . . . . . . . . . . . . . . . . . . . 100
>= . . . . . . . . . . . . . . . . . . . . . . . . . . . . . . . . . . . 100

## [

[ . . . . . . . . . . . . . . . . . . . . . . . . . . . . . . . . . . . . . 34

## ]

] . . . . . . . . . . . . . . . . . . . . . . . . . . . . . . . . . . . . . 34

## ^

^ . . . . . . . . . . . . . . . . . . . . . . . . . . . . . . . . . . . . . 99

## \

\ . . . . . . . . . . . . . . . . . . . . . . . . . . . . . . . . . . . . . 98
\ continuation marker . . . . . . . . . . . . . 123

## |

| . . . . . . . . . . . . . . . . . . . . . . . . . . . . . . . . . . . 101
|| . . . . . . . . . . . . . . . . . . . . . . . . . . . . . . . . . . . 102

## ~

~ . . . . . . . . . . . . . . . . . . . . . . . . . . . . . . . . . . . 101
~/.octaverc . . . . . . . . . . . . . . . . . . . . . . . . 20
~= . . . . . . . . . . . . . . . . . . . . . . . . . . . . . . . . . . 100

## A

abs . . . . . . . . . . . . . . . . . . . . . . . . . . . . . . . 258
accumarray . . . . . . . . . . . . . . . . . . . . . . . 263
acknowledgements . . . . . . . . . . . . . . . . . . 4
acos . . . . . . . . . . . . . . . . . . . . . . . . . . . . . 259
acosd . . . . . . . . . . . . . . . . . . . . . . . . . . . . 262
acosh . . . . . . . . . . . . . . . . . . . . . . . . . . . . 260
acot . . . . . . . . . . . . . . . . . . . . . . . . . . . . . 260
acotd . . . . . . . . . . . . . . . . . . . . . . . . . . . . 262
acoth . . . . . . . . . . . . . . . . . . . . . . . . . . . . 260
acsc . . . . . . . . . . . . . . . . . . . . . . . . . . . . . 260
acscd . . . . . . . . . . . . . . . . . . . . . . . . . . . . 262
acsch . . . . . . . . . . . . . . . . . . . . . . . . . . . . 260
addition . . . . . . . . . . . . . . . . . . . . . . . . . . 98
addpath . . . . . . . . . . . . . . . . . . . . . . . . . 134
airy . . . . . . . . . . . . . . . . . . . . . . . . . . . . . 265
all . . . . . . . . . . . . . . . . . . . . . . . . . . . . . . 235
ancestor . . . . . . . . . . . . . . . . . . . . . . . . . 219
and operator . . . . . . . . . . . . . . . . . . . . . 101
angle . . . . . . . . . . . . . . . . . . . . . . . . . . . . 258
anonymous functions . . . . . . . . . . . . . . 140
anova . . . . . . . . . . . . . . . . . . . . . . . . . . . . 359
ans . . . . . . . . . . . . . . . . . . . . . . . . . . . . . 161
answers, incorrect . . . . . . . . . . . . 506, 507
any . . . . . . . . . . . . . . . . . . . . . . . . . . . . . 235
any key . . . . . . . . . . . . . . . . . . . . . . . . . . 13
arch_fit . . . . . . . . . . . . . . . . . . . . . . . . . 420
arch_rnd . . . . . . . . . . . . . . . . . . . . . . . . 420
arch_test . . . . . . . . . . . . . . . . . . . . . . . . 420
archive functions . . . . . . . . . . . . . . . . . . 454
area . . . . . . . . . . . . . . . . . . . . . . . . . . . . . 205
arg . . . . . . . . . . . . . . . . . . . . . . . . . . . . . 258
argnames . . . . . . . . . . . . . . . . . . . . . . . . 142

# Index

arguments in function call .......... 95
argv ................................. 19
arithmetic operators ................. 98
arma_rnd ............................ 421
arrayfun ............................ 244
asctime ............................. 441
asec ................................ 260
asecd ............................... 262
asech ............................... 260
asin ................................ 259
asind ............................... 261
asinh ............................... 260
assert .............................. 491
assignin ............................ 111
assignment expressions .............. 103
assignment operators ................ 103
atan ................................ 259
atan2 ............................... 261
atand ............................... 262
atanh ............................... 260
atexit ............................... 21
audio processing .................... 435
autocor ............................. 421
autocov ............................. 421
autoload ............................ 137
autoloading ......................... 136
autoreg_matrix ...................... 421
autumn .............................. 431
axes ................................ 219
axis ................................ 205

## B

balance ............................. 271
bar ................................. 197
barh ................................ 197
bartlett ............................ 421
bartlett_test ....................... 359
base2dec ............................. 59
beep ................................ 149
beep_on_error .................. 89, 149
benchmarking ........................ 439
besselh ............................. 264
besseli ............................. 264
besselj ............................. 264
besselk ............................. 264
bessely ............................. 264
beta ................................ 265
betacdf ............................. 368

betainc ............................. 265
betainv ............................. 368
betaln .............................. 266
betapdf ............................. 368
betarnd ............................. 376
bicubic ............................. 401
bin2dec .............................. 57
binary i/o .......................... 186
bincoeff ............................ 266
binocdf ............................. 369
binoinv ............................. 369
binopdf ............................. 369
binornd ............................. 376
bitand ............................... 42
bitcmp ............................... 42
bitget ............................... 41
bitmax ............................... 42
bitor ................................ 42
bitset ............................... 41
bitshift ............................. 43
bitxor ............................... 42
blackman ............................ 421
blanks ............................... 48
blkdiag ............................. 243
body of a loop ...................... 117
bone ................................ 431
boolean expressions ................. 101
boolean operators ................... 101
box ................................. 212
break statement ..................... 120
breakpoints, debugging .............. 158
bsxfun .............................. 244
bug criteria ........................ 506
bug report mailing lists ............ 506
bug_report .................... 505, 506
bugs ................................ 505
bugs, investigating ................. 508
bugs, known ......................... 505
bugs, reporting ............... 506, 507
built-in data types .................. 29
built-in function .................... 15
built-in variables ................... 89
builtin ............................. 137
bunzip2 ............................. 454

## C

calendar ............................ 447
call by name ........................ 109

| | |
|---|---|
| call by value | 96 |
| canonicalize_file_name | 453 |
| cart2pol | 268 |
| cart2sph | 268 |
| case statement | 115 |
| cast | 29 |
| cat | 239 |
| catch | 123 |
| catching errors | 149 |
| cauchy_cdf | 369 |
| cauchy_inv | 369 |
| cauchy_pdf | 369 |
| cauchy_rnd | 377 |
| caxis | 207 |
| ccolamd | 302 |
| cd | 15, 463 |
| ceil | 253 |
| cell | 72 |
| cell arrays | 31, 71 |
| cell2mat | 77 |
| cell2struct | 77 |
| celldisp | 72 |
| cellfun | 76 |
| cellidx | 75 |
| cellstr | 75 |
| center | 355 |
| char | 48 |
| character strings | 30, 47 |
| chdir | 15, 463 |
| chi2cdf | 369 |
| chi2inv | 369 |
| chi2rnd | 377 |
| chisquare_pdf | 369 |
| chisquare_test_homogeneity | 360 |
| chisquare_test_independence | 360 |
| chol | 276 |
| chol2inv | 276 |
| Cholesky factorization | 276 |
| cholinv | 276 |
| circshift | 240 |
| class | 29 |
| clear | 87 |
| clearing the screen | 478 |
| clf | 221 |
| clock | 443 |
| cloglog | 357 |
| close | 222 |
| closereq | 222 |
| coding standards | 495 |
| colamd | 304 |
| colloc | 325 |
| colon | 38 |
| color conversion | 434 |
| colormap | 430 |
| colors | 231 |
| colperm | 305 |
| columns | 31 |
| comma separated lists | 78 |
| command and output logs | 484 |
| command completion | 479 |
| command descriptions | 15 |
| command echoing | 484 |
| command history | 480 |
| command options | 17 |
| command-line editing | 23 |
| commands, defining | 143 |
| comments | 26 |
| common_size | 237 |
| commutation_matrix | 268 |
| compan | 386 |
| comparison expressions | 100 |
| complement | 382 |
| completion_append_char | 89, 480 |
| completion_matches | 480 |
| complex | 33 |
| complex arithmetic | 258 |
| complex-conjugate transpose | 98 |
| computer | 466 |
| cond | 272 |
| condest | 309 |
| confirm_recursive_rmdir | 449 |
| conj | 259 |
| constantsm, mathematical | 269 |
| containers | 63 |
| continuation lines | 123 |
| continue statement | 121 |
| contour | 200 |
| contourc | 200 |
| contributing to Octave | 6 |
| contributors | 3 |
| conv | 387 |
| conv2 | 387 |
| conversion specifications (printf) | 177 |
| conversion specifications (scanf) | 183 |
| conversions, string | 57 |
| convhull | 412 |
| convhulln | 412 |
| cool | 431 |

# Index

543

coordinate transformations ......... 268
copper ............................ 431
copyfile .......................... 453
copyright ......................... 527
cor ............................... 353
cor_test .......................... 360
core dump ......................... 506
corrcoef .......................... 354
cos ............................... 259
cosd .............................. 261
cosh .............................. 260
cot ............................... 259
cotd .............................. 261
coth .............................. 260
cov ............................... 353
cplxpair .......................... 253
cputime ........................... 444
crash_dumps_octave_core ...... 89, 171
create_set ........................ 381
cross ............................. 268
csc ............................... 259
cscd .............................. 261
csch .............................. 260
csymamd ........................... 305
ctime ............................. 440
cumprod ........................... 263
cumsum ............................ 263
cumtrapz .......................... 325
customizing the prompt ............ 483
cut ............................... 357

## D

DAE ............................... 329
daspk ............................. 332
daspk_options ..................... 333
dasrt ............................. 335
dasrt_options ..................... 337
data structures ............... 31, 63
data types ........................ 29
data types, built-in .............. 29
date .............................. 443
datestr ........................... 445
datevec ........................... 447
dbclear ........................... 158
dbstatus .......................... 158
dbstop ............................ 158
dbtype ............................ 159
dbwhere ........................... 159

deal .............................. 131
deblank ........................... 52
debug_on_error .................... 157
debug_on_interrupt ................ 157
debug_on_warning .................. 157
debugging ......................... 157
dec2base .......................... 58
dec2bin ........................... 57
dec2hex ........................... 58
deconv ............................ 387
decrement operator ................ 105
default arguments ................. 132
default_save_options ......... 89, 170
defining functions ................ 125
del2 .............................. 253
delaunay .......................... 403
delaunay3 ......................... 404
delaunayn ......................... 404
delete ............................ 221
demo .............................. 492
derivative, polynomial ............ 389
description format ................ 14
descriptive statistics ............ 351
det ............................... 272
detrend ........................... 418
diag .............................. 245
diary ............................. 484
diary of commands and output ..... 484
diff .............................. 236
Differential Equations ............ 329
diffpara .......................... 422
diffs, submitting ................. 509
dir ............................... 463
discrete_cdf ...................... 369
discrete_inv ...................... 370
discrete_pdf ...................... 370
discrete_rnd ...................... 377
disp .............................. 161
dispatch .......................... 136
distribution of Octave ............ 6
distributions, statistical ........ 368
division .......................... 98
dmperm ............................ 306
dmult ............................. 272
do-until statement ................ 118
do_string_escapes ................. 60
do_what_i_mean_not_what_i_say .... 16
document .......................... 88
documentation notation ............ 13

documenting Octave programs....... 26
dos................................ 458
dot................................ 272
double............................. 33
drawnow............................ 220
dsearch............................ 408
dsearchn........................... 408
Dulmage-Mendelsohn decomposition
................................... 306
dup2............................... 460
duplication_matrix................. 268
durbinlevinson..................... 422

# E

e.................................. 270
echo............................... 484
echo_executing_commands............ 485
echoing executing commands....... 484
edit_history....................... 481
editing the command line........... 23
EDITOR..................... 89, 91, 482
eig................................ 272
element-by-element evaluation...... 101
else statement..................... 113
elseif statement................... 113
Emacs TAGS files................... 524
empirical_cdf...................... 370
empirical_inv...................... 370
empirical_pdf...................... 370
empirical_rnd...................... 377
end of file........................ 190
end statement...................... 113
end_try_catch...................... 123
end_unwind_protect................. 122
endfor statement................... 118
endfunction statement.............. 125
endgrent........................... 465
endif statement.................... 113
endpwent........................... 464
endswitch statement................ 115
endwhile statement................. 117
environment variables, accessing.... 462
eomday............................. 447
eps................................ 270
equality operator.................. 100
equality, tests for................ 100
equations, nonlinear............... 283
erf................................ 266

erfc............................... 266
erfinv............................. 266
errno.............................. 151
errno_list......................... 151
erroneous messages................. 506
erroneous results............. 506, 507
error.............................. 147
error message notation.............. 14
error messages..................... 24
error messages, incorrect.......... 506
errorbar........................... 201
errors............................. 147
errors, debugging.................. 157
escape sequence notation........... 47
etime.............................. 444
etree.............................. 297
etreeplot.......................... 297
eval............................... 109
evalin............................. 111
evaluation notation................ 13
example............................ 493
exceptions, see errors............. 147
exec............................... 460
EXEC_PATH..................... 89, 459
executable scripts................. 25
execution speed.................... 495
exist.............................. 86
exit............................... 21
exiting octave.................. 9, 21
exp................................ 254
expcdf............................. 370
expinv............................. 370
expm............................... 282
exponentiation..................... 98
exppdf............................. 370
expression, range.................. 38
expressions........................ 93
expressions, assignment............ 103
expressions, boolean............... 101
expressions, comparison............ 100
expressions, logical............... 101
exprnd............................. 377
eye................................ 244

# F

F_DUPFD............................ 461
F_GETFD............................ 461
F_GETFL............................ 461

| | |
|---|---|
| F_SETFD | 461 |
| F_SETFL | 461 |
| f_test_regression | 361 |
| factor | 254 |
| factorial | 254 |
| factorial function | 97 |
| factorization, matrices | 276 |
| fail | 491 |
| false | 44 |
| fast fourier transform | 415 |
| fatal signal | 506 |
| fcdf | 370 |
| fclose | 175 |
| fcntl | 461 |
| fdisp | 171 |
| feof | 190 |
| ferror | 190 |
| feval | 110 |
| fflush | 165 |
| fft | 415 |
| fft2 | 415, 416 |
| fftconv | 416 |
| fftfilt | 416 |
| fftn | 416 |
| fftshift | 422 |
| fftw | 417 |
| fgetl | 176 |
| fgets | 176 |
| fieldnames | 69 |
| figure | 214 |
| file i/o | 167 |
| file_in_loadpath | 135 |
| file_in_path | 452 |
| fileattrib | 451 |
| fileparts | 453 |
| filesep | 454 |
| filesystem functions | 448 |
| filter | 418 |
| filter2 | 419 |
| find | 236 |
| findstr | 52 |
| finite | 236 |
| finv | 370 |
| fix | 254 |
| fixed_point_format | 36, 90 |
| flag character (printf) | 178 |
| flag character (scanf) | 184 |
| flipdim | 238 |
| fliplr | 237 |
| flipud | 237 |
| floor | 254 |
| flying high and fast | 81 |
| fmod | 254 |
| fnmatch | 452 |
| fonts | 13 |
| foo | 14 |
| fopen | 174 |
| for statement | 118 |
| Fordyce, A. P. | 110 |
| fork | 459 |
| format | 162 |
| formatted input | 182 |
| formatted output | 176 |
| formula | 142 |
| fpdf | 370 |
| fplot | 195 |
| fprintf | 176 |
| fputs | 175 |
| fractdiff | 422 |
| fread | 186 |
| freport | 190 |
| freqz | 419 |
| freqz_plot | 419 |
| frewind | 191 |
| frnd | 377 |
| Frobenius norm | 275 |
| fscanf | 182 |
| fseek | 191 |
| fsolve | 283 |
| fsolve_options | 283 |
| ftell | 191 |
| full | 292 |
| fullfile | 454 |
| func2str | 141 |
| function descriptions | 14 |
| function file | 15, 133 |
| function statement | 125 |
| *function_name* | 22 |
| functions | 141 |
| functions, user-defined | 125 |
| funding Octave development | 6 |
| fwrite | 189 |

## G

| | |
|---|---|
| gamcdf | 371 |
| gaminv | 371 |
| gamma | 266 |

gammainc . . . . . . . . . . . . . . . . . . . . . . . . . 267
gammaln . . . . . . . . . . . . . . . . . . . . . . . . . . 267
gampdf . . . . . . . . . . . . . . . . . . . . . . . . . . . 371
gamrnd . . . . . . . . . . . . . . . . . . . . . . . . . . . 378
gaussian quadrature . . . . . . . . . . . . . . . 323
gca . . . . . . . . . . . . . . . . . . . . . . . . . . . . . . . 218
gcd . . . . . . . . . . . . . . . . . . . . . . . . . . . . . . . 254
gcf . . . . . . . . . . . . . . . . . . . . . . . . . . . . . . . 218
genpath . . . . . . . . . . . . . . . . . . . . . . . . . . 135
geocdf . . . . . . . . . . . . . . . . . . . . . . . . . . . 371
geoinv . . . . . . . . . . . . . . . . . . . . . . . . . . . 371
geometry . . . . . . . . . . . . . . . . . . . . . . . . . 403
geopdf . . . . . . . . . . . . . . . . . . . . . . . . . . . 371
geornd . . . . . . . . . . . . . . . . . . . . . . . . . . . 378
get . . . . . . . . . . . . . . . . . . . . . . . . . . . . . . . 219
getegid . . . . . . . . . . . . . . . . . . . . . . . . . . 462
getenv . . . . . . . . . . . . . . . . . . . . . . . . . . . 462
geteuid . . . . . . . . . . . . . . . . . . . . . . . . . . 462
getfield . . . . . . . . . . . . . . . . . . . . . . . . . . . 69
getgid . . . . . . . . . . . . . . . . . . . . . . . . . . . 462
getgrent . . . . . . . . . . . . . . . . . . . . . . . . . 465
getgrgid . . . . . . . . . . . . . . . . . . . . . . . . . 465
getgrnam . . . . . . . . . . . . . . . . . . . . . . . . 465
getpgrp . . . . . . . . . . . . . . . . . . . . . . . . . . 462
getpid . . . . . . . . . . . . . . . . . . . . . . . . . . . 462
getppid . . . . . . . . . . . . . . . . . . . . . . . . . 462
getpwent . . . . . . . . . . . . . . . . . . . . . . . . 464
getpwnam . . . . . . . . . . . . . . . . . . . . . . . 464
getpwuid . . . . . . . . . . . . . . . . . . . . . . . . 464
getrusage . . . . . . . . . . . . . . . . . . . . . . . . 467
getting a good job . . . . . . . . . . . . . . . . . 81
getuid . . . . . . . . . . . . . . . . . . . . . . . . . . . 462
givens . . . . . . . . . . . . . . . . . . . . . . . . . . . 273
glob . . . . . . . . . . . . . . . . . . . . . . . . . . . . . 452
global statement . . . . . . . . . . . . . . . . . . 82
global variables . . . . . . . . . . . . . . . . . . . 82
glpk . . . . . . . . . . . . . . . . . . . . . . . . . . . . . 339
gls . . . . . . . . . . . . . . . . . . . . . . . . . . . . . . 349
gmap40 . . . . . . . . . . . . . . . . . . . . . . . . . 433
gmtime . . . . . . . . . . . . . . . . . . . . . . . . . . 440
gnuplot_binary . . . . . . . . . . . . . . 90, 232
gnuplot_use_title_option . . . . . . . . . 233
gplot . . . . . . . . . . . . . . . . . . . . . . . . . . . . 296
gradient . . . . . . . . . . . . . . . . . . . . . . . . . 255
graphics . . . . . . . . . . . . . . . . . . . . . . . . . 193
graphics object properties . . . . . . . . . 222
graphics objects . . . . . . . . . . . . . . . . . . 217
gray . . . . . . . . . . . . . . . . . . . . . . . . . . . . . 431
gray2ind . . . . . . . . . . . . . . . . . . . . . . . . . 430

greater than operator . . . . . . . . . . . . . . 100
grid . . . . . . . . . . . . . . . . . . . . . . . . . . . . . 212
griddata . . . . . . . . . . . . . . . . . . . . . . . . . 413
griddata3 . . . . . . . . . . . . . . . . . . . . . . . . 413
griddatan . . . . . . . . . . . . . . . . . . . . . . . . 414
gunzip . . . . . . . . . . . . . . . . . . . . . . . . . . 454
gzip . . . . . . . . . . . . . . . . . . . . . . . . . . . . 454

# H

hadamard . . . . . . . . . . . . . . . . . . . . . . . 249
hamming . . . . . . . . . . . . . . . . . . . . . . . . 422
handle, function handles . . . . . . . . . . . 140
hankel . . . . . . . . . . . . . . . . . . . . . . . . . . 249
hanning . . . . . . . . . . . . . . . . . . . . . . . . . 423
hash functions . . . . . . . . . . . . . . . . . . . . 468
header comments . . . . . . . . . . . . . . . . . 497
help . . . . . . . . . . . . . . . . . . . . . . . . . . . . . 22
help, on-line . . . . . . . . . . . . . . . . . . . . . . 22
help, where to find . . . . . . . . . . . . . . . . 510
Hermitian operator . . . . . . . . . . . . . . . . 98
hess . . . . . . . . . . . . . . . . . . . . . . . . . . . . 276
Hessenberg decomposition . . . . . . . . . 276
hex2dec . . . . . . . . . . . . . . . . . . . . . . . . . . 58
hidden . . . . . . . . . . . . . . . . . . . . . . . . . . 209
hilb . . . . . . . . . . . . . . . . . . . . . . . . . . . . . 249
hist . . . . . . . . . . . . . . . . . . . . . . . . . . . . . 198
history . . . . . . . . . . . . . . . . . . . . . . . . . . . . 3
history . . . . . . . . . . . . . . . . . . . . . . . . . . 481
history of commands . . . . . . . . . . . . . . 480
history_file . . . . . . . . . . . . . . . . . . . 90, 482
history_size . . . . . . . . . . . . . . . . . . 90, 482
history_timestamp_format_string
. . . . . . . . . . . . . . . . . . . . . . . . . . . 482
hold . . . . . . . . . . . . . . . . . . . . . . . . . . . . 221
horzcat . . . . . . . . . . . . . . . . . . . . . . . . . 239
hot . . . . . . . . . . . . . . . . . . . . . . . . . . . . . 431
hotelling_test . . . . . . . . . . . . . . . . . . . . 361
hotelling_test_2 . . . . . . . . . . . . . . . . . . 361
housh . . . . . . . . . . . . . . . . . . . . . . . . . . . 281
hsv . . . . . . . . . . . . . . . . . . . . . . . . . . . . . 431
hsv2rgb . . . . . . . . . . . . . . . . . . . . . . . . . 434
hull, convex . . . . . . . . . . . . . . . . . . . . . . 412
hurst . . . . . . . . . . . . . . . . . . . . . . . . . . . 423
hygecdf . . . . . . . . . . . . . . . . . . . . . . . . . 371
hygeinv . . . . . . . . . . . . . . . . . . . . . . . . . 371
hygepdf . . . . . . . . . . . . . . . . . . . . . . . . . 372
hygernd . . . . . . . . . . . . . . . . . . . . . . . . . 378

## Index

## I

| | |
|---|---|
| I | 269 |
| identity matrix | 244 |
| if statement | 113 |
| ifft | 415 |
| ifftn | 416 |
| ifftshift | 422 |
| ignore_function_time_stamp | 90, 134 |
| imag | 259 |
| image | 428 |
| IMAGE_PATH | 428 |
| image_viewer | 429 |
| imagesc | 429 |
| improving Octave | 506, 509 |
| imshow | 428 |
| incorrect error messages | 506 |
| incorrect output | 506, 507 |
| incorrect results | 506, 507 |
| increment operator | 105 |
| ind2gray | 430 |
| ind2rgb | 430 |
| ind2sub | 95 |
| index | 53 |
| index expressions | 93 |
| Inf | 269 |
| infinity norm | 275 |
| info_file | 23 |
| info_program | 23 |
| initialization | 20 |
| inline | 142 |
| inline, inline functions | 140 |
| inpolygon | 411 |
| input | 166 |
| input and output | 161 |
| input conversions, for scanf | 184 |
| input history | 480 |
| input, formatted | 182 |
| inputname | 127 |
| installation trouble | 505 |
| installing Octave | 511 |
| int16 | 39 |
| int2str | 50 |
| int32 | 39 |
| int64 | 39 |
| int8 | 39 |
| integral, polynomial | 389 |
| integration, numeric | 323 |
| interp1 | 393 |
| interp2 | 398 |
| interp3 | 399 |
| interpft | 396 |
| interpn | 399 |
| interpolation | 393 |
| interpolation, multidimensional | 398 |
| interpolation, scattered data | 413 |
| intersect | 382 |
| intmax | 39 |
| intmin | 40 |
| introduction | 9 |
| inv | 273 |
| invalid input | 506 |
| inverse | 273 |
| invhilb | 250 |
| ipermute | 240 |
| iqr | 357 |
| is_duplicate_entry | 236 |
| is_leap_year | 444 |
| isa | 29 |
| isalnum | 61 |
| isalpha | 61 |
| isascii | 61 |
| iscell | 73 |
| iscellstr | 75 |
| ischar | 49 |
| iscntrl | 61 |
| iscommand | 143 |
| iscomplex | 45 |
| isdefinite | 45 |
| isdigit | 61 |
| isdir | 452 |
| isempty | 32 |
| isequal | 100 |
| isequalwithequalnans | 100 |
| isfield | 69 |
| isfigure | 217 |
| isglobal | 83 |
| isgraph | 61 |
| ishandle | 217 |
| ishold | 221 |
| isieee | 466 |
| isinf | 236 |
| isinteger | 39 |
| isletter | 61, 62 |
| islogical | 45 |
| islower | 62 |
| ismatrix | 45 |
| ismember | 381 |
| isna | 30 |

| | |
|---|---|
| isnan | 236 |
| isnumeric | 44 |
| ispc | 466 |
| isprime | 46 |
| isprint | 62 |
| ispunct | 62 |
| israwcommand | 144 |
| isreal | 45 |
| isscalar | 45 |
| isspace | 62 |
| issparse | 293 |
| issquare | 45 |
| isstruct | 68 |
| issymmetric | 45 |
| isunix | 466 |
| isupper | 62 |
| isvarname | 81 |
| isvector | 45 |
| isxdigit | 62 |

## J

| | |
|---|---|
| jet | 432 |
| job hunting | 81 |

## K

| | |
|---|---|
| kbhit | 166 |
| kendall | 357 |
| keyboard | 159 |
| kill | 462 |
| known causes of trouble | 505 |
| kolmogorov_smirnov_cdf | 372 |
| kolmogorov_smirnov_test | 362 |
| kolmogorov_smirnov_test_2 | 362 |
| kron | 282 |
| kruskal_wallis_test | 362 |
| krylov | 281 |
| kurtosis | 354 |

## L

| | |
|---|---|
| laplace_cdf | 372 |
| laplace_inv | 372 |
| laplace_pdf | 372 |
| laplace_rnd | 378 |
| lasterr | 151 |
| lasterror | 150 |
| lastwarn | 153 |

| | |
|---|---|
| lcm | 255 |
| legend | 211 |
| legendre | 267 |
| length | 31 |
| less than operator | 100 |
| lgamma | 267 |
| lin2mu | 435 |
| line | 220 |
| line-oriented input | 176 |
| linear algebra | 271 |
| link | 448 |
| linspace | 246 |
| load | 169 |
| loadable function | 15 |
| loadaudio | 435 |
| loadimage | 427 |
| loading data | 167 |
| lobatto rule, quadrature | 323 |
| localtime | 441 |
| locking, of functions | 137 |
| log | 255 |
| log10 | 256 |
| log2 | 256 |
| logging commands and output | 484 |
| logical | 44 |
| logical expressions | 101 |
| logical operators | 101 |
| logistic_cdf | 372 |
| logistic_inv | 372 |
| logistic_pdf | 372 |
| logistic_regression | 367 |
| logistic_rnd | 378 |
| logit | 356 |
| loglog | 196 |
| loglogerr | 203 |
| logm | 282 |
| logncdf | 372 |
| logninv | 373 |
| lognpdf | 373 |
| lognrnd | 378 |
| logspace | 246 |
| lookfor | 22 |
| lookup | 397 |
| loop | 117 |
| looping over structure elements | 120 |
| LP | 339 |
| ls | 463 |
| ls_command | 463 |
| lsode | 329 |

Index 549

lsode_options .................. 330
lstat ...................... 449, 451
lu................................ 277
LU decomposition ........ 277, 316
luinc ............................. 316
lvalue ........................... 103

## M

magic ............................ 250
mahalanobis .................... 355
makeinfo_program .............. 23
manova .......................... 363
mapping function ................ 15
mark_as_command ............... 143
mark_as_rawcommand ........... 143
mat2cell ........................ 73
mat2str ......................... 50
matching failure, in scanf .... 183
mathematical constants ........ 269
matrices ........................ 34
matrices, empty ................ 37
matrices, standard ............. 244
matrix manipulation ........... 235
matrix multiplication ......... 98
matrix, functions of ........... 282
matrix_type .................... 273
max .............................. 256
max_recursion_depth ....... 90, 97
maximum field width (scanf) ... 184
mcnemar_test ................... 363
md5sum .......................... 468
mean ............................ 351
meansq .......................... 352
median .......................... 352
menu ............................ 166
mesh ............................ 208
meshc ........................... 208
meshgrid ........................ 209
messages, error ................. 24
mfilename ...................... 134
min ............................. 256
minimum field width (printf) .. 178
mislocked ...................... 139
missing data .................... 30
mkdir ........................... 449
mkfifo .......................... 449
mkpp ............................ 391
mkstemp ......................... 189

mktime .......................... 441
mlock ........................... 138
mod ............................. 257
mode ............................ 353
moment .......................... 354
more ............................ 165
movefile ........................ 453
mu2lin .......................... 435
multiple plots .................. 213
multiple return values ......... 128
multiplication .................. 98
munlock ......................... 138

## N

NA ............................... 30
NaN ............................. 269
nargchk ......................... 129
nargin .......................... 127
nargout ......................... 129
native_float_format ........... 171
nbincdf ......................... 373
nbininv ......................... 373
nbinpdf ......................... 373
nbinrnd ......................... 379
nchoosek .................. 355, 503
ndgrid .......................... 209
ndims ............................ 31
negation ........................ 98
network functions .............. 456
newplot ......................... 221
newtroot ........................ 110
nextpow2 ........................ 257
NLP ............................. 339
nnz ............................. 293
nonlinear equations ............ 283
nonlinear programming ......... 339
nonzeros ........................ 293
norm ............................ 274
normcdf ......................... 373
normest ......................... 309
norminv ......................... 373
normpdf ......................... 373
normrnd ......................... 379
not operator .................... 101
now ............................. 440
nthroot ......................... 257
ntsc2rgb ........................ 434
null ............................ 275

num2cell ........................... 73
num2str ............................ 50
numel .............................. 31
numeric constant ................ 30, 33
numeric value.................... 30, 33
nzmax ............................. 293

## O

O_APPEND........................... 461
O_ASYNC............................ 461
O_CREAT............................ 461
O_NONBLOCK......................... 461
O_RDONLY........................... 461
O_RDWR............................. 461
O_SYNC............................. 461
O_WRONLY........................... 461
ocean ............................. 432
Octave command options ............ 17
octave-tags ....................... 524
octave_config_info................. 467
octave_core_file_limit ........... 172
octave_core_file_name ............ 172
octave_core_file_options ......... 172
OCTAVE_EXEC_PATH................... 91
OCTAVE_HISTFILE.................... 92
OCTAVE_HISTSIZE.................... 92
OCTAVE_HOME.................... 89, 466
OCTAVE_INFO_FILE................... 92
OCTAVE_INFO_PROGRAM................ 92
OCTAVE_PATH........................ 91
OCTAVE_VERSION.................... 467
ODE................................ 329
ols ............................... 348
on-line help ....................... 22
ones .............................. 245
operator precedence ............... 106
operators, arithmetic .............. 98
operators, assignment ............. 103
operators, boolean................. 101
operators, decrement .............. 105
operators, increment............... 105
operators, logical................. 101
operators, relational ............. 100
optimization ...................... 339
options, Octave command ........... 17
or operator........................ 101
orderfields........................ 69
oregonator ........................ 331

orient ............................ 216
orth .............................. 275
orthogonal collocation ............ 325
otherwise statement ............... 115
output conversions, for printf ..... 179
output, formatted ................. 176
output_max_field_width......... 35, 90
output_precision .............. 35, 90
overloading........................ 136

## P

P_tmpdir .......................... 454
pack .............................. 455
packages .......................... 469
page_output_immediately .......... 165
page_screen_output ............ 90, 165
PAGER.......................... 89, 165
PAGER_FLAGS ...................... 165
parseparams ....................... 130
pascal ............................ 250
patch ............................. 220
patches, submitting ............... 509
path .............................. 135
pathdef ........................... 135
pathsep ........................... 135
pause ............................. 445
pcg ............................... 312
pchip ............................. 423
pclose ............................ 458
pcolor ............................ 205
pcr ............................... 314
peaks ............................. 216
periodogram ....................... 423
perms ............................. 355
permute ........................... 240
persistent statement .............. 83
persistent variables................ 83
pi................................. 269
pie ............................... 204
pink .............................. 432
pinv .............................. 275
pipe .............................. 460
pkg ............................... 469
playaudio.......................... 436
plot .............................. 194
plot3 ............................. 210
plots, printing.................... 214
plots, statistical ................ 358

# Index

plotting .......................... 193
poisscdf ......................... 374
poissinv ......................... 374
poisspdf ......................... 374
poissrnd ......................... 379
pol2cart ......................... 268
polar ............................ 203
polar coordinates ................ 268
poly ............................. 392
polyarea ......................... 410
polyder .......................... 390
polyderiv ........................ 389
polyfit .......................... 390
polygcd .......................... 387
polyint .......................... 390
polynomial manipulations ......... 385
polynomial, interpolation ........ 390
polyout .......................... 392
polyreduce ....................... 392
polyval .......................... 385
polyvalm ......................... 385
popen ............................ 458
popen2 ........................... 458
position, in file ................ 191
postpad .......................... 243
pow2 ............................. 257
ppplot ........................... 358
ppval ............................ 391
precedence, of operators ......... 106
precision (printf) ............... 179
predicates, numeric ............... 44
prepad ........................... 243
primes ........................... 257
print ............................ 214
print_answer_id_name ......... 90, 164
print_empty_dimensions ........ 37, 90
print_usage ...................... 149
printf ........................... 176
printing notation ................. 14
prism ............................ 432
probit ........................... 356
prod ............................. 262
program, self contained ........... 25
program_name ...................... 19
programs .......................... 26
prompt customization ............. 483
prop_test_2 ...................... 363
PS1 .......................... 89, 483
PS2 .......................... 89, 484

PS4 .......................... 89, 484
putenv ........................... 462
puts ............................. 175
pwd .............................. 464

## Q

qp ............................... 345
QP ............................... 339
qqplot ........................... 358
qr ............................... 277
QR factorization ................. 277
quad ............................. 323
quad_options ..................... 324
quadl ............................ 324
quadratic programming ............ 339
quadrature ....................... 323
quit .............................. 21
quitting octave ................ 9, 21
quiver ........................... 204
quotient .......................... 98
qz ............................... 278
qzhess ........................... 279

## R

rainbow .......................... 432
raising errors ................... 147
rand ............................. 246
rande ............................ 247
randg ............................ 248
randn ............................ 247
random number distributions ...... 376
randp ............................ 248
randperm ......................... 249
range ............................ 356
range expressions ................. 38
rank ............................. 275
ranks ............................ 356
rat .............................. 172
rational approximation ........... 172
rats ............................. 173
read_readline_init_file .......... 482
readdir .......................... 448
readlink ......................... 448
real ............................. 259
realmax .......................... 270
realmin .......................... 270
record ........................... 436

rectangle_lw ...................... 423
rectangle_sw ...................... 423
recursion .......................... 97
regexp ............................. 54
regexpi ............................ 56
regexprep .......................... 56
regular expressions ................ 52
rehash ............................ 135
relational operators .............. 100
rem ............................... 258
rename ............................ 448
repmat ............................ 245
reporting bugs ............... 505, 506
reshape ........................... 240
residue ........................... 388
results, incorrect ........... 506, 507
rethrow ........................... 151
return ............................ 132
return values, multiple ........... 128
return_last_computed_value ........ 90
rgb2hsv ........................... 434
rgb2ind ........................... 430
rgb2ntsc .......................... 434
rindex ............................. 53
rmdir ............................. 449
rmfield ............................ 68
rmpath ............................ 135
roots ............................. 386
rosser ............................ 250
rot90 ............................. 238
rotdim ............................ 238
round ............................. 258
rows ............................... 31
rref .............................. 276
run ............................... 111
run_count ......................... 356
run_history ....................... 481
run_test .......................... 364

## S

save .............................. 167
save_header_format_string ........ 170
save_precision ............... 90, 170
saveaudio ......................... 436
saveimage ......................... 427
savepath .......................... 135
saving data ....................... 167
saving_history ............... 91, 482

schur ............................. 279
Schur decomposition ............... 279
scoping ........................... 111
script files ................. 125, 139
scripts ............................ 25
sec ............................... 259
secd .............................. 261
sech .............................. 260
SEEK_CUR .......................... 191
SEEK_END .......................... 191
SEEK_SET .......................... 191
self contained programs ............ 25
semilogx .......................... 196
semilogxerr ....................... 202
semilogy .......................... 196
semilogyerr ....................... 202
set ............................... 219
setaudio .......................... 436
setdiff ........................... 382
setenv ............................ 462
setfield ........................... 69
setgrent .......................... 465
setpwent .......................... 464
sets .............................. 381
setxor ............................ 383
shading ........................... 210
shg ............................... 221
shift ............................. 241
shiftdim .......................... 241
short-circuit evaluation .......... 102
side effect ....................... 103
SIG ............................... 462
sighup_dumps_octave_core ..... 91, 171
sign .............................. 258
sign_test ......................... 364
signal processing ................. 415
sigterm_dumps_octave_core .... 91, 171
silent_functions .............. 91, 127
sin ............................... 259
sinc .............................. 419
sind .............................. 261
sinetone .......................... 423
sinewave .......................... 424
single ............................. 33
singular value decomposition ...... 280
sinh .............................. 260
size ............................... 32
size_equal ......................... 32
sizeof ............................. 32

# Index

| | |
|---|---|
| skewness | 354 |
| sleep | 445 |
| sombrero | 216 |
| sort | 241 |
| sortrows | 242 |
| sound files | 435 |
| source | 140 |
| spalloc | 291 |
| sparse | 290 |
| sparse matrices | 287 |
| sparse_auto_mutate | 299 |
| spconvert | 291 |
| spdiag | 289 |
| spdiags | 289 |
| spearman | 356 |
| special functions | 264 |
| spectral_adf | 424 |
| spectral_xdf | 424 |
| speed | 493 |
| speedups | 495 |
| spencer | 424 |
| speye | 289 |
| spfun | 289 |
| sph2cart | 269 |
| spherical coordinates | 268 |
| spline | 397 |
| split | 54 |
| split_long_rows | 35, 91 |
| spmax | 293 |
| spmin | 294 |
| spones | 289 |
| spparms | 310 |
| sprand | 292 |
| sprandn | 292 |
| sprandsym | 292 |
| sprank | 311 |
| spring | 432 |
| sprintf | 177 |
| spstats | 294 |
| spy | 295 |
| sqp | 346 |
| sqrt | 258 |
| sqrtm | 282 |
| squeeze | 32 |
| sscanf | 183 |
| stairs | 198 |
| standards of coding style | 495 |
| startup | 20 |
| startup files | 20 |
| stat | 449 |
| statements | 113 |
| statistics | 351 |
| statistics | 354 |
| std | 352 |
| stderr | 173 |
| stdin | 173 |
| stdout | 173 |
| stem | 199 |
| stft | 424 |
| *str* | 22 |
| str2double | 59 |
| str2func | 141 |
| str2mat | 49 |
| str2num | 60 |
| strcat | 49 |
| strcmp | 51 |
| strcmpi | 51 |
| strfind | 53 |
| strftime | 441 |
| string_fill_char | 49 |
| strings | 30, 47 |
| strjust | 60 |
| strmatch | 53 |
| strncmp | 51 |
| strncmpi | 52 |
| strptime | 443 |
| strrep | 54 |
| strtok | 53 |
| strtrunc | 49 |
| struct | 68 |
| struct_levels_to_print | 64, 91 |
| struct2cell | 70 |
| structfun | 70 |
| Structural Rank | 311 |
| structure elements, looping over | 120 |
| structures | 31, 63 |
| strvcat | 49 |
| studentize | 355 |
| sub2ind | 95 |
| subfunctions | 136 |
| submitting diffs | 509 |
| submitting patches | 509 |
| subplot | 213 |
| subprocesses, controlling | 457 |
| subsasgn | 105 |
| subsref | 94 |
| substr | 54 |
| substruct | 70 |

| | |
|---|---|
| subtraction | 98 |
| suggestions | 506 |
| sum | 262 |
| summer | 432 |
| sumsq | 263 |
| suppress_verbose_help_message | 23, 91 |
| surf | 209 |
| surface | 220 |
| surfc | 209 |
| svd | 280 |
| swap | 242 |
| swapbytes | 30 |
| swapcols | 242 |
| swaprows | 242 |
| switch statement | 115 |
| syl | 282 |
| sylvester_matrix | 250 |
| symamd | 306 |
| symbfact | 311 |
| symlink | 448 |
| symrcm | 307 |
| synthesis | 425 |
| system | 457 |
| system functions | 439 |

## T

| | |
|---|---|
| t_test | 364 |
| t_test_2 | 364 |
| t_test_regression | 365 |
| table | 356 |
| TAGS | 524 |
| tan | 259 |
| tand | 261 |
| tanh | 260 |
| tar | 455 |
| tcdf | 374 |
| tempdir | 454 |
| tempname | 454 |
| temporary files | 189 |
| terminal i/o | 161 |
| test | 487 |
| test functions | 487 |
| tests for equality | 100 |
| tests, statistical | 359 |
| text | 212 |
| three-dimensional plotting | 207 |
| tic | 444 |

| | |
|---|---|
| tilde_expand | 453 |
| time | 440 |
| timing functions | 439 |
| tinv | 374 |
| tips | 495 |
| title | 211 |
| tmpfile | 189 |
| tmpnam | 190 |
| toascii | 60 |
| toc | 444 |
| toeplitz | 251 |
| tolower | 60 |
| toupper | 60 |
| tpdf | 374 |
| trace | 276 |
| transpose | 98 |
| transpose, complex-conjugate | 98 |
| trapezium rule | 323 |
| trapz | 325 |
| treeplot | 297 |
| triangle_lw | 425 |
| triangle_sw | 425 |
| triangulation, delaunay | 403 |
| trigonometric functions | 259 |
| tril | 243 |
| trimesh | 405 |
| triplot | 405 |
| triu | 243 |
| trnd | 379 |
| troubleshooting | 505 |
| true | 44 |
| try statement | 123 |
| tsearch | 407 |
| tsearchn | 407 |
| type | 88 |
| typecast | 29 |
| typeinfo | 29 |

## U

| | |
|---|---|
| u_test | 365 |
| uint16 | 39 |
| uint32 | 39 |
| uint64 | 39 |
| uint8 | 39 |
| umask | 449 |
| uname | 466 |
| unary minus | 98 |
| undefined behavior | 506 |

# Index

undefined function value .......... 506
undo_string_escapes ............... 61
unidcdf ........................... 374
unidinv ........................... 374
unidpdf ........................... 374
unidrnd ........................... 379
unifcdf ........................... 374
unifinv ........................... 375
unifpdf ........................... 375
unifrnd ........................... 379
union ............................. 382
unique ............................ 381
unlink ............................ 448
unmark_command .................... 143
unmark_rawcommand ................. 144
unmkpp ............................ 391
unpack ............................ 455
untar ............................. 455
unwind_protect statement .......... 122
unwind_protect_cleanup ............ 122
unwrap ............................ 420
unzip ............................. 455
urlread ........................... 456
urlwrite .......................... 456
usage ............................. 149
use of comments ................... 26
user-defined functions ............ 125
user-defined variables ............ 81
usleep ............................ 445
Utility Functions ......... 256, 293, 294

## V

values ............................ 355
vander ............................ 251
var ............................... 353
var_test .......................... 365
variable descriptions ............. 15
variable-length argument lists .... 129
variable-length return lists ...... 131
variables, global ................. 82
variables, persistent ............. 83
variables, user-defined ........... 81
vec ............................... 243
vech .............................. 243
vectorize ......................... 142
ver ............................... 467
version ........................... 467

vertcat ........................... 239
view .............................. 210
voronoi ........................... 409
voronoin .......................... 409

## W

waitpid ........................... 460
warning ........................... 152
warnings .......................... 147
warranty .......................... 527
wavread ........................... 436
wavwrite .......................... 437
wblcdf ............................ 375
wblinv ............................ 375
wblpdf ............................ 375
wblrnd ............................ 380
weekday ........................... 447
welch_test ........................ 366
which ............................. 88
while statement ................... 117
white ............................. 432
who ............................... 85
whos .............................. 85
whos_line_format .................. 85
wienrnd ........................... 380
wilcoxon_test ..................... 366
wilkinson ......................... 251
winter ............................ 433
wrong answers ............... 506, 507

## X

xlabel ............................ 212
xor ............................... 235

## Y

ylabel ............................ 212
yulewalker ........................ 425

## Z

z_test ............................ 366
z_test_2 .......................... 367
zeros ............................. 245
zip ............................... 455
zlabel ............................ 212